Ramarao Poduri, Gaurav Joshi, Mayank, Asim Kumar (Eds.)

Drug Repurposing

Also of interest

Phytochemicals in Medicinal Plants.
Biodiversity, Bioactivity and Drug Discovery
Edited by: Charu Arora, Dakeshwar Kumar Verma, Jeenat Aslam and Pramod Kumar Mahish, 2023
ISBN 9783110791761, e-ISBN (PDF) 9783110791891

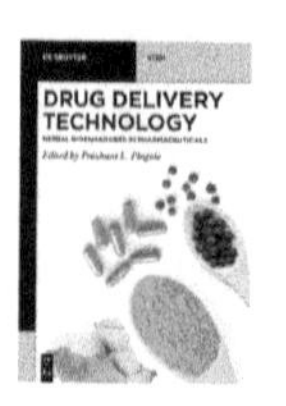

Drug Delivery Technology.
Herbal Bioenhancers in Pharmaceuticals
Edited by: Prashant L. Pingale, 2022
ISBN 9783110746792, e-ISBN (PDF) 9783110746808

Natural Poisons and Venoms.
Plant Toxins: Terpenes and Steroids
Eberhard Teuscher and Ulrike Lindequist, 2023
ISBN 9783110724721, e-ISBN (PDF) 9783110724738

Chemistry of Natural Products.
Phytochemistry and Pharmacognosy of Medicinal Plants
Edited by: Mayuri Napagoda and Lalith Jayasinghe, 2022
ISBN 9783110595895, e-ISBN (PDF) 9783110595949

Biopharmaceutical Manufacturing.
Principles, Processes, and Practices
Gary Gilleskie, Charles Rutter and Becky McCuen, 2021
ISBN 9783110616873, e-ISBN (PDF) 9783110616880

Drug Repurposing

Edited by
Ramarao Poduri, Gaurav Joshi, Mayank, Asim Kumar

DE GRUYTER

Editors
Dr. Ramaro Poduri
GITAM Institute of Pharmacy
GITAM University
Gandhi Nagar 530045, Rushikonda
Visakhapatnam
Andhra Pradesh
India
rpoduri@gitam.edu

Dr. Gaurav Joshi
Department of Pharmaceutical Sciences
Hemvati Nandan Bahuguna Garhwal (A Central)
University Srinagar – 246174
Dist. Garhwal (Uttarakhand)
India
garvpharma29@gmail.com

Dr. Mayank
University College of Pharmacy
Guru Kashi University
Talwandi Sabo
Punjab 151302
India
mayank6103@gmail.com

Dr. Asim Kumar
Amity Institute of Pharmacy
Amity University
Gurugram 122413
Haryana
India
akumar13@ggn.amity.edu

ISBN 978-3-11-079114-3
e-ISBN (PDF) 978-3-11-079115-0
e-ISBN (EPUB) 978-3-11-079152-5

Library of Congress Control Number: 2023939213

Bibliographic information published by the Deutsche Nationalbibliothek
The Deutsche Nationalbibliothek lists this publication in the Deutsche Nationalbibliografie; detailed bibliographic data are available on the internet at http://dnb.dnb.de.

Cover image: metamorworks/iStock/Getty Images Plus
Typesetting: Integra Software Services Pvt. Ltd.
Printing and binding: CPI books GmbH, Leck

www.degruyter.com

Preface

Drug discovery and development is a complex and time-consuming process, often spanning several years and requiring significant financial investments. However, in recent years, drug repurposing has emerged as a promising strategy to accelerate the identification of new therapeutic options. By leveraging existing drugs and exploring their potential in different disease contexts, researchers and pharmaceutical companies can tap into a vast reservoir of knowledge and resources. This approach not only offers potential cost savings but also provides the possibility of rapidly bringing safe and effective therapies to patients.

"Drug Repurposing text book" is a comprehensive and timely compilation of insights and research findings that shed light on the contemporary trends, challenges, and opportunities in the field of discovering new drug indications. This book combines the expertise of leading experts from pharmaceutical sciences, offering a multidimensional perspective on this rapidly evolving field.

Chapter 1 delves into the contemporary trends in drug repurposing, focusing on identifying new targets for existing drugs. This chapter highlights the importance of computational methods, high-throughput screening, and innovative data analysis techniques in uncovering novel therapeutic opportunities. Intellectual property and regulatory considerations are critical aspects that influence the landscape of drug repurposing. Chapter 2 explores these factors in-depth, discussing the challenges and potential solutions in navigating the legal and regulatory frameworks associated with repurposed drugs. Chapter 3 offers a unique viewpoint by presenting drug repurposing from both academic and industrial perspectives. It examines the diverse strategies employed by researchers in academia and pharmaceutical companies, showcasing their collaborative efforts and the unique insights they bring to the table. The implications of drug repurposing in specific therapeutic areas are explored in subsequent chapters. Chapter 4 focuses on identifying drugs for neurological disorders, an area of immense unmet medical need. Similarly, Chapters 5 and 6 examine the potential for repurposing drugs as antiviral and antibacterial agents to combat infectious diseases. Gastrointestinal disorders and renal disorders, which affect a significant portion of the global population, are the focal points of Chapters 7 and 8, respectively. These chapters highlight the exciting advancements made in repurposing existing drugs to target these challenging medical conditions. Chapter 9 delves into the application of drug repurposing strategies in the identification of drugs for cardiovascular disorders. With cardiovascular diseases being a leading cause of mortality worldwide, exploring repurposed drugs offers hope for improved treatment options. While exploring the realm of drug repurposing, it is essential to recognize successful examples that have shaped this field. Chapter 10 provides a detailed analysis of blockbuster drugs that have undergone repurposing, showcasing the transformative potential of this approach. Natural and marine sources have long been reservoirs of bioactive compounds with therapeutic potential. Chapter 11 explores the repurposing of drugs

https://doi.org/10.1515/9783110791150-202

derived from these sources, highlighting the wealth of opportunities within the natural world. Chapter 12 focuses on repurposing COX-2 inhibitors and their applications in various diseases. By examining the potential of these drugs beyond their conventional usage, this chapter showcases the versatility of repurposed therapeutics.

Lastly, Chapter 13 comprehensively analyses the advantages, opportunities, and challenges associated with drug repurposing. It explores the economic considerations, regulatory complexities, and ethical dimensions that shape the landscape of this evolving field.

"Drug Repurposing text book" serves as a valuable resource for undergraduates, postgraduates researchers, clinicians, pharmaceutical professionals, and policymakers interested in exploring innovative approaches to drug discovery and development. It provides a roadmap for leveraging existing knowledge and resources to address unmet medical needs effectively. We hope that this book inspires further exploration, collaboration, and breakthroughs in the fascinating field of drug repurposing.

Editors

Contents

List of authors

Arshad J. Ansari
Department of Pharmacy,
School of Chemical Sciences and Pharmacy,
Central University of Rajasthan,
Bandarsindri, Ajmer 305817, Rajasthan,
India
arshadpharm@hotmail.com
Chapters 7, 10

Shikha Asthana
Department of Pharmacology and Toxicology,
National Institute of Pharmaceutical Education
and Research,
Raebareli, Transit Campus,
Bijnor-Sisendi Road, Sarojini Nagar,
Lucknow 226002, Uttar Pradesh,
India
shikhasthana111@gmail.com
Chapter 4

Akanksha Bhatt
Graphic Era Hill University,
Dehradun, Uttarakhand
India
akanksha.rinki@gmail.com
Chapter 5

Soumili Biswas
School of Biological Sciences,
Indian Association for the Cultivation of Science
(IACS),
Jadavpur, Kolkata 700032, West Bengal,
India
erzaayona@gmail.com
Chapter 12

Asit K. Chakraborti
School of Chemical Sciences,
Indian Association for the Cultivation of Science
(IACS),
Jadavpur, Kolkata 700032, West Bengal,
India
ocakc@iacs.res.in
asitkumarchakraborti@gmail.com
Chapter 12

Yirivinti Hayagreeva Dinakar
Department of Pharmaceutics,
JSS College of Pharmacy,
SS Nagara, Bannimantap, Mysuru 570015,
Karnataka,
India
yhayagreeva@gmail.com
Chapter 10

Leif A. Eriksson
Department of Chemistry and Molecular Biology,
University of Gothenburg,
405 30 Göteborg,
Sweden
leif.eriksson@chem.gu.se
Chapter 6

Urooj Fatima
Center of Interdisciplinary Research in Basic
Sciences,
Jamia Millia Islamia,
New Delhi,
India
ufatima@jmi.ac.in
Chapter 7

Ankit Ganeshpurkar
Department of Pharmaceutical Chemistry,
Poona College of Pharmacy,
Bharati Vidyapeeth (Deemed to be University),
Pune, Maharashtra 411038,
India
ankitganeshpurkar@gmail.com
Chapter 3

Avtar Singh Gautam
Department of Pharmacology and Toxicology,
National Institute of Pharmaceutical Education
and Research,
Raebareli, Transit Campus,
Bijnor-Sisendi Road, Sarojini Nagar,
Lucknow 226002, Uttar Pradesh,
India
singh1997avi@gmail.com
Chapter 4

https://doi.org/10.1515/9783110791150-204

Sumeet Gupta
MM College of Pharmacy,
Maharishi Markandeshwar (Deemed to Be University),
Mullana, Ambala 133207, Haryana,
India
sumeetgupta25@mmumullana.org
Chapter 8

Vibhu Jha
Department of Chemistry and Molecular Biology,
University of Gothenburg,
405 30 Göteborg,
Sweden
vibhu.jha@gu.se
Chapter 6

Mayank Joshi
MM College of Pharmacy,
Maharishi Markandeshwar (Deemed to Be University),
Mullana, Ambala 133207, Haryana,
India
and
Department of Chemical Sciences,
Indian Institute of Science Education and Research,
Sector 81, Knowledge City, SAS Nagar,
Manauli PO,
Mohali 140306, Punjab,
India
mayankjoshidj@gmail.com
Chapters 2, 8

Iliyas Khan
Department of Pharmacy,
School of Chemical Sciences and Pharmacy,
Central University of Rajasthan,
Bandarsindri, Ajmer 305817, Rajasthan,
India
iliyaspharm@hotmail.com
Chapters 7, 10

Kiran
Department of Chemistry,
COBS&H,
CCS Haryana Agricultural University,
Hisar 125004, Haryana
India
kiranbhaber@gmail.com
Chapter 1

Asim Kumar
Amity Institute of Pharmacy (AIP),
Amity University Haryana (AUH),
Amity Education Valley,
Panchgaon, Manesar 122413, Haryana,
India
asimniper02@gmail.com
akumar13@ggn.amity.edu
Chapters 12, 13

Bhupinder Kumar
Department of Pharmaceutical Sciences,
HNB Garhwal University,
Chauras Campus,
Srinagar, Garhwal 246174, Uttarakhand,
India
bhupinderkumar25@gmail.com
Chapter 11

Devendra Kumar
School of Pharmacy and Technology Management,
SVKM's NMIMS University,
Mukesh Patel Technology Park,
Shirpur 425405,
Maharashtra,
India
devendrak.phe@gmail.com
Chapter 3

Hitesh Kumar
Faculty of Pharmacy,
Kalinga University,
Kotni, Near Mantralaya, Naya Raipur 492101
(CG),
India
and
Department of Pharmaceutics,
JSS College of Pharmacy,
SS Nagara, Bannimantap, Mysuru 570015,
Karnataka,
India
hitesh.sahu1921@gmail.com
hitesh.kumar@kalingauniversity.ac.in
Chapter 10

Vaibhav Lasure
Department of Pharmacology and Toxicology,
National Institute of Pharmaceutical Education
and Research,
Raebareli, Transit Campus,
Bijnor-Sisendi Road, Sarojini Nagar,
Lucknow 226002, Uttar Pradesh,
India
vaibhavlasure25@gmail.com
Chapter 4

Mubashir H. Masoodi
Department of Pharmaceutical Sciences,
School of Applied Sciences and Technology,
University of Kashmir,
Hazratbal, Srinagar 190006, J & K,
India
mubashir@kashmiruniversity.ac.in
Chapter 9

Mayank
Department of Pharmacy,
Guru Kashi University, Talwandi Sabo,
Bathinda, Punjab,
India
mayank6103@gmail.com
Chapter 2,8

Ravi K Mittal
Galgotia College of Pharmacy,
Greater Noida,
Uttar Pradesh,
India
ravimittalniper@gmail.com
Chapter 5

Shivam Kumar Pandey
Department of Pharmacology and Toxicology,
National Institute of Pharmaceutical Education
and Research,
Raebareli, Transit Campus,
Bijnor-Sisendi Road, Sarojini Nagar,
Lucknow 226002, Uttar Pradesh,
India
shivamkumarpandey067@gmail.com
Chapter 4

Priyanka Verma
Faculty of Pharmacy,
Shri Ram Murti Smarak College of Engineering
and Technology,
Bareilly, Uttar Pradesh,
India
priyankapharma331993@gmail.com
Chapter 7

Priyank Purohit
Graphic Era Hill University,
Dehradun, Uttarakhand,
India
priyank.niper@gmail.com
prpurohit@gehu.co.in
Chapter 5

Naresh Kumar Rangra
Department of Pharmaceutical Chemistry and
Analysis,
ISF College of Pharmacy,
GT Road, NH-95, Ghal Kalan 142001,
Punjab,
India
nareshrangra@gmail.com
Chapter 11

Sandeep Rathor
MM College of Pharmacy,
Maharishi Markandeshwar (Deemed to Be University),
Mullana, Ambala 133207, Haryana,
India
rathor.s347@gmail.com
Chapter 8

Debajyoti Roy
Department of Pharmacognosy,
ISF College of Pharmacy,
GT Road, NH-95, Ghal Kalan 142001,
Punjab,
India
droy4896@gmail.com
Chapter 11

Nirjhar Saha
School of Chemical Sciences,
Indian Association for the Cultivation of Science (IACS),
Jadavpur, Kolkata 700032, West Bengal,
India
nirjharsaha91@gmail.com
Chapter 12

Kirti Sharma
Amity Institute of Pharmacy (AIP),
Amity University Haryana,
Amity Education Valley,
Panchgaon, Manesar 122413, Haryana,
India
and
GD Goenka University,
GD Goenka Educational City,
Sohna, Gurgaon 122103, Haryana,
India
sharmakirti260@gmail.com
Chapter 13

Himanshu Shrivasatava
Amity Institute of Pharmacy (AIP),
Amity University Haryana,
Panchgaon, Manesar 122413, Haryana,
India
himanshu26012000@gmail.com
Chapter 13

Jayant Sindhu
Department of Chemistry,
COBS&H,
CCS Haryana Agricultural University,
Hisar 125004, Haryana,
India
jayantchem@gmail.com
jayantchem1402@hau.ac.in
Chapter 1

Rakesh Kumar Singh
Department of Pharmacology and Toxicology,
National Institute of Pharmaceutical Education and Research,
Raebareli, Transit Campus,
Bijnor-Sisendi road, Sarojini Nagar,
Lucknow 226002, Uttar Pradesh,
India
rakesh.singh@niperraebareli.edu.in
Chapter 4

Firdoos Ahmad Sofi
Department of Pharmaceutical Sciences,
School of Applied Sciences and Technology,
University of Kashmir,
Hazratbal, Srinagar 190006, J & K,
India
firdous.phscpdf@kashmiruniversity.net
Chapter 9

Vinita Pandey
MIT College of Pharmacy,
MIT Campus Moradabad,
Moradabad 244001, Uttar Pradesh
India
vineetapandey33@gmail.com
Chapter 7

Kiran, Jayant Sindhu

1 Contemporary trends in drug repurposing: Identifying new targets for existing drugs

Abstract: Drug repositioning or drug repurposing has evolved as a successful, well-liked alternative strategy in contemporary drug discovery to find older medications for new targets economically and dynamically. The world's most difficult healthcare issue of the century, the Covid-19 pandemic, gives this idea a tremendous boost. In this method, researchers look for new applications and low-risk clinical uses of medications that have already been pharmacologically proven and licensed. With the rapid advancement of technology, numerous new drug-target diseases are discovered, and as a result, a wealth of information is now available in various databases, including Drug Bank, OMIM, ChemBank, KEGG, Pubmed and Genecard. The idea of medication repurposing along with its effects on companies, academia and society has been discussed in this chapter. Moreover, the chapter covers the methods and strategies used in drug repositioning for identifying drugs for different diseases and disorders. The current chapter also takes into account the most recent market research and updates on the costs of drug repurposing-based drug discovery program, its comparability to conventional drug discovery methodologies, difficulties associated with drug repurposing and future views.

Keywords: Drug repositioning, Drug repurposing, Case studies, Polypharmacology, In silico approaches, Pharmacological approaches

1.1 Introduction

Therapeutic switching is an alternative and contemporary drug discovery method that determines the function of earlier drugs for newer targets and offers robust and less expensive drug development [1]. The first instance of drug repurposing goes back to the 1920s when unexpected events resulted into drug development and history is rich with such tales. Therefore, if a medicine was discovered to have an effect that was either newly identified or off-target, it was typically carried forward to capitalize on it [2]. According to Ashburn and Thor, repurposing is the process of finding new targets for the available drugs, many of which are generic and may or may not be in the public domain [3]. This definition has evolved greatly over time and now includes active ingredients from medications that failed at the clinical level as a result of their extreme toxicity and unintended side effects. The substances that fall under the heading of "selective optimization of side activities" (SOSA) are all excluded by this broadened definition. In SOSA approach, failed or withdrawn drugs are chemically modified with the help of chemical modification strategy to be repositioned for a new target [4].

https://doi.org/10.1515/9783110791150-001

According to one of the USFDA studies, fewer new drugs have been approved by the agency since 1995. The reports claims that out of total approved vaccines and drugs, 30% are from therapeutic switching [5]. Blockbuster drugs like sildenafil, thalidomide, minoxidil, aspirin and methotrexate were established by drug repurposing strategy. Pfizer repurposes sildenafil, a hypertension-related medicine for the treatment of erectile dysfunction-related ailments. It has been sold as Viagra and had impressive global sales of almost US$2.05 billion. Another example is thalidomide, which was first sold in 1957 to treat nausea in pregnant women. Due to its teratogenic impact, it was discontinued from the market as its associated side effects related to bone deformities in newborns. Thalidomide, however, was later effectively repurposed for the treatment of multiple myeloma and erythema nodosum leprosum (ENL) [6]. Drug repurposing often relies on the fact that all medications approved or rejected by regulatory authorities have already completed comprehensive testing and safety evaluations. As a result, their repositioning is simple and does not cost as much as it would to do the exhaustive studies [7].

In contrast, traditionally, the presence of "*polypharmacology*" or the biological activity of a drug on several receptors, was considered to be negative when looking for a promiscuous drug candidate. However, if the disease pathophysiology is extremely complicated, polypharmacology currently offers larger scope for identifying off-targets [8]. The field of medication repurposing has become even more invaded as a result of integration of system biology into polypharmacology [6]. Since the method is less dangerous, cost-effective and time-saving, repurposing medications is often considered as attractive. Drug repurposing is an important and popular topic in contemporary drug technology since it quickly expands the economy. In traditional drug discovery, there are normally five lengthy steps: (i) Lead discovery followed by preclinical research; (ii) Safety review; (iii) Clinical trials; (iv) Review by the USFDA and (v) Post-market surveillance and long-term safety assessment. However, compared to traditional drug discovery, drug repurposing requires only four steps.

The major steps include: (i) Compound identification from the approved leads; (ii) Acquisition of lead compounds; (iii) Validation of new target and (iv) FDA approval and market surveillance for safety parameters. Traditional de novo drug development requires a lot of time, money and risk from drug identification to drug commercialization, which makes it less appealing to investors of pharmaceutical sector. Drug launching using drug repositioning requires less time and money than de novo drug development approach. In this instance, the "in vitro" and "in vivo" screening, validation and efficacy investigations involve less time and money and do not require phase 1 clinical trials, which increases investor interests.

Repositioned compounds have undergone numerous safety and pharmacokinetics investigations for medication development with 100% accuracy and zero error. It has been revealed that repositioning medications results in cost savings and market earnings of about 40% [9, 10]. Repurposing of drugs requires three to four years to complete significant clinical trials and an investment of roughly US$1.65 billion, in comparison to the typical requirement of US$12 billion for original drug research [11]. Drug repurposing is being used to treat orphan, rare or neglected diseases, where funding is extremely limited, as well as pandemics, which necessitate the quick finding of potential treatments [12]. Despite a lot of benefits associated with drug repurposing, it also faces a lot of challenges. The crucial one entails intellectual property and data accessibility [13]. To safeguard the repurposed category medicine in the pharmaceutical market, a patent application may be filed. However, patenting repurposed medicine may be subject to legal restrictions in some countries where the original drug was first patented. Therefore, the data is protected for at least eight years in the European Union and five years in the United States, taking into account the original patent (obtained for first use). The grant is further extended by one year in the European Union and by three years in the United States if the second use is discovered during these years (8 or 5 years). This brief period (1 and 3 years) offers a very limited window of opportunity to generate considerable returns in lieu of the initial investment. As a result, the pharmaceutical industry frequently slows down the acceleration of such breakthroughs. A further obstacle to drug repositioning is the availability of data and compounds. The goal of drug repurposing may be hampered by the inaccessibility of secondary clinical trial data, disclosure of failed pharmaceuticals and selective pharmaceutical drug development company relationships. Additionally, the loss of generic active pharmaceutical ingredients (API) from the global market could seriously impede scientist's access to that chemical from reputable vendors, which would have a negative impact on their capacity to conduct repurposing research [13–15]. Nevertheless, despite these negative aspects, there has been a marked increase in the efforts to locate previously used substances for new uses. The effective use of this innovation may be attributed to the pandemic Covid-19, where initial reorganization of the pharmaceutical candidates was carried out. Few medications that were rearranged also got USFDA emergency approval; however, these approvals were quickly cancelled due to toxicity in Covid-19 patients. The main drugs that repurposed were hydroxychloroquine, tocilizumab, remdesivir and dexamethasone [16]. Recently, the Drugs Controller General of India approved the use of drug 2-DG medicine (2-Deoxy-D-Glucose) as an adjuvant therapy for the treatment of Covid-19. Dr. Reddy's Laboratory and DRDO India collaborated to create 2-DG as a joint venture. The therapeutic effects of 2-DG against cancer, cardiovascular disease and Alzheimer's disease are well established and promising. However, due to the host cell's high energy need (glucose) for viral growth in the form of glucose, 2-DG selectively accumulates in the infected cells (host) for its effects against SARS-CoV-2. By acting as a fictitious substrate for glucose, 2-DG stops the creation of energy and interferes with the glycosylation of viral glycoprotein. By limiting the glycolysis process, this stops the spread of the virus and thus lowers the pathogenicity within the host cell [17, 18].

Further, the popularity of drug repurposing may be analyzed from the data retrieved from the Scopus database on February 12, 2023. The data was searched using two keywords, "Drug" and "Repurposing," providing 9,069 documents. The thorough analysis revealed that the first paper on drug repurposing came in 1965, with the significant boom coming after 2015. The year 2020 saw 1,502 publications, followed by

2,236 publications in 2021, 2,080 in 2022 and 201 publications in 2023 (till February 12, 2023). The publication types were mainly confined to research articles (5,956), reviews (2,194), Editorial (128), notes (121), Conference papers (180) and book chapters (283). Considering the country-wise research scenario, drug repurposing is focused primarily in the United States (2,898 publications), followed by India (1,304 publications). Moreover, the Scopus database also revealed a total of 2,744 patent results with major accounting in the United States Patent & Trademark Office (2,383); European Patent Office (157); Japan Patent Office (109); World Intellectual Property Organization (82) and United Kingdom Intellectual Property Office (13).

The goal of this chapter is to clarify the function of medication repurposing from the viewpoint of the academic and industrial scientific communities. In addition, information on methods and strategies to find medications for repositioning and their use in identifying drugs for different diseases and disorders is covered in this chapter. The current review also anticipates recent market analysis updates on the costs of drug research and development that entail drug repurposing compared to traditional drug discovery, difficulties associated with drug repurposing and future views.

1.2 Tools and techniques for drug repurposing

In general, there are two efficient methods for drug repositioning. The first is "*on target*" medication repositioning, in which an existing pharmacological mechanism is implemented to the repurposed medicine and the second is "*off target*" drug repurposing, in which the same drug-receptor is targeted but for other disorders [19]. Minoxidil, which was initially used for its vasodilator effect and antihypertensive qualities, is an illustration of "*on target*" therapeutic repurposing. However, when it was applied to hair, it also made it easier for blood vessels to expand and supply the follicles, so later on, its use in hair development became apparent. Aspirin is a prime example of a medicine that has been repurposed *off-target*. Aspirin was first used to treat inflammation, but now we know that it has anticoagulant properties. Aspirin has a different method for treating inflammation and avoiding blood clots despite sharing a comparable drug-receptor interaction (inhibition of platelet aggregation). The two primary approaches for medication repurposing are: (i) Activity- or profile-based approaches & (ii) In silico approaches, which are further divided into two sub-integrated approaches, namely network-based approaches and data-based approaches [2]. In the activity or profile-based repurposing procedure, experimental screening is carried out via "in vivo" and "in vitro" assays. Though time-consuming, this strategy often has a lesser likelihood of discovering false hits (Figure 1.1). The second strategy uses a computational method that primarily uses target protein structure data for high-throughput screening of possible hits. Despite being rapid, this strategy typically results in a high percentage of false hits during screening [20].

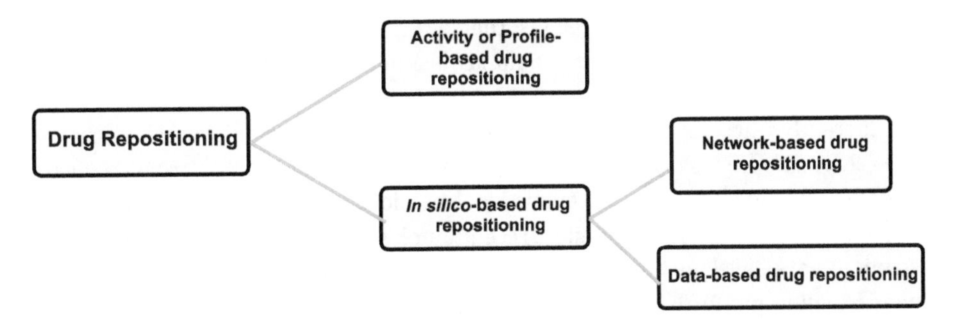

Figure 1.1: Various approaches to achieve drug repurposing.

The major challenge in repurposing includes finding the off-patent compounds from the authorized and failed medicine categories. According to sources, the DrugBank database has been expanded to include 7,800 drug candidates, out of which 2,254 have received regulatory approval. Nearly 80% of drug repositioning uses "*on target*" strategy. The success of the Human Genome Project, which provides over 6,000 druggable targets to support the repurposing endeavors, has further increased the repositioning ventures [9]. Several in silico drug repurposing methodologies are currently available and they may be further supported by activity- or profile-based approaches to hasten the repositioning of therapeutic candidates. The critical strategies might not be confined to network-based, semantics-based or data-based drug repositioning.

Network-based methods build drug-disease networks that incorporate data from many different sources, including gene expression, protein–protein interactions and many more. Interactions are directly or indirectly extracted by computational algorithms from data in existing databases. There are two different kinds of network-based methods used to investigate core disease-associated genes that are already known. These are two network-based approaches: one for clustering and one for dissemination. In order to find modules that could be utilized to find new drug-disease or drug-target associations, network-based clustering algorithms are used to identify biological entities from drug-disease, drug-drug or drug-target links in a comparable module of networks. The Hedgehog signaling pathway is inhibited by the medication Vismodegib, which was developed to treat Gorlin syndrome. On the other hand, Iloperidone, a typical antipsychotic used to treat schizophrenia, was found to be a unique potential treatment for hypertension.

Another approach was network-based propagation, popularly used to find prospective candidate drugs. Using a random walk propagation algorithm linked to a specific disease, the workflow of these approaches was extended to the genes sharing the neighbors in a protein–protein or gene–gene interaction network. Based on how they propagate, these methods can be divided into two groups: global and local perspectives. According to this method, the medications methotrexate, gabapentin, donepezil, risperidone and cisplatin now have a new indication for treating Parkinson's disease,

Crohn's disease and anxiety-related disorders [21, 22]. Semantics-based techniques are frequently used in image retrieval, data acquisition and other related field of drug development.

Currently, these approaches used for the drug repurposing process consist of three steps. The first is to create a semantic network by using the data from the medical database to identify the associations between biological entities. After that, semantic networks are built using the most recent ontology networks. Additionally, mining techniques are developed to anticipate future semantic network interactions. According to a notion put forth by Guillermo et al., similar medications are associated with similar targets and outcomes and greatly aid in drug-target connections. The results demonstrate that the suggested strategy correctly predicted the interactions between drugs and their targets [23]. Using medications, targets, proteins and diseases as the basis, Chen et al., created a semantic network with 2,90,000 nodes and 7,20,000 edges [24].

Further, there are numerous difficult issues associated with the extraction of useful and novel biological entity association from the body of existing research. Text mining (TM) techniques have been widely used to meet this challenge. This method helps extract new data from current literature and reveal the connections between biological concepts or entities. The four steps in the biological TM pipeline are: IR (information retrieval), BNER (biological name entity recognition), BIE (biological information extraction) and BKD (biological knowledge discovery).

In the first phase, related documents are retrieved from the literature. These related documents must then be filtered because they contain some unrelated concepts. Controlled vocabularies are then used to identify key biological concepts in the BENR stage. In two more phases known as the BIE and BKD phases, pertinent data is extracted in order to discover and expand knowledge about the biological idea. Additionally, it was shown that there may be a correlation between the knowledge of drug-disease and drug-target relationships. Table 1.1 compiles the various databases needed for an "in silico" method.

Table 1.1: Databases employed for drug repurposing.

Data	Database use
Drug chemical structure and activities	PubChem, CheEMBL, STITCH
Structure of proteins	PDB, ProBis, PLIP
Clinical data	repoDB, repurposeDB
Structure of protein, transcriptional profile, and drug's activities	Open Targets, Drug Repurposing Hub, Drug Target Commons
Drug-induced transcriptional response	Connectivity Map, LINCS

1.3 Case studies of blockbuster drugs identified using drug repurposing

As discussed, drug repositioning is mostly used to find treatments for orphan, rare or neglected diseases due to shortage of funding. Additionally, this strategy makes sense in pandemics where it is important to find potential treatments quickly. According to estimates, 90% of rare diseases are incurable, making drug repositioning extremely important in these situations. The best examples of drug repositioning and drug discovery have been accidental discoveries [25–27]. Here, some significant blockbuster repurposed medications have been discussed.

1.3.1 Aspirin

One of the pharmaceuticals that is most commonly used worldwide is aspirin. Aspirin, one of the oldest known medications, was originally introduced to the market by Bayer in 1899 as an analgesic. In modest doses, the same medication was used as an antiplatelet aggregation treatment [28]. In 1971, Vane and colleagues made the landmark discovery that aspirin prevented the formation of prostaglandins by inhibiting cyclooxygenase (COX). Dr. Lawrence first noticed that aspirin, which was initially prescribed as an analgesic for patients undergoing tonsillectomy, had increased bleeding as a side effect. Further research revealed that aspirin prevents platelet aggregation by permanently inhibiting the COX-1 enzyme, which is necessary for the synthesis of the prostaglandin analogue thromboxane [29, 30]. Aspirin primarily binds to COX-1 at low dosages, i.e., less than 300 mg/day and inhibits COX-2 at higher levels. This COX-1 enzyme also contributes in the production of prostaglandins, which protect the gastrointestinal system by cytoprotection [31]. Additionally, it is anticipated that aspirin may soon be used for its effects in oncology to treat a variety of malignant illnesses. Aspirin slows the growth of colorectal cancer and numerous other cancer types also showed similar results [32, 33].

1.3.2 Thalidomide

Thalidomide, an antiemetic medicine used in the early stages of pregnancy, was banned by the World Health Organization (WHO) in 1962 because it degraded the transcription factor SALL4, which was linked to the development. Additionally, thalidomide prevents the production of pro-inflammatory cytokines, a key indicator of leprosy complications. In 1998, thalidomide was repurposed as an orphan medication for the leprosy problem [34]. Only rare disorders allows the approval of a medicine with teratogenic harm. Similar to aspirin, thalidomide also has an antiangiogenic action, a particular characteristic of cancer treatment medications. With the same precautions, i.e., not to administer during pregnancy, thalidomide is now used as a first-line treatment for multiple myeloma because of its antiangiogenic, immunomodulatory, antiproliferative and proapoptotic actions [35].

1.3.3 Sildenafil

The first medication to be repurposed before it was released into the market, Sildenafil, is the most successful drug repurposing product in the world. It was initially created by Pfizer in 1990 for cardiovascular conditions linked to hypertension and angina [36]. It caused vasodilation and prevented platelet aggregation by blocking PDE-5, an enzyme from the phosphodiesterase family that can break down cGMP and cause vasoconstriction. Clinical investigations carried out in the United Kingdom have revealed a surprising adverse effect in the form of penile erection. This adverse reaction was the result of vasodilation, which breaks down cGMP and ultimately leads to increases in production of nitric oxide (NO) [37]. In order to cure erectile dysfunction, Pfizer introduced sildenafil under the brand name "Viagra" in 1998 [38]. Pfizer is still working on PDE-5 inhibitors and taking advantage of its vasodilator effects. They discovered that a dose equivalent to one-fifth of the dosage used to treat erectile dysfunction, can treat pulmonary arterial hypertension [39]. Pfizer received USFDA permission for second sildenafil "Revatio" in 2005 for pulmonary hypertension, a rare illness [40]. It is possible that sildenafil will soon be repositioned for its effects against cancer. There is currently sufficient evidence to support its therapeutic benefit in the treatment of cancer, whether used alone or in combination with other anticancer medications. The main mechanism for sildenafil's anticancer activity appears to be its nitric oxide (NO)/PDE-5 dependent proapoptotic effect, which is accompanied by autophagy activation and modification of the immune system's response to tumors [41].

1.3.4 Amantadine

It was initially used to treat Influenza A virus due to its antiviral action against avian influenza. Later, it was shown that it can lessen akinesia, extrapyramidal disorders and Parkinson's disease. It was authorized in 1969 [42].

1.3.5 Bupropion

Another instance of repurposing a medicine is bupropion. It was initially used as antidepressants. Lately, it has been employed as a smoking treatment since it aids participants in quitting smoking in clinical trials. The study discovered that bupropion aids in changing the brain's chemical signals, lessening the impact of nicotine withdrawal. In 1997, it got authorization [43].

1.3.6 Finasteride

It aids in shrinking an enlarged prostate. Finasteride is sold under the brand name Proscal. It was discovered to aid in the treatment of male pattern baldness later in 1997. It aids in preventing testosterone from being converted to dihydrotestosterone [44]. The complete list of medications that have been repurposed, together with their original uses, is presented in Table 1.2 [7]. The relevant chemical structures have been represented in Figure 1.2.

Table 1.2: List of reported repurposed drugs along with their original use of indication.

Drug	Brand Name	Originally discovered for	Repurposed use	Chemical structure No.
Aspirin	Aspir 81	Inflammation, analgesia	Antiplatelet agent and colorectal cancer	S1
Arsenic trioxide	Trisenox	Syphilis	Cancer (Leukemia)	S2
Amphotericin	Fungizone	Fungal infections caused by fungi	Anti-platelet	S3
Amantadine	Gocovri, Symadine,	Influenza flu	Parkinson's disorder	S4
Allopurinol	Zyloprim	Cancer	Gout	S5
Bromocriptine	Parlodel	Parkinson disorder	Diabetes mellitus	S7
Bleomycin	Bleocip	Cancer	Pleural effusion (lungs edema)	S8
Bimatoprost	Lumigan	Glaucoma	Promoting eyelash growth	S9
Azathoprine	Azasan and Imuran	Rheumatoid arthritis	Kidney transplant	S10
Atomoxetine	Strattera	Parkinson disorder	Attention deficit hyperactivity disorder	S6
Bupropion	Aplenzin	Depression	Smoking cessation	S11
Celecoxib	Celebrex	Inflammation and pain	Adenomatous polyposis	S12
Colchicine	Colcrys	Gout	Recurrent pericarditis	S13
Colesevelam	Welchol	Hyperlipidemia	Diabetes mellitus	S14

Table 1.2 (continued)

Drug	Brand Name	Originally discovered for	Repurposed use	Chemical structure No.
Cycloserine	Seromycin	Urinary tract infection	Tuberculosis	S15
Cyclosporine	Gengraf	Rheumatoid arthritis	Transplant rejection	S16
Dapoxetine	Priligy	Analgesia and depression	Premature ejaculation	S17
Dapsone	Aczone	Leprosy	Malaria	S18
Disulfiram	Antabuse	Alcoholism	Melanoma	S19
Doxepin	Silenor	Depressive disorder	Antipruritic	S20
Duloxetine	Cymbatla	Depression	Stress urinary incontinence	S21
Eflornithine	Vaniqa	Depression	Attention deficit hyperactivity disorder	S22
Everolimus	Afinitor	Renal Cancer	Renal transplant	S23
Fingolimod	Gilenya	Organ transplant rejection	Multiple sclerosis	S24
Gabapentin	Gabarone	Seizures	Neuropathic pain	S25
Gemcitabine	Infugem	Antiviral	Cancer	S26
Histrelin	Vantas	Prostate cancer	Precocious puberty	S27
Hydroxychloroquine	Plaquenil	Malaria	Covid-19	S28
Ketoconazole	Nizoral	Fungal infections	Cushing syndrome	S29
Lomitapide	Juxtapid	Hyperlipidemia	Familial hypercholesterolemia	S30
Methotrexate	Trexall	Cancer treatment	Rheumatoid arthritis and Psoriasis	S31
Miltefosine	Impavido	Cancer treatment	Visceral leishmaniosis	S32
Minoxidil	Loniten	Hypertension	Alopecia	S33
Naltrexone	Vivitrol	Opioid addiction	Alcohol withdrawal	S34
Naproxen	Anaprox	Inflammation, pain	Alzheimer disorder	S35
Nortriptyline	Pamelor	Depression	Neuropathic pain	S36
Pemetrexed	Alimta	Mesothelioma	Lung cancer treatment	S37
Propranolol	Inderal	Hypertension	Migraine control and treatment	S38

Table 1.2 (continued)

Drug	Brand Name	Originally discovered for	Repurposed use	Chemical structure No.
Raloxifene	Evista	Osteoporosis, contraceptive	Breast cancer	S39
Retinoic acid	Aberela	Acne	Acute promyelocytic leukemia	S40
Zileuton	Zyflo	Asthma	Acne	S41
Zidovudine	Retrovir	Cancer	HIV/AIDS	S42
Topiramate	Topamax	Epilepsy	Obesity	S43
Thalidomide	Thalomid	Morning sickness	Leprosy, multiple myeloma	S44
Sildenafil	Viagra	Angina	Erectile dysfunction, pulmonary hypertension	S45
Ropinirole	Requip	Parkinson's disease	Restless leg syndrome	S46

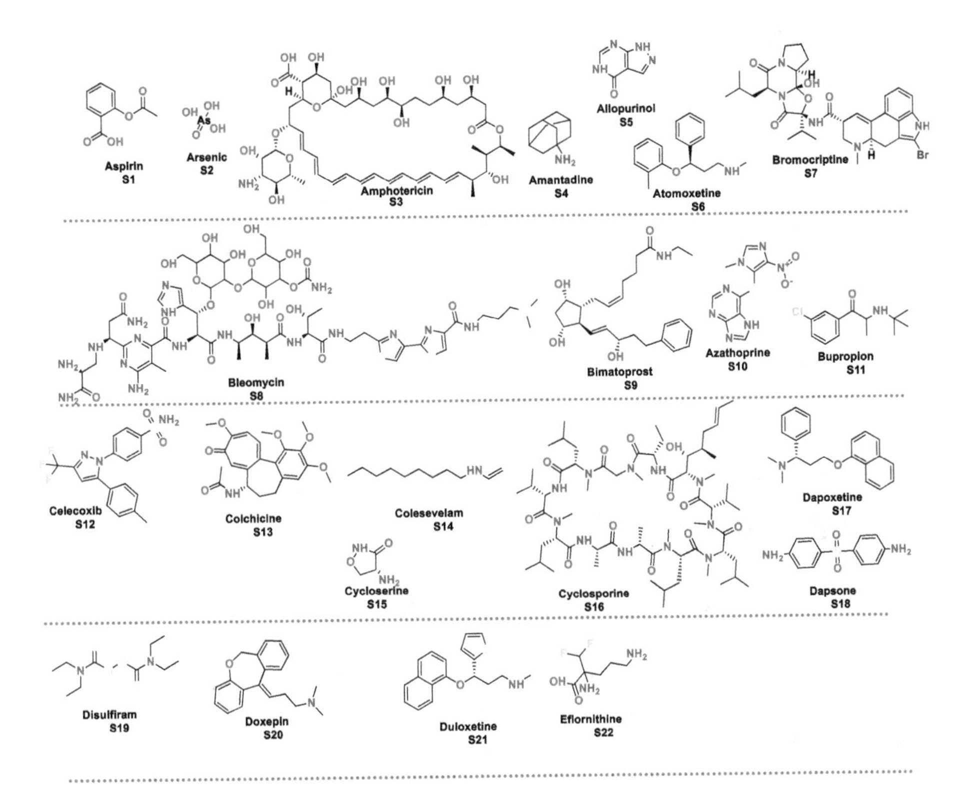

Figure 1.2: Chemical structures of reported repurposed drugs.

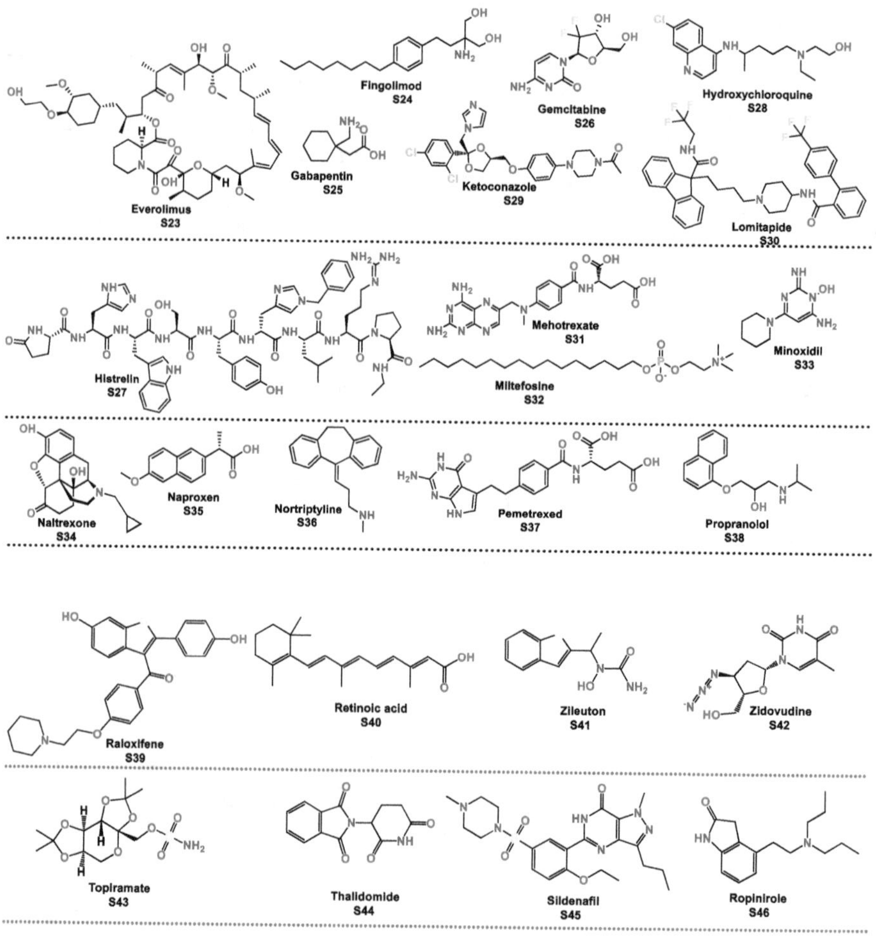

Everolimus S23
Fingolimod S24
Gabapentin S25
Gemcitabine S26
Histrelin S27
Hydroxychloroquine S28
Ketoconazole S29
Lomitapide S30
Mehotrexate S31
Miltefosine S32
Minoxidil S33
Naltrexone S34
Naproxen S35
Nortriptyline S36
Pemetrexed S37
Propranolol S38
Raloxifene S39
Retinoic acid S40
Zileuton S41
Zidovudine S42
Topiramate S43
Thalidomide S44
Sildenafil S45
Ropinirole S46

Figure 1.2 (continued)

1.4 Drug repurposing from an academic and industrial view

The pursuit of translational research in academia involves adjusting the processes and objectives of drug discovery and integrating traditional academic practices (Figure 1.3) with the industry. Drug discovery landscape and ambit are rapidly evolving [27]. For a successful medication research and development process, the pharmaceutical sector and academia should collaborate. The primary instance of both the crucial pillars of drug research and development coming together favorably is drug repurposing [2, 3]. In the past ten years, it has been noted that the hard-core academic institutions have taken over drug research initiatives from major pharmaceutical market players. It is imperative to have an effective strategy in terms of drug repurposing for battling the rising number of neglected and orphan diseases, orphan disease targets and communicable diseases [3]. The drug repurposing technique has taken center stage as a result of various flaws in the de novo drug development process. De novo drug discovery faces several significant challenges, including the chemical space being already saturated, the pharmaceutical industry's innovation gap, the higher attrition rate of newer scaffolds, the enormous costs associated with the development of new drugs, the slow pace of drug discovery and growing environmental concerns [45]. Up to 25% of yearly pharmaceutical revenue is generated by repurposed pharmaceuticals and more than one-third of recently authorized drugs are truly repurposed drugs [13]. Programs to promote drug repurposing have been supported by various government and nonprofitable groups. The National Center for Advancing Translational Sciences at the National Institutes of Health and the Bill & Melinda Gates foundation are two notable donors to this initiative. Drug repurposing approach has been incorporated by pharmaceutical companies into the life cycle management (LCM) of pharmaceutical goods [46].

Figure 1.3: Interplay of academia and industry in drug repurposing strategy.

With the growing concern for environmental degradation, greener and sustainable development approaches are largely being encouraged across the various sectors. Pharmaceutical industries are now taking up and adopting the novel and greener approaches developed by academic setup for drug discovery and development [47, 48]. Drug repurposing strategy is relatively greener as compared to the de novo drug discovery programs because it does not require generation of a library of new chemical entity (NCE), which in turn would require lots of chemical reagents and solvents.

1.5 Integration of pharmaceutical industry and clinical studies

The availability of well-established safety data in preclinical models as well as recognized pharmacokinetic and manufacturing data leads to a decrease in the time frame for drug discovery and development. Therefore, unless a serious drug-disease interaction is discovered, these drug candidates are frequently less likely to be unsuccessful on account of safety concerns [49, 50]. Based on the previously standardized pharmacology and their off-level side effects, a multitude of gangbuster repurposed medications, such as valproic acid, protease inhibitors, aspirin, hydroxy chloroquine and propranolol have emerged [51, 52]. Drug repurposing has evolved into a systemic, robust, data-driven and reliable practice with the introduction of computer-aided drug design, in silico investigations and improved proteomic data [53]. The enormous increase in publications over the years provides insight into the significance of pharmacological repurposing [19]. Figure 1.4 represents the benefits of drug repositioning methodology. The academic community's increased emphasis on interdisciplinary and translational research which benefitted the drug discovery process. A productive partnership between research facilities, small drug discovery start-ups and academic institutions could have a big impact on the drug discovery process. As a result, emphatic drug screening libraries would be produced, which might speed up the process of finding candidates for repurposing.

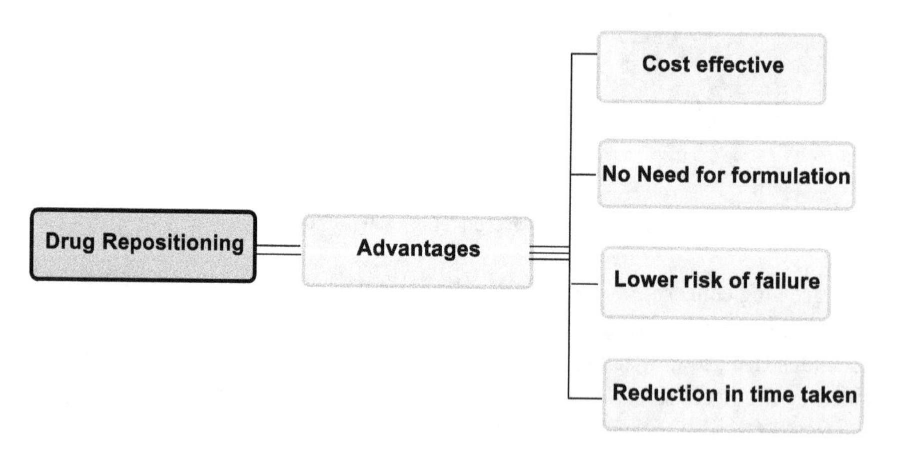

Figure 1.4: Advantages of drug repurposing.

Drug repurposing requires information from a range of sources and involves multiple systematic steps. It includes "in silico" screening, literature mining and phenotype screening and needs pre-clinical, clinical and genomics data [54]. One of the most important steps in this process is choosing an appropriate strategy for the drug's repurposing. Any of the three scenarios – new target but old therapeutic indication, new target and new therapeutic indication and old target but new therapeutic indication – could serve as the basis for the repurposing [20]. After appropriately completing these steps, a candidate for drug repurposing is obtained. The pharmaceutical sector will then get ready to produce the medication in huge quantities under urgent circumstances. This shortens the amount of money and time needed to initiate a de novo drug discovery [15]. Recently, the approach of using an old target but a novel therapeutic indication was used for the treatment of SARS-CoV-2. RdRP inhibitors and protease inhibitors were two kinds of medicines that were thoroughly researched for COVID-19 therapy. Viral protease inhibitors function by preventing downstream proteolytic processes and protein translation. Numerous protease inhibitors, including indinavir, lopinavir, sequinavir and ritonavir, were investigated as potential repurposed treatments for SARS-CoV-2 [16]. Various protease inhibitors that were used to treat COVID-19 are shown in Figure 1.5.

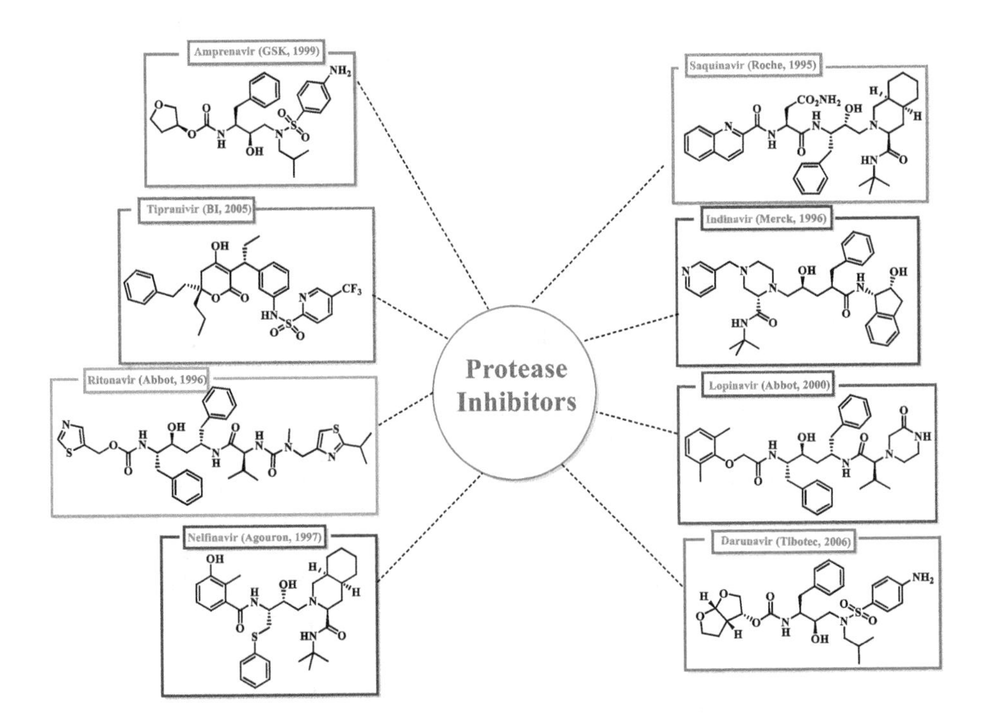

Figure 1.5: Various protease inhibitors repurposed for SARS-CoV-2.

A replication complex that includes the RNA-dependent RNA polymerase (RdRp) mediates the replication of the viral genome. One such RdRP inhibitor developed by Gilead Sciences in 2015 for HCV and also under consideration for the treatment of the Ebola outbreak is remdesivir. Remdesivir is a phosphoramidate prodrug that is transformed into remdesivir triphosphate, an active NTP (nucleoside triphosphate) analogue. Figure 1.6 shows the mechanism of remdesivir [16, 55].

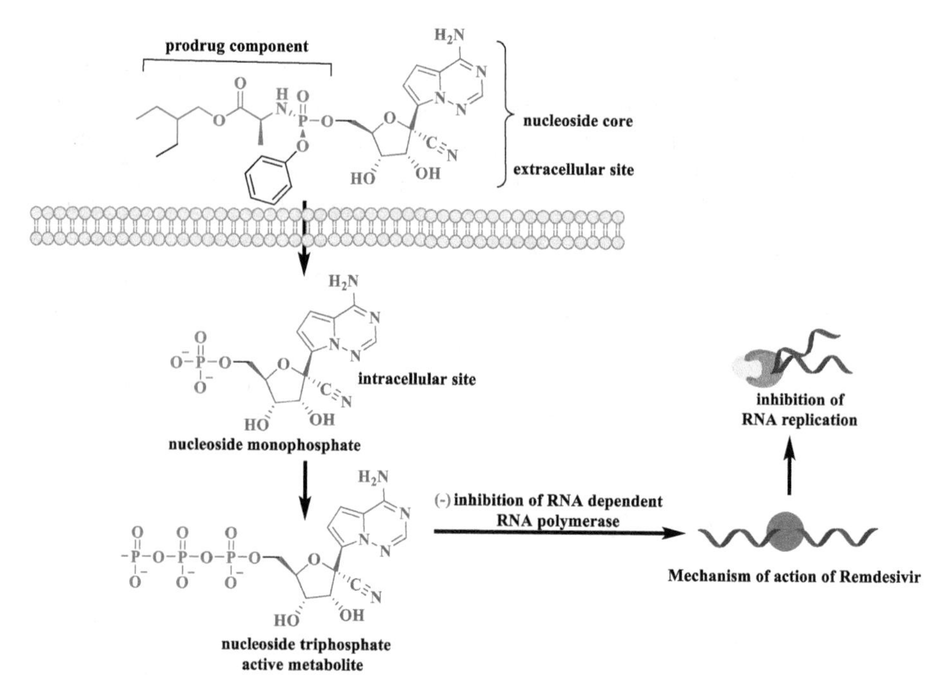

Figure 1.6: Mechanism of action of Ramdesivir.

A similar scenario of the medication tocilizumab being repurposed for the same target but new indications was investigated [56]. Hoffman-La-Roche and Chugai Pharmaceutical developed the immunosuppressive monoclonal antibody (mAb), tocilizumab, for the treatment of cytokine release syndrome in rheumatoid arthritis in 2003. However, in 2010, the USFDA gave it the green light for the treatment of moderately to severely active rheumatoid arthritis. It was the first biological therapy that was FDA-approved for the treatment of interstitial lung disease linked to systemic sclerosis [57]. It works by blocking the IL-6 receptor. In SARS-CoV-2 infected individuals, it was discovered to inhibit both free and membrane-bound IL-6 receptors produced by the pro-inflammatory cascade pathway [58].

The third scenario of drug repurposing entails creating a brand-new molecular or pharmacological target for an existing medication. Drugs with well-known mechanisms of action, such as minoxidil, thalidomide, tamoxifen, celebrex, pentostatin, dapsone, rapamycin and sildenafil, have been successfully repurposed for fresh indications with unique molecular targets [20, 36, 49].

One such medicine that has been repurposed for SARS-CoV-2 is hydroxychloroquine (HCQ). Different modes of action against COVID-19 infection have been proposed for HCQ, including lowering immune cell activity in infected individuals, preventing SARS-CoV-2 viral entry into human cells and preventing SARS-CoV-2 virus fusion [59].

Another instance is the development of 2-DG, which functions as an antimetabolite for natural glucose and shuts off the energy source in host cells, affecting the spread of viruses [17, 18].

1.6 Intellectual coverage and knowledge transfer

Medication repurposing is a relatively new topic for academics and industry; the greatest disadvantage is the dearth of legal expertise in this area. For off-patent pharmaceuticals, there are a few intellectual property protection mechanisms [13]. Due to the fact that the medicine is being used for a new indication, it has already been successfully developed for a number of other indications and is thoroughly documented in the specialized literature. For off-patent medications, a patent for the new indication may be secured but may be contested because it makes use of strength and dose form that are already widely accepted. The ideal situation was one in which the medicine being repurposed had a fresh molecular mechanism for a newer indication in addition to having a distinctive formulation. However, it should be remembered that derivatizing a well-known medication resulted in the production of a new molecule; therefore, it would not be regarded as a repurposing tactic. Lack of coordination between multiple parties involved in the supply chain of a medicine is another factor related to the challenge of applying the patents for a repurposed drug (second indication) [60–62].

1.7 Regulatory process involved

Various regulatory agencies have their own laid down procedures and established methods for granting market exclusivity right to an NCE. European Union (EU) provides a data protection time period of eight years during which no other generic company would be able to make use of the originator's data. EU also provides two years of market exclusivity to the originator. However, if the originator devised a new indication for the same drug during the eight year's period of exclusivity, one more year of protection would be given. It is evident that during the two years of market exclusivity to the originator, any generic company would not be able to market a product based on the data it collected. On the other hand, the United States grants five years of market exclusivity to an NCE, which is extendable by three more years in case of finding a new indication. This extended period of exclusivity in case of nongeneric drug constitutes an appropriate time to get an acceptable return of investment [63].

1.8 Comparing the cost involved in drug repurposing versus traditional drug development

Modern medicinal chemist's tool box is well-versed with techniques like bioinformatics and cheminformatics, in silico research, artificial intelligence, the availability of biological and structural databases and structure-based drug design, which has made drug repurposing strategy more feasible than ever. Comparing the drug repurposing technique to de novo drug discovery, there are many advantages. The cost involved is one of the main benefits. Table 1.3 contrasts a few factors between drug repurposing techniques and conventional drug development. According to recent assessments, developing a new medicine will cost somewhere around 2.6–2.8 billion US dollars [64]. On the other hand, drug repurposing only needs 1–1.6 billion US dollars [65].

Table 1.3: Comparison of traditional drug discovery and drug repurposing strategy.

S. no.	Parameters	De novo drug discovery	Drug repurposing strategy
1.	Cost involved	To the tune of 2.8 billion USD	1.6 billion USD
2.	Time	Up to 12 years	3 years
3.	Duration of clinical trial	6 years	1–2 years
4.	Success rate	1 out of 10,000 (0.01%)	3 out of 10 (30%)
5.	Regulatory requirements	Stringent	Somewhat relaxed

1.9 Challenges and future perspectives

Despite being a promising technique, the medication repurposing approach has its own drawbacks and challenges. It is imperative to streamline drug repurposing and repositioning strategy in light of the expanding epidemics around the world, the high costs associated with conventional drug development and the identification of orphan diseases. The lack of coordination between basic and clinical sciences, which is frequently referred to as the "valley of death" in the drug discovery community, needs to be addressed. Therefore, a more coordinated strategy between the numerous parties involved in drug discovery and regulatory bodies is required to prevent this. Governmental incentives and business backing for academia might have a huge impact here. The drug repurposing and repositioning process would be accelerated by a more robust technology transfer framework, better protection of intellectual property rights, a collaborative industry–academia strategy and improved regulatory elements.

References

[1] Pushpakom S, et al. Drug repurposing: Progress, challenges and recommendations. Nat Rev Drug Discov. 2018;18:41–58.
[2] Xue H, Li J, Xie H, Wang Y. Review of drug repositioning approaches and resources. Int J Biol Sci. 2018;14:1232.
[3] Ashburn TT, Thor KB. Drug repositioning: Identifying and developing new uses for existing drugs. Nat Rev Drug Discov. 2004;3:673–683.
[4] Wermuth CG. Selective optimization of side activities: The SOSA approach. Drug Discov Today. 2006;11:160–164.
[5] Aggarwal S, Verma SS, Aggarwal S, Gupta SC. Drug repurposing for breast cancer therapy: Old weapon for new battle. In Seminars in Cancer Biology, Vol. 68. Elsevier, 2021, 8–20.
[6] Jourdan J-P, Bureau R, Rochais C, Dallemagne P. Drug repositioning: A brief overview. J Pharm Pharmacol. 2020;72:1145–1151.
[7] Ko Y. Computational drug repositioning: Current progress and challenges. Appl Sci. 2020;10:5076.
[8] Liu X, et al. Predicting targeted polypharmacology for drug repositioning and multi-target drug discovery. Curr Med Chem. 2013;20:1646–1661.
[9] Placchi M, Phillips R. The Benefits and pitfalls of repurposing drugs. 2018.
[10] Elder D, Tindall S. The many advantages of repurposing existing drugs. Eur Pharm Rev. 2020;25:34–37.
[11] Ojezele MO, Mordi J, Adedapo EA. Drug repurposing: Cost effectiveness and impact on emerging and neglected diseases. J Cameroon Acad Sci. 2020;16:3–17.
[12] Muthyala R. Orphan/rare drug discovery through drug repositioning. Drug Discov Today Ther Strateg. 2011;8:71–76.
[13] Talevi A, Bellera CL. Challenges and opportunities with drug repurposing: Finding strategies to find alternative uses of therapeutics. Expert Opin Drug Discov. 2020;15:397–401.
[14] Agrawal P. Advantages and challenges in drug re-profiling. J Pharmacovigil S. 2015;2(2).
[15] Cha Y, et al. Drug repurposing from the perspective of pharmaceutical companies. Br J Pharmacol. 2018;175:168–180.
[16] Poduri R, Joshi G, Jagadeesh G. Drugs targeting various stages of the SARS-CoV-2 life cycle: Exploring promising drugs for the treatment of Covid-19. Cell Signal. 2020;74(109721).
[17] Mesri EA, Lampidis TJ. 2-Deoxy-D-glucose exploits increased glucose metabolism in cancer and viral-infected cells: Relevance to its use in India against SARS-CoV-2. IUBMB Life. 2021.
[18] DCGI approves anti-COVID drug developed by DRDO for emergency use.
[19] Rudrapal M, Khairnar JS, Jadhav GA. Drug repurposing (DR): An emerging approach in drug discovery. Drug Repurpos Hypothesis Mol Asp Ther Appl. 2020.
[20] Sahoo BM, et al. Drug Repurposing Strategy (DRS): Emerging approach to identify potential therapeutics for treatment of novel coronavirus infection. Front Mol Biosci. 2021;8:35.
[21] Lotfi Shahreza M, Ghadiri N, Mousavi SR, Varshosaz J, Green JR. A review of network-based approaches to drug repositioning. Brief Bioinform. 2018;19:878–892.
[22] Alaimo S, Pulvirenti A. Network-based drug repositioning: Approaches, resources and research directions. In Computational Methods for Drug Repurposing. Springer, 2019, 97–113.
[23] Palma G, Vidal M-E, Raschid L. Drug-target interaction prediction using semantic similarity and edge partitioning, Vol. 8796, 2014.
[24] Chen B, Ding Y, Wild DJ. Assessing drug target association using semantic linked data. PLoS Comput Biol. 2012;8(e1002574).
[25] Tanoli Z, et al. Exploration of databases and methods supporting drug repurposing: A comprehensive survey. Brief Bioinform. 2021;22:1656–1678.

[26] Issa NT, Stathias V, Schürer S, Dakshanamurthy S. Machine and deep learning approaches for cancer drug repurposing. In Seminars in Cancer Biology, Vol. 68. Elsevier, 2021, 132–142.
[27] Oprea TI, et al. Drug repurposing from an academic perspective. Drug Discov Today Ther Strateg. 2011;8:61–69.
[28] Bohuon C, Monneret C. Fabuleux Hasards: Histoire de la découverte de médicaments. EDP sciences, 2012.
[29] Vane JR. Inhibition of prostaglandin synthesis as a mechanism of action for aspirin-like drugs. Nat New Biol. 1971;231:232–235.
[30] Vane JR, Botting RM. The mechanism of action of aspirin. Thromb Res. 2003;110:255–258.
[31] Cadavid AP. Aspirin: The mechanism of action revisited in the context of pregnancy complications. Front Immunol. 2017;8(261).
[32] Mills EJ, et al. Low-dose aspirin and cancer mortality: A meta-analysis of randomized trials. Am J Med. 2012;125:560–567.
[33] Rothwell PM, et al. Effect of daily aspirin on long-term risk of death due to cancer: Analysis of individual patient data from randomised trials. Lancet. 2011;377:31–41.
[34] Raje N, Anderson K. Thalidomide – a revival story, 1999.
[35] Teo SK, Stirling DI, Zeldis JB. Thalidomide as a novel therapeutic agent: New uses for an old product. Drug Discov Today. 2005;10:107–114.
[36] Kauppi DM. Therapeutic drug repurposing, repositioning and rescue. Drug Discov. 2015;16(16).
[37] Brazil R. Repurposing viagra: The 'little blue pill' for all ills. Pathophysiology. 2018;14(20).
[38] Lexchin J. Bigger and better: How Pfizer redefined erectile dysfunction. PLoS Med. 2006;3(e132).
[39] Ghofrani HA, Osterloh IH, Grimminger F. Sildenafil: From angina to erectile dysfunction to pulmonary hypertension and beyond. Nat Rev Drug Discov. 2006;5:689–702.
[40] Dudley J, Berliocchi L. Drug repositioning: Approaches and applications for neurotherapeutics. CRC press, 2017.
[41] Iratni R, Ayoub MA. Sildenafil in combination therapy against cancer: A literature review. Curr Med Chem. 2021;28:2248–2259.
[42] Butterworth RF. Potential for the repurposing of adamantane antivirals for COVID-19. Drugs R D. 2021:1–6.
[43] Huecker MR, Smiley A, Saadabadi A. Bupropion, 2017.
[44] Chavez-Dozal AA, Lown L, Jahng M, Walraven CJ, Lee SA. In vitro analysis of finasteride activity against Candida albicans urinary biofilm formation and filamentation. Antimicrob Agents Chemother. 2014;58:5855–5862.
[45] Mouchlis VD, et al. Advances in de novo drug design: From conventional to machine learning methods. Int J Mol Sci. 2021;22:1676.
[46] Global drug repurposing service providers market – a look at repurposed drugs in the fight against COVID-19itle, 2020.
[47] Anastas P, Eghbali N. Green chemistry: Principles and practice. Chem Soc Rev. 2010;39:301–312.
[48] Jadhavar PS, et al. Sustainable approaches towards the synthesis of quinoxalines. In Green Chemistry: Synthesis of bioactive heterocycles. Springer, 2014, 37–67.
[49] Schcolnik-Cabrera A, Juárez-López D, Duenas-Gonzalez A. Perspectives on drug repurposing. Curr Med Chem. 2021.
[50] Baker NC, Ekins S, Williams AJ, Tropsha A. A bibliometric review of drug repurposing. Drug Discov Today. 2018;23:661–672.
[51] Talevi A. Drug repositioning: Current approaches and their implications in the precision medicine era. Expert Rev Precis Med Drug Dev. 2018;3:49–61.
[52] DeMonaco HJ, Ali A, Von Hippel E. The major role of clinicians in the discovery of off-label drug therapies. Pharmacother J Hum Pharmacol Drug Ther. 2006;26:323–332.

[53] Gns HS, Saraswathy GR, Murahari M, Krishnamurthy M. An update on drug repurposing: Re-written saga of the drug's fate. Biomed Pharmacother. 2019;110:700–716.

[54] Corsello SM, et al. The drug repurposing hub: A next-generation drug library and information resource. Nat Med. 2017;23:405–408.

[55] Tian L, et al. RNA-dependent RNA polymerase (RdRp) inhibitors: The current landscape and repurposing for the COVID-19 pandemic. Eur J Med Chem. 2021:113201.

[56] Denton CP, Khanna D. Rational repurposing of tocilizumab for treatment of lung fibrosis in systemic sclerosis. Lancet Rheumatol. 2021;3:e321–e323.

[57] Actemra FDA Approval History.

[58] Samaee H, Mohsenzadegan M, Ala S, Maroufi SS, Moradimajd P. Tocilizumab for treatment patients with COVID-19: Recommended medication for novel disease. Int Immunopharmacol. 2020;107018.

[59] Joshi G, Thakur S, Poduri R. Exploring insights of hydroxychloroquine, a controversial drug in Covid-19: An update. Food Chem Toxicol. 2021;151:112106.

[60] Fetro C, Scherman D. Drug repurposing in rare diseases: Myths and reality. Therapies. 2020;75:157–160.

[61] Smith RB. Repositioned drugs: Integrating intellectual property and regulatory strategies. Drug Discov Today Ther Strateg. 2011;8:131–137.

[62] Halabi SF. The drug repurposing ecosystem: Intellectual property incentives, market exclusivity, and the future of new medicines. Yale JL Tech. 2018;20:1.

[63] Breckenridge A, Jacob R. Overcoming the legal and regulatory barriers to drug repurposing. Nat Rev Drug Discov. 2019;18:1–2.

[64] Wouters OJ, McKee M, Luyten J. Estimated research and development investment needed to bring a new medicine to market, 2009–2018. Jama. 2020;323:844–853.

[65] DiMasi JA, Grabowski HG, Hansen RW. Innovation in the pharmaceutical industry: New estimates of R&D costs. J Health Econ. 2016;47:20–33.

Mayank Joshi, Mayank

2 Intellectual property and regulatory considerations in drug repurposing

Abstract: Drug repurposing has grown to be a crucial tool in rediscovering new drug uses for the medication compound. Drug repurposing has been suggested as a method for creating novel therapeutics that carries fewer risks, costs less money, and takes less time to produce than creating entirely new medications. An appealing strategy that can help reduce major time and financial investments made throughout the drug development process is the repositioning or "repurposing" of current treatments into different disease indications. However, because of legal and regulatory restrictions, the promise of this technique has not been as fully realized as envisioned. According to the situation, it is possible to reduce the length of the regulatory approval procedure while also saving money and time. Even though there are many aspects to consider when pricing a pharmaceutical redevelopment project commercially, obtaining intellectual property for the new application often linked with a new formula, which may include a unique route of administration, will always be a vital aspect. Here, we draw attention to these obstacles and discuss potential solutions. We conclude by outlining the methods the pharmaceutical sector has previously used to successfully circumvent regulatory concerns and launch their therapeutic products. To the greatest extent possible, both patients and drug developers must be taken into account when repurposing current medications for brand-new indications.

Keywords: IPR, drug poisoning, regulatory bodies, challenges in IPR

2.1 Intellectual property rights

Intellectual property rights (IPR) is defined as an idea, invention, and artistic expression based on a community's readiness to grant the status of the property. The inventor or creator has some exclusive rights to that particular invention or creation under intellectual property rights. These rights allow them to earn commercial profits from their creativity, hard work, or status. Several types of intellectual property shields include patents, trademarks, and copyrights. The patent is an appreciation for an invention with specific criteria like global novelty, nonobviousness, and industrial applicability. IPR has become the prerequisite for an idea, innovation, or creativity for their better credentials, planning, commercialization and rendering. Industries should grow individually with their particular IPR strategies, management style, and policies based on their area of specialization [1].

https://doi.org/10.1515/9783110791150-002

Intellectual property (IP) applies to any innovative construct of the human brain, for example technology, art, literature, and science. IPR implies that anyone can protect his/her invention or creative product for a specific time period under legal rights provided to the inventor or creator. The legal constitutional rights confirm a special right assigned to the originator, maker, or assignee to exploit his origination or creation for a specified period entirely. IP has become an essential part of the modern economy [2]. All companies prioritize the creative and intelligent workforce that can improve the new ideas and products to the market for the public interest and their profit. There have been new technologies in the market with massive investment in research and development, which has increased exponentially these days. The total price share of the technology of new drug developers has grown immensely. Subsequently, this knowledge must be protected from illicit use, for a particular time. IPR would safeguard the regaining of the all the expenditures in research and development and associated expenses and satisfactory earnings for constant inflow of money for research and development. Therefore, IPR helps a country's economic development by encouraging healthy competition and inspiring industrial expansion and financial growth [3].

The patent grants started in Europe in the fourteenth century, and they initiated law and administrative procedures concerning IPR. England has advanced technology and attracted artists with special terms in comparison of all other European nations. The first acknowledged copyright was granted in Italy. Most of the legal framework, law, and systems of the IPs were made in Venice and followed by other countries; this was the first effort in the world. The Patents Act was introduced in various countries and was based on the British patent system, which was amended multiple times by adding several actions into it [4].

Formerly, the industrial property covered only patents, trademarks, and industrial designs, but now it has been upgraded to the much broader term "Intellectual Property." IPRs can advance technology and offer a mechanism for managing infringement, piracy, and unauthorized use of original creatures. IP provides knowledge to the general public in the published form, except for trade secrets. IP protection can be pursued for a diversity of scholarly works, as shown in Figure 2.1.

2.2 Patents

Patents can be granted for products and processes of an invention that fulfils some criteria as global novelty, nonobviousness, and industrial or commercial applicability. The idea of an invention is not disclosed and protected under most of the laws meant for the defense of inventions. Several nations express inventions as innovative resolutions to technical difficulties. These issues may be ancient or contemporary, but the answer must be new to merit the invention's name. Discovering something that is al-

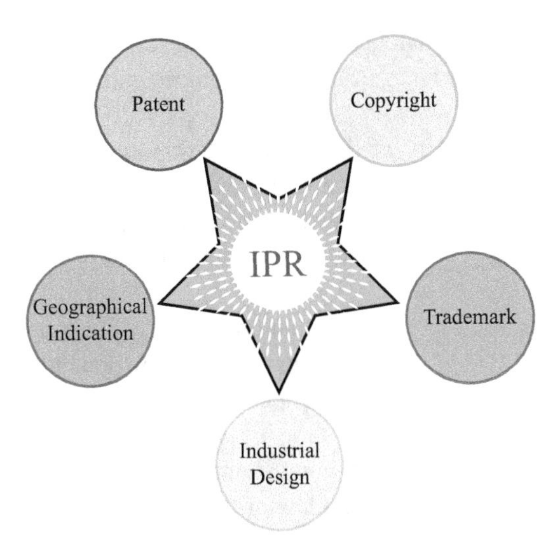

Figure 2.1: Types of intellectual property rights (IPR).

ready present in nature like an unknown plant variety is not an invention. Some efforts and involvement must be present; so we can call it a new invention if we extract a novel molecule from that plant. It is not necessary that invention should always be a complex process. For instance, a safety pin was a simple creation, which is used to resolve present issues. New solutions and ideas are protected as such. Therefore, the protection of creations under patent law does not involve invention signified in a physical embodiment [5].

Generally, patents for any invention mean the protection of rights of an inventor. These rights under patent are granted to a creator by a government or a local office serving many states. These license the inventor to eliminate anyone else from taking commercial advantage of his invention for a specific time period, usually 20 years. Patents deliver incentives to inventors, presenting credit for the originality and reward for its commercial creation only after granting an exclusive right. These incentives inspire innovation, which sequentially contributes towards improving the quality of life of human beings. In response to their sole ownership, the inventor must satisfactorily reveal their invention to the general public, so that others can increase the innovative information and additionally enhance the technology. The disclosure of new invention is a mandatory part of any patent grant. The patent system is planned to ensure equilibrium between inventors' benefits and the general public's interests [6].

The government authority allots the relevant documents for patent. The person or the firm submits an application to the national or regional patent for their invention. The inventor describes the details of his invention in an application and compares it with the prior knowledge or technology available in public domain for the novelty.

Every creation or invention does not have patentability. As per the law, an innovation needs to accomplish certain conditions recognized as the requirements of patentability, and these are as follows:

- **Novelty:** Novelty represents the originality of the work and this knowledge should not be present in the public domain on its technical basis.
- **Nonobviousness:** It indicates the involvement of creative efforts made by the inventor and that which cannot be replicated by an ordinary person without having technical knowledge.
- **Industrial applicability**: The invention must be such that one can produce it on industrial scale and for mass use. If the invention is novel and nonobvious but does not have the ability for industrial application, we cannot protect it by a patent.
- **Patentable subject matter:** The invention or creation must present scope of patentable fields or process and it is determined by the law of the country. It may vary from nation to nation. Many fields such as theories of science, mathematical methods, animal and plant varieties, discoveries of natural ingredients rational for medical treatment and any invention with public health interest where we need to prevent commercial exploitation for ethical reasons are excluded from patentability by different countries.

The date on which the application of patent is filed it should fulfil the conditions of novelty and nonobviousness. The Paris Convention protects the industrial IPs and regulates the priority of the inventor but this is an exemption to the rule. This exclusion is only for the members nations of the Paris Convention. The priority is given under right of priority to that person who has applied in one member country of the Paris Convention; now the applicant or his successor becomes eligible for the protection of his/her invention in all other member nations for a certain time period. If he/she files patent for another member country, the date of first application will be considered for the priority. It works on the principle of first come, first served. For instance, it is sufficient that the conditions of nonobviousness existed at the time the French application was filed if an inventor first files for patent protection in France and then submits a second application for the identical invention in Germany. This means that any applications for the same invention submitted by other applicants between the dates of the inventor's first and second applications and the later, German application retain priority. The time between the two dates must not exceed 12 months in order for this to apply [7].

We need to differentiate between the product and process patent. For example, an invention that includes formation of a new alloy is a product patent but the development of a new method or process for manufacturing a known or unknown alloy is a process patent. These patents are considered for product inventions and process inventions, respectively.

The patentee is a person to whom a patent is granted, and we can acknowledge him/her as a patent owner or patent holder. Once the patent has been granted to a person in a country, he can give approval to other persons who wish to exploit the invention commercially in the same nation. Technically, if anyone uses a patented work or creation without the authorization of patentee, it will be considered as an illegal act. The patent for an invention is approved in most countries for 20 years. The innovation enters the public domain immediately after a patent's expiration date, and the holder's commercial protection ceases. The patentee no longer embraces exclusive rights to the invention, which then becomes available for commercial use by others [8].

The rights deliberated by a patent are not defined in the patent itself. Those rights are described in the patent law of the country where the patent is granted. The patent owner's exclusive rights generally consist of the right to prevent third parties from manufacturing, using, offering for sale, selling or importing these products and the process patent without the holder's permission. The patentee is not given a legislative right to exploit his invention but a legal right to prevent others from commercially misusing it. He may provide authorization or grant a license to other parties to practice the invention on mutually decided terms. The patentee may also sell his right to the invention to someone else, who will become the patent's new title-holder [9].

The patent can be used without the patent's owner's consent in certain situations. These situations allow for a sense of balance between the legitimate interests of the patentee and the general public. Patent laws may provide for cases wherein a patented invention may be exploited without the patentee's authorization, e.g., in the broader public interest by or on behalf of the government or based on a compulsory license [10]. A compulsory license is a permission to exploit the innovation provided by a governmental authority. As per the defined law, it is usually allotted in exceptional cases only and where the entity desiring to use the patented invention is incapable of getting approval from the patent owner. The decision to grant a compulsory license must deliver adequate compensation to the patentee and may be the subject of an appeal [11].

2.3 Industrial designs

It is related to structures of any outline of design, arrangement, surface pattern, and the composition of lines and colors applied to an object, whether 2D or 3D, e.g., textile toothbrush. An industrial design is the ornamental or aesthetic feature of a helpful article. This characteristic may be subject to the item's shape, pattern, and color. The design must have a visual appeal and execute its envisioned function capably. Additionally, it must be able to be replicated by industrial means; this is the essential purpose of the design, and that is why the design is called industrial.

In a legal sense, industrial design refers to the right granted in several countries, according to a registration system, to defend the original, ornamental, and nonfunctional features of a product that result from design activity. Visual appeal is one of the main features that influence customers in their preference for one product over another. Consumers will choose based on price and artistic appeal when the technical performance of a product presented by different manufacturers is comparatively alike. Thus, in registering their industrial designs, manufacturers protect one of the distinguishing elements that govern market success [12].

By rewarding creators for their determination to make new industrial designs, this legal protection also helps as a motivation to invest capital in design activities. One of the primary intentions of industrial design protection is to inspire the design element of production. That is why industrial design laws generally protect only those designs that can be used in industry or can have large-scale production. This condition of usefulness is a noteworthy difference between industrial design protection and copyright, since the latter is only concerned with artistic creations.

Industrial designs can generally be protected if they are novel or original. These designs may not be mentioned as new or unique if they cannot create a significant difference from known techniques or their blends. Most industrial design laws exclude structures, which are dictated uniquely by the function of an article, from protection. If several manufacturers produce the design for an assay, for example, a screw, it is dictated mainly by the role of the screw. In this case, protection for that design would have the consequence of not including all other constructors from producing objects projected to perform the same function. Such exclusion is unacceptable without the design being adequately novel and ingeniously suitable for patent protection. This protection does not prevent other manufacturers from producing or dealing in similar articles or goods, provided these do not symbolize or replicate the protected design [13].

Industrial design registration protects against unauthorized exploitation of the design in industrial articles. It allows the proprietor of the design the exclusive right to make, import, sell, hire, or offer for sale articles to which the design is applied or in which the design is embodied.

The tenure for an industrial design right differs from country to country. The average maximum term is 10–25 years, frequently divided into terms requiring the owner to renew the registration to obtain an extension of time. The comparatively short design protection period may be associated with more general fashion styles, mainly in highly fashion-conscious zones, for example, clothing or footwear [14].

2.4 Trademarks

This is related to any logo, mark, or name, which encompasses trade that is directed for the service or product, and it is an identification of the producer or the service provider. We can get, buy, and sell the license for trademarks. A trademark is a symbolic goodwill for the product or service. A trademark is a sign or a mixture of symbols that discriminates one enterprise's goods or services from another [15].

Such signs of trademarks include words, letters, numbers, images, shapes, and colors, along with any combination of these. A growing count of countries also permit the registration of fewer old-style formats of trademark, for example the Coca-Cola bottle and Toblerone chocolate bar has three-dimensional marks; audio symbols and sounds, like MGM production films has the roar of the lion at the starting of movie; or olfactory signs, for example aromas, like perfumes. But there are predetermined parameters in many countries for the products that may be registered under the umbrella of a trademark, usually permitting only visually observable symptoms or those can be denoted graphically. A trademark is considered as a logo used on goods or regarding the promotion of products. The trademark can be used on the article itself, but it may be printed on the container or packaging of the goods in which these are vended. Trademarks are directly related to marketing and are, therefore, frequently used for advertisements in newspapers, television, or window showcases of shops.

In the various categories of similar goods, we can identify the source of products and services by trademarks. A federation owns collective marks, and its followers use the mark to identify and determine the quality level of goods. Certification marks, for example, Woolmark, are specified for acquiescence with well-defined values but these are unrestricted to any integration. A trademark can be used to relate the service facilities and is called a service mark. The service marks are brought into service for hotels, cafeterias, commercial airline, travel agencies, car rental agencies, washing services, and cleaners. All these conditions for trademarks also apply for service marks [16].

Largely, there are four main roles of trademarks related to distinguishing marked services or goods, their corporate source, quality, and advancement in the marketplace. These functions are described as follows:

- We can discriminate the services or products of one venture from other concerns. Trademarks simplify the making of choices by the purchaser when purchasing specific goods or services. The trademark assists the customer in recognizing a service or product introduced by advertisement. We can assess the distinguishing qualities of products and services according to their trademarks; for instance, even though this is a feature of their computers, the term "Apple" and a representation of an apple cannot be used to distinguish between apples. Trademarks set items and services apart from one another and from one another in reference to the company that they are associated with.

- We can refer a particular firm to the customers for their products and services even if it is unknown to the client. Hence, trademarks differentiate goods or services from their source compared to their competitors in the same field. It gives the scope of protection of trademarks and is an essential function.
- We can pertain to a specific standard of the service or product for which the trademark is used, so it provides customers with consistency in quality. This function, we can say, is usually mentioned as the assured function of trademarks. More than one enterprise may use a trademark if the owner of a particular trademark grants certification to other enterprises. These licenses granted to other enterprises must follow the trademark owner's quality standards. Likewise, trading ventures frequently use trademarks for goods they obtain from several sources. In these instances, the owner of the trademark is not accountable for manufacturing the goods but for picking those that match his quality values and requirements. It can be understood that the trademark holder is the producer of a specific product. Still, the trademark owner may regularly use ingredients that have not been manufactured by him but only selected.
- We can endorse the sale and marketing of products and services. Trademarks can be used to differentiate or endorse a particular brand for quality and enhance sales. A trademark must appeal to the buyer, produce curiosity, and motivate a sense of assurance. Hence, we can call it the appeal function.

The trademark provides the prerogative of using an inscribed trademark to the owner. It protects the general public from misleading unlicensed parties and confusing between similar marks. One can renew the trademark by paying the corresponding fee after the protection period. The courts enforce trademark protection and has the authority to block trademark infringement [17].

2.5 Copyright

All art, publishing, music, cinematography, dramatic work, audio tapes, and computer software are the expression of ideas and are protected as copyrights in these material forms. Copyright is recognized as an author's right for most European languages except English. The copyright is a central act created by the author or by his authorization regarding literary and artistic creations. That action is the making of replicas of academic or creative work, for example, a book, painting, sculpture, photograph, or motion picture. The author's rights deal with the individual who is the creator of the artistic work or author; he has certain precise rights over his creation, like the right to prevent a partial reproduction. Only he can exercise this, whereas additional rights allow him to make copies with another person with consent, such as a publisher who has obtained a license to this effect from the author [18].

2.6 Geographical indications

This is a symbol provided for belongings of a specific geographical origin and holding qualities or status due to that place of origin or a region or locality. Farming products naturally have qualities that originate from their place of creation, and these are affected by definite local influences such as climate and soil. National law and the customer perception decide the critical functions for an indication. Many agricultural products are typically registered as geographical indications, for example, "Tuscany" for olive oil produced in a specific area of Italy and "Roquefort" for cheese made in a particular region of France [19].

The practice of geographical indications is not restricted to agricultural products only. This may be used for highlighting the specific potential of a product, which may arise due to human influences present at the origin, i.e., specific manufacturing skills and traditions. The place of origin might be a village, town, area, or country. The word "Swiss" is widely used as a geographical indication for products like watches manufactured in Switzerland. The idea of geographical indication includes appellations of origin. Illustrations of titles of origin protected in states party to the Lisbon Agreement for the Protection of Appellations of Origin and their International registration comprise "Habana" for tobacco cultivated in the Havana region of Cuba; and "Tequila" for spirits manufactured in certain zones of Mexico.

Geographical indications are protected under national laws under a wide range of perceptions, such as laws against unfair competition, consumer protection laws, laws for the protection of certification marks, or special regulations for the protection of geographical indications or appellations of origin. To conclude, unlicensed products may not use geographical indications if they misinform the public regarding the product's true origin. The honorable court may apply sanctions to prevent unauthorized use, ranging from court bans to compensations and fines and imprisonment in severe cases [20].

Drugs and pharmaceutical products have attracted global attention more than any other technology for IPRs. A lot of money is invested in introducing a new drug – around $0.3–1 billion. High risk is also associated with the progression stage, so none of the firms like to take the risk for IP and make it free, cutting out satisfactory earnings. Along with the fund-raising IP creation, obtaining protection and management have become part of corporate action [21].

Globally, pharmaceutical companies are competitive mainly due to scientific knowledge and research and development efforts rather than manufacturing expertise. Hence, a higher percentage of investments in the drug industry is made in research and development, about 15% of total sales. A high cost involves the risk of failure in medicinal research and development along with the discovery of potential medications, which may not fulfill the rigorous safety standards. Sometimes, afterward an investment of many years, it may be terminated. Due to hurdles in the drug development of many molecules, it takes 8–10 years to be synthesized. Drug companies need to change their focus on the product patent rather than process development to develop a new chemical entity. One must satisfy various the regulatory authorities' requirements in the international market.

In the last two decades, numerous documents have been submitted to regulatory authorities and the count has improved almost three times. Furthermore, nowadays, these regulatory bodies take much longer time to accept an innovative drug. Subsequently, the duration of patent protection is reduced, which increases the need for extra efforts to make great returns. The condition may be further difficult for medications invented using the biotechnological way, particularly for the drugs related to the gene therapy. The commercial world is expected to initiate obtaining extended drug patent protection speedily. Government bodies put in efforts to make the drug price controllable to meet the expectations of the general public. Hence, the companies also need to reduce their costs for development of the drug, production, and marketing. Alternatively, proper forecasting is needed for reduced profit margins if cost recovery is expected over an extended time period [22].

2.7 What is drug repurposing?

Drug repurposing is rediscovering novel therapeutic uses for the existing drug molecules. It encompasses the drug molecules that were already approved, withdrawn, abandoned, and are under investigation for other therapeutics. The definition of drug repositioning can be: "finding new uses outside the scope of the original medical indication for existing drugs." [23] Drug reprofiling or repurposing is an approach to recognize unconventional uses for the drugs that are already approved or existing medicines approved by FDA. The elementary purpose behind the repurposing of the drug molecule is to find the diversified action of already clinically failed or used drugs in the drug discovery pipeline [24]. The drug development method is costly, lengthy, and time-consuming. It is associated with a very selective drug mechanism, making it highly risky. The advantage of FDA-approved repurposed drugs over the drug development process is the absence of requirements for human trials due to their low risk. Additionally, these drugs have low failure risk, high success rate, short development time frames, and are less expensive [25]. These repurposed drugs are broadly used to treat severe disorders, i.e., cancer, psychiatric issues, gastrointestinal disease, respiratory infections, etc. [26].

Drug repositioning is the discovery of a new medical sign for a medication. The drug might be approved for other uses, withdrawn due to adverse effects, or rejected because of lack of efficacy [27, 28]. At the same time, drug repurposing may be discovered in some cases from serendipity by the discovery of side effects of drugs. It opens up new ways to develop groundbreaking medicines for rare and neglected molecules or compounds with the already existing safety, pharmacokinetic, and manufacturing data. Recent initiatives have been taken to explore drug repurposing due to its importance and availability in the market [29]. About 33% of approved drugs are examples of drug repurposing. Many approved drugs have been repurposed for treating other disease conditions, for example, Aspirin, Zidovudine, sildenafil, and thalidomide, initially used as nonsteroidal anti-inflammatory agents, anticancer agents, and to treat angina and sickness were repurposed later as antithrombotic agents, anti-HIV agents, and to treat erectile dysfunction and leprosy, respectively [30, 31].

Drug repurposing is envisioned to discover another use for a revolutionary drug or a drug discovered/synthesized by another person. We need to adopt new approaches to rediscover the compounds that have been rejected during the development process. The strategy fast-tracks this process, makes it economical and profitable, and provides effortless marketing [32]. Drug repositioning is a therapeutic revolution for the neglected compounds in drug discovery. After the thalidomide tragedy in Europe, this medicine had a harmful image but has initiated a novel drug indication [33].

There should be lower risk and low cost for the development of drug repositioning approaches and it has been extensively accepted to find a new clinical opportunity for ancient drugs. Hence, numerous approaches have been proposed for drug repositioning revisions. This includes mechanism-based repositioning, based on the patient's heterogeneity and complexity. Computational modelling can offer an advantage for drug repurposing with better customization of parameters with in-depth knowledge of integration experimental methods, with the mechanistic computational method to ensure high success rates of drug repositioning [34]. Virtual screening is the most systematic approach used for drug repurposing with the support of computational methods. Molecular docking can be used for virtual screening with the data available. Experimental validation is necessary after the virtual screening because the systematic techniques are not 100% truthful. The increasing consideration of epigenetic targets as a chance for repurposing drugs offers high anticipation of epi-drugs [35, 36].

2.8 Advantages of drug repurposing

Drug repositioning has various advantages over conventional methods of drug discovery. The first is that the time required for drug discovery is significantly less in drug repurposing than in the traditional research and development process. We need not go through preclinical studies like conventional procedures in drug repurposing. Instead, we can immediately start the testing and clinical trials phase. Thus, we can save 6–9 years and decrease the development cost, so the drug becomes economical. The drug discovery process searches for and designs innovative molecules with improved pharmacokinetics and pharmacodynamics responses [37]. The conventional method follows different steps, such as drug candidate discovery and design, preclinical studies, safety assessment, clinical trials, review and approval process, marketing, and post-approval safety observation. But in drug repurposing, a lot of preclinical and clinical knowledge of the drug candidate is already available because the drug has already passed through these stages. There are four phases of drug repurposing: determination of compound, procurement of combination, development, and safety observation after marketing [38, 39]. Drug repurposing is well-suited for the rapidly emerging and reemerging infectious diseases for which no existing medicines are available. Drug repurposing is an inventive and quick method for expanding treatments by manipulation of existing and accepted remedies [40].

2.9 Intellectual property rights for drug repurposing

Any original creation of human origin like scientific, technical, literary, and artistic work go through IP. The legal protection rights are given to the creator or inventor for the invention for certain time in IPRs. Patents, industrial design, trademarks, copyright and geographical indication, come under IPRs. The patent is provided for an invention that fulfils three essential conditions: novelty, nonobviousness, and industrial applicability. Any inventor can apply for IPR for the invention, so they can get enough time for improved identification, planning, commercialization, and execution of their creativity. It plays a vital role in the economy and encourages pharmaceutical companies to promote more and more inventions. The research and development cost is associated with the invention process, so intellectual property rights can help profit from the invention across the globe. The IPR is a reliable way to recover the cost of effort, investment, time, and money invested by the inventor of intellectual property [41]. Hence, it develops healthy competition between pharmaceutical companies and inventors and is associated with a country's economic development [42]. Pharmaceutical industries are evolving with time and focusing on the IPRs strategy. Drug patents are a knowledge source from the perspective of pharmacological and drug development and help in research and development. They provide knowledge regarding the applicability of the designed drug with novelty. Patent document searching becomes worthy of finding a new perspective of drug repositioning.

In order to address the standing challenges under drug development research, the process of drug repurposing provides a promising approach. The strategy generally has numerous benefits, if it compared to de novo discovery of drug. The foremost benefit of drug repurposing is termed to be intellectual property and to provide patent protection to highlight innovative indication of a present drug, which can simply be attained if the novel expression was not included in the previously obtained patent or not present in public domain knowledge [43]. Additionally, drug repurposing also supports the preservation of original IP to gain the upper hand over its competitor pharmaceutical concerns, and such intellectual property rights can be out-licensed to analyze the novel indication under the specific drug molecule. When the pharmaceutical company does not conduct clinical trials for the new expressions of the drug molecule, it can be out-licensed to a large pharmaceutical firm [44]. Furthermore, the IPRs in terms of drug repositioning provide the preliminary sign that presents a significant complication for repurposing the drug. Accomplishment of a different drug repurposing technique is a mode to get control of restrictions in order to achieve successful drug repurposing [45].

The drug and binding site recommendations for drug repurposing for international patent applications are not arbitrarily distributed but follow preferred vector patterns [46]. Secondary claims from unexpected findings can be made, like anticancer agents claimed to treat noninfectious lung diseases and cardiovascular mediators, which are meant for the treatment of neurological problems. Some of the patents offer motivating perspectives into drug repurposing, such as a COMT inhibitor entacapone, which is used in Parkinson's disease, can also inhibit dengue and West Nile virus proteases; clopidogrel, the antithrombotic medicine can be used for benign prostate hyperplasia; iopamidol and iohexol are X-ray contrast agents that can be used to treat influenza and Ebola virus diseases [47].

2.10 Intellectual property and regulatory considerations

A rational IP and regulatory approach can bring about encouraging repurposed drugs. Drug repurposing is not easily attainable due to lack of designed process or tools that can motivate drug repurposing used for current drugs. As a result, drug repurposing has the technical ability to tolerate industrial barriers even though it lowers healthcare costs when it is accompanied by logical, IP and regulatory defence. We need to consider and accomplish many diverse goals at the same time during development of drug purposing strategy, which may include therapeutic, monitoring, technical, and publicizing targets, to diminish burden, to boost profits, and to utilize the inclusive effect of accomplishing an extremely money-making and fruitful product. Approaches that comprise IP and legal involvement can transform an ostensibly nonviable drug repurposing assignment into an achievement. On the other hand, a drug repositioning strategy that manages appropriate IP and legal examination may quickly lose prospects due to expensive legal costs. Therefore, an appropriate drug repurposing scheme should be assisted by the advice of professional legal counsel [46].

There are various challenges in the IPR of drug repurposing methods along with the advantages. Significant challenges include the constant mandate to provide clinical trials and the commercialization technique associated with the grant of patent. We cannot ignore the patent document with a different format [48]. In the case of drug purposing, the "Use" patent becomes very important; it generally defends IP after such drug is presented for corporate proposal. A multiple sclerosis drug, dimethyl fumarate (BG-12), has been repurposed by Biogen and marketed by "Use" patent. The Orphan Drug Act of 1983 also gives exclusivity for seven years from the intellectual property rights to the repurposed drug if it comes under orphan medicine [49].

Drug repurposing has huge potential in drug discovery, but it is not a magic box. Several issues must be addressed before a drug is repurposed for a diverse indication. Thorough knowledge and understanding of the entire process are required for successful results. Before heavy investment in drug repurposing, we need to address IP issues [50]. The detailed knowledge of drug repurposing and its potential to provide novel indications encourages inventors to try more drug repurposing. The legality and ethical issues with repurposed use of any drug molecule need to be discussed with the regulatory authorities. IPRs are not applied to drugs rediscovered by drug repurposing simply for humanitarian reasons, so it becomes a disadvantage of IP for commercial use [51]. IPRs and regulatory ethics encompassing debates on drugs that come under off-patent and their patents on the action mechanism of drugs are a big concern for the marketing of such approved drugs for new indications [52, 53]. FDA-approved drugs are screened phenotypically for drug repurposing due to their ability proven already under the challenging trail of regulatory system and following the protocols for the development of drug manufacturing.

Drug repositioning generally offers more instant price as against a new chemical entity [54]. The regulatory approval of the repurposed drug can be faster and more cost-effective, depending on the situation. Many issues must be considered before the commercial valuation of a drug repurposing project, like regulatory approvals and IPR acquisition. IP for innovative use will constantly remain a dominant factor and can be achieved by developing a new formulation with a different administration route[55]. Drug repurposing also aims to acquire an innovative label for a current drug molecule by regulatory approval bodies; subsequently the drug repurposing and traditional drug development find their analogous ways. Monetary and intellectual property strategy, experimental and monitoring approach, and scientific equilibrium are the significant characteristics of drug repurposing and deliver an ethical justification for randomized controlled trials [56].

However some specific legal problems can impact decision-making for the optimal outcome and for justification of substantial undesirable trade-offs. For instance, a drug formulation that had earlier been accepted for a primary suggestion developed another physically diverse drug product, which was developed with an altered strength and preparation, enhanced for subsequent usage. The second indication may forecast extended timelines for development, cost of production, supplementary regulatory necessities, future launch date, and other marketing properties. A relatively healthier and long-standing approach would be opening development for the second drug indication along with the identical drug molecule accepted for the primary drug indication. All the hurdles and opportunities must be analyzed at the primary stage of development and regularly reviewed as expectations are established or contradicted and updated data made available, due to unique entity of each drug molecule. Swift approval may be supreme to accomplishment for certain drugs and all possibilities that have the potential for late approval will be extremely undesirable. Additional patent filing for the issue of drug product may impact positively on the matter of loss of exclusivity before one gets full benefit from the product, whereas delay in approval may affect it negatively. It is a fundamental consideration for the commercial benefit [50].

Thus, many strategy points are applied to the new biological entity analogously with new chemical entities. There are significant alterations, like in the United States, five years of definite exclusiveness for new chemical entities are provided against agreement of any succeeding drug invention with the identical active pharmaceutical ingredient depending on evidence from the original new drug application, while new biological entities are given 12 years of sure exclusiveness against consequent related biologics [55].

The rewards of repositioning of drugs are: we use known targets for drug; all previously available materials and data like long-term toxicology studies, etc., can be utilized and presented to regulatory authorities, and we can make this research and development significantly time- and cost-effective as compared to the classical method for bringing a new molecular compound to market. Renowned cases include drugs such as thalidomide, bupropion, sildenafil, and fluoxetine, with novel uses outside their originally approved therapeutic indications. Thalidomide was infamous for causing birth defects in the infants if administered during the first trimester of pregnancy. US Food and Drug Administration, manufacturers, and other regulatory bodies have withdrawn the drug thalidomide from the market but it can be brought back to use by drug repurposing. Still, thalidomide can be used in treatment of multiple myeloma, because this is not very common in pregnant women and the adverse consequence can be overcome by novel use [56].

References

[1] Department of science and technology Govt of I. Research and Development Statistics 2019–2020, 110016, 1–132.
[2] Delhi N. Department of Scientific and Industrial Research, Government of India. Anonymous Res Dev Ind Overv. 2002.
[3] Singh DR. Law Relating to Intellectual Property: A Complete Comprehensive Material on Intellectual Property Covering Acts, Rules, Conventions, Treaties, Agreements, Digest of Cases and Much More. Universal Law Publishing Company, 2008.
[4] Bainbridge DI. Intellectual Property. Pearson Education, 2009.
[5] Watal J. Intellectual Property Rights in the WTO and Developing Countries Kluwer Law International, Vol. 170. Hague, The Netherlands, 2001.
[6] Khurana N. Intellectual property rights-an understanding. Asian J Multidimens Res. 2015;4:1–16.
[7] Laxmi V, Inala MSR. Intellectual Property Rights. Bioentrepreneursh. Transf. Technol. Into Prod. Dev., Vol. 25. Emerald Group Publishing Limited, 2021, 95–110.
[8] Larsen R. Intellectual property law. Encycl Int Media Commun. 2003:429–449.

[9] Lelarge C. Innovation and intellectual property rights. Innov Firms A Microecon Perspect. 2009;9789264056:157–190.
[10] Hanel P. Intellectual property rights business management practices: A survey of the literature. Technovation. 2006;26:895–931.
[11] Stiglitz JE. Economic foundations of intellectual property rights. Duke Law J. 2008;57:1693–1724.
[12] Hertenstein JH, Platt MB, Veryzer RW. The impact of industrial design effectiveness on corporate financial performance. J Prod Innov Manag. 2005;22:3–21.
[13] Introduction Design and History, Routledge, 2020, 13–14.
[14] Gemsera* G, Leendersb MAAM. How integrating industrial design in the product development process impacts on company performance. J Prod Innov Manag. 2001;18:28–38.
[15] Mendonça S, Pereira TS, Godinho MM. Trademarks as an indicator of innovation and industrial change. Res Policy. 2004;33:1385–1404.
[16] Desai DR. From trademarks to brands. Fla Law Rev. 2012;64:981–1044.
[17] Barnes DW. A new economics of trademarks. SSRN Electron J. 2011:5.
[18] Breyer S. The uneasy case for copyright: A study of copyright in books, photocopies, and computer programs. Harv L Rev. 1970:84.
[19] Barjolle D, Paus M, Perret AO. Impacts of Geographical Indications-Review of Methods and Empirical Evidences, 2009.
[20] Moschini GC, Menapace L, Pick D. Geographical indications and the competitive provision of quality in agricultural markets. Am J Agric Econ. 2008;90:794–812.
[21] Angell M. The pharmaceutical industry – to whom is it accountable? N Engl J Med. 2000;342:1902–1904.
[22] Glasgow LJ. Stretching the limits of intellectual property rights: Has the pharmaceutical industry gone too far? Idea. 1997;41:227–258.
[23] Ashburn TT, Thor KB. Drug repositioning: Identifying and developing new uses for existing drugs. Nat Rev Drug Discov. 2004;3:673–683.
[24] Chong CR, Sullivan DJ. New uses for old drugs. Nature. 2007;448:645–646.
[25] Flower D. Drug Discovery: Today and Tomorrow. Bioinformation. 2020;16:1–3.
[26] Rajamuthiah R, Fuchs BB, Conery AL, Kim W, Jayamani E, Kwon B, et al. Repurposing salicylanilide anthelmintic drugs to combat drug resistant Staphylococcus aureus. PLoS One. 2015;10:e0124595.
[27] Ashburn TT, Thor KB. Drug repositioning: Identifying and developing new uses for existing drugs. Nat Rev Drug Discov. 2004;3:673–683.
[28] Chan J, Wang X, Turner JA, Baldwin NE, Gu J. Breaking the paradigm: Dr Insight empowers signature-free, enhanced drug repurposing. Bioinformatics. 2019;35:2818–2826.
[29] Law GL, Tisoncik-Go J, Korth MJ, Katze MG. Drug repurposing: A better approach for infectious disease drug discovery? Curr Opin Immunol. 2013;25:588–592.
[30] Masoudi-Sobhanzadeh Y, Omidi Y, Amanlou M, Masoudi-Nejad A. DrugR+: A comprehensive relational database for drug repurposing, combination therapy, and replacement therapy. Comput Biol Med. 2019;109:254–262.
[31] Graul AI, Cruces E, Stringer M. The year's new drugs & biologics, 2013: Part I. Drugs Today. 2014;50:51–100.
[32] Talevi A. Drug repositioning: Current approaches and their implications in the precision medicine era. Expert Rev Precis Med Drug Dev. 2018;3:49–61.
[33] Talevi A, Bellera CL. Challenges and opportunities with drug repurposing: Finding strategies to find alternative uses of therapeutics. Expert Opin Drug Discov. 2020;15:397–401.
[34] Oprea TI, Bauman JE, Bologa CG, Buranda T, Chigaev A, Edwards BS, et al. drug repurposing from an academic perspective. Drug Discov Today Ther Strateg. 2011;8:61–69.
[35] Park K. A review of computational drug repurposing. Transl Clin Pharmacol. 2019;27:59–63.

[36] Vanhaelen Q, Mamoshina P, Aliper AM, Artemov A, Lezhnina K, Ozerov I, et al. Design of efficient computational workflows for in silico drug repurposing. Drug Discov Today. 2017;22:210–222.
[37] Rudrapal M, Khairnar SJ, Jadhav AG. Drug Repurposing (DR): An emerging approach in drug discovery drug repurposing – hypothesis. Mol Asp Ther Appl. 2020.
[38] Boguski MS, Mandl KD, Sukhatme VP. Repurposing with a difference. Science. 2009;324:1394–1395.
[39] Cha Y, Erez T, Reynolds IJ, Kumar D, Ross J, Koytiger G, et al. drug repurposing from the perspective of pharmaceutical companies. Br J Pharmacol. 2018;175:168–180.
[40] Farha MA, Brown ED. Drug repurposing for antimicrobial discovery. Nat Microbiol. 2019;4:565–577.
[41] Novac N. Challenges and opportunities of drug repositioning. Trends Pharmacol Sci. 2013;34:267–272.
[42] Nath Saha C, Bhattacharya S. Intellectual property rights: An overview and implications in pharmaceutical industry. J Adv Pharm Technol Res. 2011;2:88–93.
[43] Naylor S, Kauppi DM, Schonfeld JM. Therapeutic drug repurposing, repositioning and rescue: Part II: Business review. Drug Discovery World. 2015;16:57–72.
[44] Mucke HA. Patent highlights February–March 2022. Pharm Pat Anal. 2022;11:119–126.
[45] Gaziano L, Giambartolomei C, Pereira AC, Gaulton A, Posner DC, Swanson SA, et al. Actionable druggable genome-wide Mendelian randomization identifies repurposing opportunities for COVID-19. Nat Med. 2021;27:668–676.
[46] Mucke HA. Patent Highlights June–July 2015. Pharm Pat Anal. 2015;4:423–430.
[47] Cheong JE, Ekkati A, Sun L. A patent review of IDO1 inhibitors for cancer. Expert Opin Ther Pat. 2018;28:317–330.
[48] Mucke HAM. Drug repositioning in the mirror of patenting: Surveying and mining uncharted territory. Front Pharmacol. 2017;8:927.
[49] Keating GM, Garnock-Jones KP. Alemtuzumab: A guide to its use in relapsing–remitting multiple sclerosis. Drugs Ther Perspect. 2014;30:337–341.
[50] Bloom BE. Creating new economic incentives for repurposing generic drugs for unsolved diseases using social finance. Assay Drug Dev Technol. 2015;13:606–611.
[51] Shineman DW, Alam J, Anderson M, Black SE, Carman AJ, Cummings JL, et al. Overcoming obstacles to repurposing for neurodegenerative disease. Ann Clin Transl Neurol. 2014;1:512–518.
[52] Shim JS, Liu JO. Recent advances in drug repositioning for the discovery of new anticancer drugs. Int J Biol Sci. 2014;10:654–663.
[53] Bhattarai S, Park JM, Gao B, Bian K, Lehr W. An overview of dynamic spectrum sharing: Ongoing initiatives, challenges, and a roadmap for future research. IEEE Trans Cogn Commun Netw. 2016;2:110–128.
[54] Caban A, Pisarczyk K, Kopacz K, Kapuśniak A, Toumi M, Rémuzat C, et al. Filling the gap in CNS drug development: Evaluation of the role of drug repurposing. J Mark Access Heal Policy. 2017;5:1299833.
[55] Sternitzke C. Drug repurposing and the prior art patents of competitors. Drug Discov Today. 2014;19:1841–1847.
[56] Begley CG, Ashton M, Baell J, Bettess M, Brown MP, Carter B, et al. Drug repurposing: Misconceptions, challenges, and opportunities for academic researchers. Sci Transl Med. 2021:13.

Devendra Kumar, Ankit Ganeshpurkar

3 Drug repurposing from an academic and industrial view

Abstract: Traditional drug discovery protocol may be replaced by the method of repurposing. Drug repurposing may reduce the burden of rigorous preclinical and clinical data required for traditional drug discovery. Thus, the technique will provide maximum output with minimal time and capital expense. Further, the drug repurposing method also has less risk of failure. These factors together impact the cost of the drug. The chapter includes drug repurposing in the case of neurological disorders. Moreover, drug repurposing from the industry and academic perspectives are also discussed. The tool and data bases used by these two are mentioned in the chapter. Various successful case studies adopting the method of drug repurposing are elaborated.

Keywords: Drug repurposing, neurological disorder, infliximab, azidothymidine

3.1 Introduction

Drug repurposing has the potential to complement traditional drug discovery. Primary drug discovery needs rigorous preclinical and clinical trials. Thus, huge amounts of time and capital must be spent. Further, the risk of failure is very high. These factors together impact the cost of the drug. The risk of failure is around 47%, which is associated with toxicity or safety. Additionally, the drug discovery required an average of 5–7 years. On the other hand, drug repurposing is reported to be quick and has minimal chance of failure due to documented safety and efficacy profile. Generally, molecules used for repurposing have successful records in phase I and phase II clinical trials. Thus, the molecule has validated tolerability and safety profile. These data provide an upper hand for the swift approval of the repurposed drugs by the approval authority. Drug repurposing performed by pharmaceutical industries and academics involved the cases of serendipitous discovery or mechanism of action-based discovery [1].

The use of life cycle management activities by the pharmaceutical and biotech industries is important to extend the patent on the potential candidate for repurposing. Infliximab was developed by following the same regulatory procedure and used for rheumatoid arthritis and ulcerative colitis. Sometimes, in-depth understanding of mode of action of drugs and diseases is critical for drug repurposing. Azidothymidine was developed for the chemotherapy and the drug failed in the clinical trial. However, the drug was repurposed for treating human immunodeficiency virus (HIV) and it opened a new arena of HIV therapy. It is an example of successful drug repurposing helping control the rising price of new drugs obtained from traditional drug discovery. The

https://doi.org/10.1515/9783110791150-003

field of repurposing that began with the serendipity, unintentional and constraint research adopted updated drug discovery tools. Drug repurposing research uses systematic, rational and high throughput techniques for obtaining the treatment for new diseases using existing drugs. The strategies used by the pharmaceutical companies are high-throughput screening platforms, in silico approach, omics-supported repurposing and use of real-world data.

3.2 Drug repurposing in neurological disorders

3.2.1 Traumatic brain injury

Mechanical brain shocks resulted in traumatic brain injuries (TBI). TBI is estimated to be one of the top three injuries that cause death and disability by 2030. Every year 50–60 million people are suffering from TBI [2]. Of all neurological disorders, TBI has the highest public health burden. The pathophysiology of the TBI is not fully understood, and thus, more than 30 clinical trials have failed over the past three decades [3]. However, the use of in silico drug design tools provided the pace for the development of novel drug candidates. Further, high-throughput screening facilitates the screening of a set of libraries against the target. But, repurposing of FDA-approved drugs performed well for the treatment of new disorders. Complex secondary injuries are reported after primary injury in TBI.

Mitochondrial membrane potential imbalance, glutamate excitotoxicity, ionic imbalance and microvascular disruption collectively generate free radicals and induce oxidative stress. Oxidative stress was found to be involved in secondary injuries that lead to neurodegeneration and death. Edaravone, a potent antioxidant was found to improve the neurological function and symptoms associated with acute ischemic stroke. Unlike other antioxidants, Edaravone, a small molecule readily crosses the Blood-Brain Barrier (BBB) and scavenges free radicals in the brain [4]. The drug is considered as multi-targeting due to its constructive action on inflammation, matrix metalloproteinases, nitric oxide production and apoptotic cell death. These neuroprotective effects became the basis for the approval of Edaravone for the treatment of Amyotrophic lateral sclerosis (ALS) in Japan and Korea (2015), the United States (2017) and Canada (2018). Besides the multi-target effect of Edaravone on TBI, a detailed clinical trail is needed to evaluate the safety and efficacy of the drug. Rhabdomyolysis, renal toxicity, thrombocytopenia, fulminant hepatitis and acute lung injury have been reported with the drug [5].

Cerebral oedema is found to be fatal in the secondary TBI. Many transporters and channels are involved in the cerebral oedema. However, SUR1-TRPM4 was found to be notably upregulated in such cases. Glyburide is reported to interact with SUR1, which is a type of sulfonylurea receptor. However, the effect of drug varies with tissues. It depends on the co-assembled proteins with the SUR1. Glyburide is found to interact with

SUR1-Kir6.2 complex, which is an ATP-sensitive potassium channel reported on the β-islet cells of the pancreas. Glyburide, the weakly acidic drug has low solubility at physiological pH. Further, the drug is also not reported to cross the BBB in normal conditions. However, TBI and ischemic brain damage lead to compromise the BBB, resulting in the decrease of pH in the extracellular fluid. Thus, the pH condition allows the drug to cross the BBB and concentrates in the neuron, astrocytes, glia and vascular endothelial cells [6, 7]. Further, the drug interacts with SUR1-TRPM4-AQP4 and SUR1-TRPM4 complexes and promotes prolonged shutdown of the cationic channel. Thus, it decreases cytopathic cellular enlargement to restrict cerebral oedema. Glyburide was also found to retain the integrity of BBB [8–10]. However, most of the studies associated with glyburide are from preclinical rat model [11, 12]. The drug candidates used in repositioning for TBI are compiled in Table 3.1.

3.2.2 Parkinson's disease

Parkinson's disease (PD) is the second most frequent neurodegenerative disorder. Nigrostriatal dopaminergic neurones degeneration is the primary pathological marker. The degeneration is primarily responsible for the motor symptoms. However, dysregulation of other brain regions is responsible for the non-motor symptoms, viz. cognition deficit and anxiety. The available therapies, viz. dopamine agonist and dopamine precursor cannot reverse or slowdown the disease. The prevalence of PD is accelerated as compared to other neurological disorders due to ageing population and trends of smoking [13]. Unfortunately, the progress of PD pathophysiology is very slow. Thus, the development of new drug candidates and completion of clinical trials are difficult, as treatment under trail should show effect of slow decline in memory or prolonged treatment regimen or large size of cohort. Thus, the drug development for the PD is greatly harnessing the possibility of drug repurposing. Amantadine, the antiviral drug was the first one in the category repurposed for the PD. The repurposing was based on the experience of PD patients during early 1960s. The PD patients undergoing influenza therapy experienced relief in motor symptoms. Thus, amantadine become part of PD treatment regimen and relieves the symptoms of dyskinesia induced by levodopa treatment. This type of drug repurposing is exceptional and based on the widely available drug interaction records. However, the repurposing is mainly driven by robust hypothesis preparation and its execution. The amantadine drug was found to antagonize the NMDA receptor and block the $K_{ir}2.1K^{+}$ channel [14]. Similarly, zonisamide, an anticonvulsant drug was clinically approved in Japan, based on observation in PD patients. Zonisamide is reported to improve the dyskinesia induced by levodopa [15, 16].

Type II diabetes mellitus (T2DM) was found to be the potential risk factor for the PD. Norwegian Prescription (NorP) database and UK Clinical Practice Research Datalink (CPRD) database were independently analysed to assess the risk of PD in T2DM patients. The effect of glitazones on the comorbidity was assessed in comparison with

Table 3.1: Drug candidates for repurposing to the TBI.

Drug	Primary use	Primary mechanism of action	Result of preclinical TBI model	Chemical structure
Progesterone	Birth control & hormonal therapy	Downstream gene modulation through by binding with PR receptor	Neuroprotective; decreases vasogenic oedema	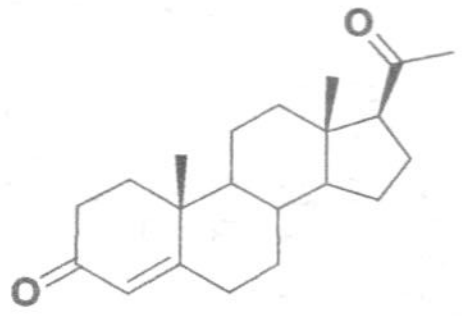 O, O
Edaravone	ALS & Stroke	Free radical scavenger	Decreases oxidative stress and lipid peroxidation	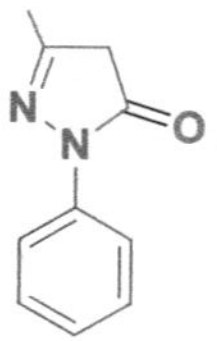 N, N, O
Levetiracetam	Epilepsy	Binds SV2A membrane protein and regulates neurotransmitter release	Control seizures by maintaining excitation/inhibition imbalance	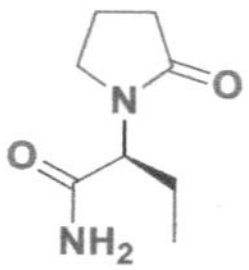 N, O, O, NH_2

Glyburide	Diabetes mellitus II	Initiates the release of insulin by antagonizing SUR1-Kir6.2 complex	Decreases cerebral oedema and maintains BBB integrity	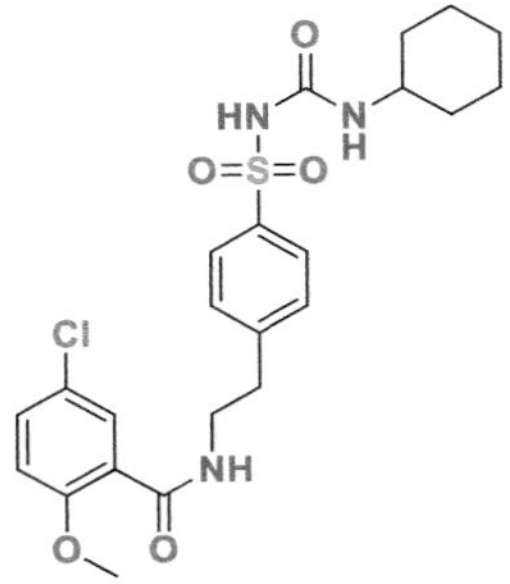
Ceftriaxone	Antibacterial	Inhibits bacterial cell wall synthesis	Improves GLT-1 expression and decreases Astrogliosis	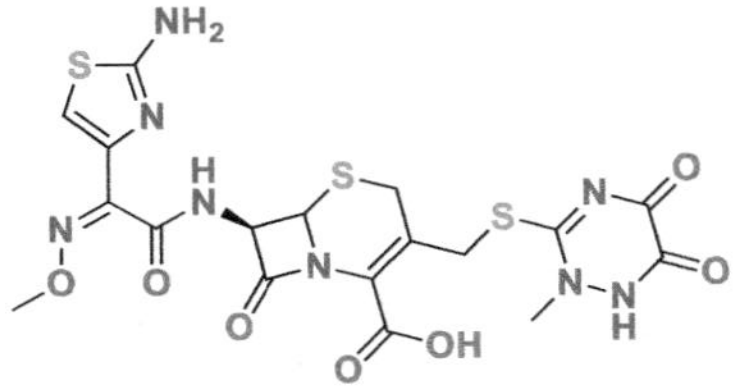

metformin or sulphonylureas [17–19]. However, the health improvement network (THIN) database assessment showed no significant difference between PD prevalence in T2DM and in control on glitazones treatment [20]. But, treatment with GLP-1 agonist (exenatide) displayed decreased incidence of PD. Preclinical studies of exenatide showed neuroprotective effect of GLP1 receptor in a dose-dependent manner [21–23]. Further, open label trial was performed with randomized controlled trial showing improvement in motor score [24, 25]. Multi-centre Phase III clinical trial is in process to assess the neuroprotective effect of exenatide in PD patients with comorbidity of T2DM (ClinicalTrials.gov Identifier: NCT04232969). The brief flowchart is pictorially represented in Figure 3.1.

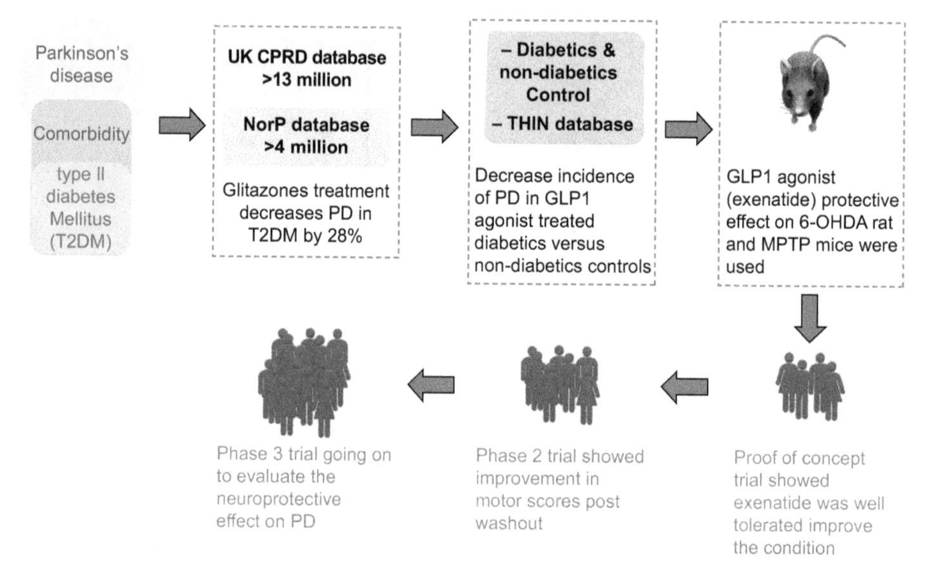

Figure 3.1: Exenatide repurposing journey for the GLP1 against PD.

3.2.3 Neuroinflammation

Analgesic therapies were repurposed for neuropathic pain. The major criteria that should be accomplished by the analgesic drugs for the repurposing are: clinical confirmation for the analgesic effect, drug tolerability, therapeutic range of the analgesic dose, drug interaction in obvious comorbidity and lack of long term effect. Reuptake inhibitor of the serotonin-norepinephrine duloxetine was developed for depression. Duloxetine is recommended as one of the first-line drugs for the treatment of neuropathic pain [26]. However, tricyclic antidepressant amitriptyline is also considered as first-line treatment for neuropathic pain. Likewise, many antidepressants having analgesic properties have been identified. Gabapentin and pregabalin are the established as standard therapeutic drugs for the treatment of neuropathic pain. However, their efficacy is found to be lesser than duloxetine and amitriptyline [26]. Many drugs have been repurposed for the management of spinal and sensory neuron-mediated neuroinflammation.

Decreasing neuroinflammation and increase of neuronal activity are the viable therapeutic approaches to treat persistent pain. Neuroinflammation shares the mechanism with the pain state. Further, neuroinflammation interferes with neurological disorders like Parkinson's disease and multiple sclerosis. Thus, it is interesting to repurpose the drug for these diseases. Various classes of the drug have anti-inflammatory properties, but few of them may act on neuroinflammation. Some antibiotics, viz. demeclocycline, ceftriaxone, chlortetracycline and minocycline, along with fingolimod, the sphingolipid modulator, as well as ibudilast, a phosphodiesterase inhibitor and antidiabetic pioglitazone are repurposed (Table 3.2).

Table 3.2: Repurposed drugs for neuroinflammation.

Drug and class	Model	Mechanism and reference
Biomarker-mediated therapy	Diagnostic biomarkers to assess the pain mechanism	Analgesic molecules; pyridoxine, SC-560, haloperidol, methylergometrine [27]
β-lactam antibiotic, Ceftriaxone	Trigeminal neuropathic pain (TNP) model in rat	Restore GLT1 in the area of dorsal horn, thus maintain hyperalgesia and hypersensitivity [28]
	Formalin-based rat model	Formalin-based flinching behaviour reduces on systemic and local administration [29]
	Formalin-based spinal nerve ligation	Delayed formalin-mediated nocifensive action, mechanical hypersensitivity unaffected but reduces thermal hyperalgesia [30]
Tetracyclines viz. Demeclocycline Chlortetracycline	Complete Freund's adjuvant (CFA) animal model, capsaicin-induced model, formalin mice model	EphB1 activation inhibition due to binding of Chlortetracycline, prevented phosphorylation in the area of dorsal root ganglion, brain and spinal cord [31]
Sphingosin-1-phosphate (S1P) modulator (Fingolimod)	Neuropathic pain (Bortezomib-induced)	Increased production of IL10, IL4 and decrease level of TNFα, IL1β in astrocytes [32]
	Mice neuropathic pain model; Chronic constriction injury (CCI)	Prolonged fingolimod mediated effect [33]
Phosphodiesterase 4 inhibitor (Ibudilast)	Rat sciatic neuronal injury	Hypersensitivity mediated by mechanical injury is rescued in sciatic neuronal injury [34]

3.3 Drug repurposing with the perspective of pharmaceutical industries and academia

3.3.1 Major contributors in the field of drug repurposing

The major contributors included pharmaceutical industries, research institute, academia and technology companies. Mostly, they worked on the principle of existing drugs having new indications, expertise of the institution, capital potential and business model. Academia is mainly focused on the development of novel molecules using the techniques of high-throughput screening (HTS) and in silico screening. The approach is not focused on the commercial potential but works for the soundness of the scientific approach. The development will attract government funding, incubate the scientific skill in the students and collaborate with the for-profit organizations. Repurposing technology industries are driven by their business model. This model primarily depends on the vision and capability of the companies. These models primarily consist of a consulting services provider, drug database platform, drug pipeline and screening facilities. Commonly, mixed model is used for maximizing the output and business success. Advanced technologies have been used by the repurposing industries. However, the access to the preclinical and clinical resources and compliance to the regulatory bodies are main hurdles. Most of the pharmaceutical companies are dependent on the life cycle management (LCM) data as part of post-marketing and late stage development. Hence, collaboration between small and large pharmaceutical companies is required for deep expertise in drug development. A successful case is the development of a drug combination of innovative memantine with donepezil for the management of Alzheimer's disease. The original proprietor of memantine is the Forrest Labs that has exclusive rights [35].

3.3.2 Screening approach

Identification of the hit molecules for drug repurposing and drug discovery is performed. These hit molecules are notably different in their outcome and uses. The screening for the de novo candidates is performed using the HTS method. The compounds libraries mentioned in the Table 3.3 are commonly used. Repurposing requires in-depth knowledge of the molecules. The molecules should be approved if failed in the clinical trial, so that their safety and toxicity profile are known. Typically, 500–2,000 molecules are approved and a similar number of compounds are failed or still in the clinical trial for the approval. The sources to access these compounds have been mentioned in Table 3.3. Further, the destiny of the hit molecule is entirely different in drug repurposing and drug discovery. The hit molecules may be promoted for the clinical trial. However, the hit molecules identified must undergo a series of hit refinement process before preclinical and clinical trial.

Cell-based and cell-free screening techniques are adopted for HTS. These are either phenotypical screens or target-focused screens. The phenotypical screens are based on cell growth, death-like cell behaviour in the presence and absence of test molecules. However, target-focused screens are based on the relationship between activity and mode of action. Phenotypical screening is reported to be more productive [36]. The method uses the signals important for the biology of the disease. In vitro screening used is questioned for its validity to mimic human diseases. Fruit fly (*Drosophila melanogaster*), worms (*Caenorhabditis elegans*) and zebrafish (*Danio rerio*) are the multicellular modelled species [35]. These are introduced with disease-causing mutation or human genes. Such species provided compelling methodology for the phenotypical screening [37]. The limitation is the extent to which the human disease is represented in these models. This model has the potential to clearly identify the drug for repurposing in the early stage.

Various laboratories working independently repurposed the molecules promoting myelin repair. Myelin remodelling includes increased differentiation of oligodendrocyte precursor (OP) to the oligodendrocytes, which promotes axon remyelination. The zebrafish model showed OP cell migration and enhanced production of the myelin protein. Benztropine identified for the primary rat optic nerve was found as the muscarinic antagonist. The compound benztropine acts as an anti-cholinergic drug used in the treatment of Parkinson's disease (PD) and Alzheimer's disease (AD). The drug was found to inhibit the B^0AT1 (SLC6A19), which is transporter of neutral amino acid. Thus, the drug may be repurposed for Type 2 diabetes [38]. The main requirement for the myelin repair is the property of the drug to cross the BBB [39]. Clemastine is histamine and muscarinic receptor antagonist that easily penetrates the BBB when used with benztropine. Further, miconazole, clobetasol and quetiapine are also reported to be muscarinic receptor antagonists and improve remyelination in the rat OP cell differentiation study [40].

3.3.3 In silico-based repurposing

Sophisticated analytical methods are used to identify the new potential association between drug and disease. The methods are mainly divided into two categories, viz. molecular approach and real-world data (RWD) approach. The molecular approach is based on the understanding of the drug activity and pathophysiology that is assisted by large "omics data". The omics data included transcriptomic, genomic, proteomic, and data based on drug structure and action. RWD approach is based on the unexpected relationship between drug and target or the symptom.

Table 3.3: Online library used for the drug repurposing by pharmaceutical companies and academia.

Repurposing Library	Small molecules	Remark	References
Drug repurposing hub	>5,000 preclinical and launched candidates	Collaboration between Map group and Broad institute cancer program	https://clue.io/repurposing
FDA-approved anticancer drug	FDA-approved anticancer drugs	Current version, AODVII contains more than 129 anticancer drugs	https://dtp.cancer.gov/organization/dscb/obtaining/available_plates.htm
NIH small Molecule Repository	Contains bioactive compounds and toxins	NIH library contains data for HTS	https://mlsmr.evotec.com/
UD Drug database collection: The microsource spectrum collection	4,000 molecules	These molecules are marketed in Europe and Asia	http://www.msdiscovery.com/intdrug.html
John S. Dunn Gulf Coast Consortium	>125 compounds	57% are used in clinic and 37% are in the trials	http://www.gulfcoastconsortia.org/
LOPAC1280	1,280 molecules	Inhibitors, receptors ligands, developed tools and approved drugs	http://www.sigmaaldrich.com/life-science/cell-biology/bioactive small-molecules/lopac1280-navigator.html
SCREEN-WELL FDA-approved drug library V2	>770 molecules	FDA-approved drug collection	http://www.enzolifesciences.com/BML-2843/screen-well-fda-approveddrug-library-v2
FDA-approved molecules library	1,447 drugs	FDA-approved drug collection	http://www.selleckchem.com/screening/fda-approved-druglibrary.html
Teva Screening Set	640 FDA and non FDA approved drugs	Teva marketed drugs	Not available

3.3.3.1 Molecular approach

Identifying the action mechanism of the drug and correlating it with the new diseases is the core of drug repurposing. Accessibility to the large amount of data in the form of omics library and availability of the detailed mechanism of action of the drug facilitate rug repurposing [41, 42]. Genomics and transcriptomics are the two databases frequently used for drug repurposing due to availability of data on drug and diseases and reproducibility of data. Gene expression microarray/RNA sequence, used to quantify the RNA, is commonly used in transcriptomics. The use of transcriptomics is based on the fact that regulating gene expression may translate to the drug. Gene regulation data was used for the repurposing of mTOR inhibitor sirolimus. Sirolimus was found to reverse the gene expression of lymphoblastic leukaemia resistant to dexamethasone [43]. Further, in silico and in vitro results found that sirolimus along with dexamethasone significantly decreases the cell viability when compared with dexamethasone treatment. Sirolimus-dexamethasone combination reinforced the transcriptomics-based approach and was reported to be a durable treatment for acute cases of xenografted lymphoblastic leukaemia [44].

3.4 Conclusion

The chapter discussed drug repurposing in the case of neurological disorders. Moreover, drug repurposing from the industry and academia perspectives are also discussed. The tool and databases used by these two are also mentioned in the chapter. Various successful case studies adopting the method of drug repurposing are elaborated.

References

[1] Ashburn TT, Thor KB. Drug repositioning: Identifying and developing new uses for existing drugs. Nat Rev Drug Discov. 2004;3(8):673–683.
[2] Maas AI, Menon DK, Manley GT, Abrams M, Åkerlund C, Andelic N, Aries M, Bashford T, Bell MJ, Bodien YG. Traumatic brain injury: Progress and challenges in prevention, clinical care, and research. Lancet Neurol. 2022.
[3] Ng SY, Lee AYW. Traumatic brain injuries: Pathophysiology and potential therapeutic targets. Front Cell Neurosci. 2019;13:528.
[4] Cruz MP. Edaravone (Radicava): A novel neuroprotective agent for the treatment of amyotrophic lateral sclerosis. Pharm Ther. 2018;43(1):25.
[5] Lapchak PA. A critical assessment of edaravone acute ischemic stroke efficacy trials: Is edaravone an effective neuroprotective therapy?. Expert Opin Pharmacother. 2010;11(10):1753–1763.
[6] Wu N, Yang X, Song L, Wei J, Liu Z. Effect of Tianqi antitremor granules on behavioral manifestations and expression of G protein-coupled receptor kinase 6 and β-arrestin1 in levodopa-induced dyskinesia in a rat model of Parkinson's disease.. Drug Des Devel Ther. 2013:1481–1489.

[7] Hersh D, Simard J, Eisenberg H. The application of glibenclamide in traumatic brain injury. In New Therapeutics for Traumatic Brain Injury. Elsevier, 2017, 95–107.
[8] Woo SK, Tsymbalyuk N, Tsymbalyuk O, Ivanova S, Gerzanich V, Simard JM. SUR1-TRPM4 channels, not KATP, mediate brain swelling following cerebral ischemia. Neurosci Lett. 2020;718:134729.
[9] Khalili H, Derakhshan N, Niakan A, Ghaffarpasand F, Salehi M, Eshraghian H, Shakibafard A, Zahabi B. Effects of oral glibenclamide on brain contusion volume and functional outcome of patients with moderate and severe traumatic brain injuries: A randomized double-blind placebo-controlled clinical trial. World Neurosurg. 2017;101:130–136.
[10] Xu Z-M, Yuan F, Liu Y-L, Ding J, Tian H-L. Glibenclamide attenuates blood–brain barrier disruption in adult mice after traumatic brain injury. J Neurotrauma. 2017;34(4):925–933.
[11] Kochanek PM, Bramlett HM, Dixon CE, Dietrich WD, Mondello S, Wang KK, Hayes RL, Lafrenaye A, Povlishock JT, Tortella FC. Operation brain trauma therapy: 2016 update.. Mil Med. 2018;183 (suppl_1):303–312.
[12] Yadav R, Weng H-R. EZH2 regulates spinal neuroinflammation in rats with neuropathic pain. Neurosci. 2017;349:106–117.
[13] Rossi A, Berger K, Chen H, Leslie D, Mailman RB, Huang X. Projection of the prevalence of Parkinson's disease in the coming decades: Revisited. Mov Disord. 2018;33(1):156–159.
[14] Shen W, Ren W, Zhai S, Yang B, Vanoye CG, Mitra A, George AL, Surmeier DJ. Striatal Kir2 K+ channel inhibition mediates the antidyskinetic effects of amantadine. J Clin Invest. 2020;130(5):2593–2601.
[15] Yang LP, Perry CM. Zonisamide: In Parkinson's disease. CNS Drugs. 2009;23:703–711.
[16] Murata M, Horiuchi E, Kanazawa I. Zonisamide has beneficial effects on Parkinson's disease patients. Neurosci Res. 2001;41(4):397–399.
[17] Brauer R, Bhaskaran K, Chaturvedi N, Dexter DT, Smeeth L, Douglas I. Glitazone treatment and incidence of Parkinson's disease among people with diabetes: A retrospective cohort study. PLoS Med. 2015;12(7):e1001854.
[18] The Norwegian prescription database. The Norwegian Institute of Public Health wwwnorpdno.
[19] Brakedal B, Flønes I, Reiter SF, Torkildsen Ø, Dölle C, Assmus J, Haugarvoll K, Tzoulis C. Glitazone use associated with reduced risk of Parkinson's disease. Mov Disord. 2017;32(11):1594–1599.
[20] Brauer R, Wei L, Ma T, Athauda D, Girges C, Vijiaratnam N, Auld G, Whittlesea C, Wong I, Foltynie T. Diabetes medications and risk of Parkinson's disease: A cohort study of patients with diabetes. Brain. 2020;143(10):3067–3076.
[21] Harkavyi A, Abuirmeileh A, Lever R, Kingsbury AE, Biggs CS, Whitton PS. Glucagon-like peptide 1 receptor stimulation reverses key deficits in distinct rodent models of Parkinson's disease. J Neuroinflammation. 2008;5:1–9.
[22] Li Y, Perry T, Kindy MS, Harvey BK, Tweedie D, Holloway HW, Powers K, Shen H, Egan JM, Sambamurti K. GLP-1 receptor stimulation preserves primary cortical and dopaminergic neurons in cellular and rodent models of stroke and Parkinsonism. Proc Natl Acad Sci. 2009;106(4):1285–1290.
[23] Harkavyi A, Whitton PS. Glucagon-like peptide 1 receptor stimulation as a means of neuroprotection. Br J Pharmacol. 2010;159(3):495–501.
[24] Aviles-Olmos I, Dickson J, Kefalopoulou Z, Djamshidian A, Ell P, Soderlund T, Whitton P, Wyse R, Isaacs T, Lees A. Exenatide and the treatment of patients with Parkinson's disease. J Clin Invest. 2013;123(6):2730–2736.
[25] Athauda D, Maclagan K, Skene SS, Bajwa-Joseph M, Letchford D, Chowdhury K, Hibbert S, Budnik N, Zampedri L, Dickson J. Exenatide once weekly versus placebo in Parkinson's disease: A randomised, double-blind, placebo-controlled trial. Lancet. 2017;390(10103):1664–1675.
[26] Finnerup NB, Attal N, Haroutounian S, McNicol E, Baron R, Dworkin RH, Gilron I, Haanpää M, Hansson P, Jensen TS. Pharmacotherapy for neuropathic pain in adults: A systematic review and meta-analysis. Lancet Neurol. 2015;14(2):162–173.

[27] Niculescu A, Le-Niculescu H, Levey D, Roseberry K, Soe K, Rogers J, Khan F, Jones T, Judd S, McCormick M. Towards precision medicine for pain: Diagnostic biomarkers and repurposed drugs. Mol Psychiatry. 2019;24(4):501–522.
[28] Luo X, He T, Wang Y, Wang J-L, Yan X-B, Zhou H-C, Wang -R-R, Du R, Wang X-L, Chen J. Ceftriaxone relieves trigeminal neuropathic pain through suppression of spatiotemporal synaptic plasticity via restoration of glutamate transporter 1 in the medullary dorsal horn. Front Cell Neurosci. 2020;14:199.
[29] Baeza-Flores GDC, Rodríguez-Palma EJ, Reyes-Pérez V, Guzmán-Priego CG, Torres-López JE. Antinociceptive effects of ceftriaxone in formalin-induced nociception. Drug Dev Res. 2020;81 (6):728–735.
[30] Eljaja L, Bjerrum OJ, Honoré PH, Abrahamsen B. Effects of the excitatory amino acid transporter subtype 2 (EAAT-2) inducer ceftriaxone on different pain modalities in rat. Scand J Pain. 2011;2 (3):132–136.
[31] Ahmed MS, Wang P, Nguyen NUN, Nakada Y, Menendez-Montes I, Ismail M, Bachoo R, Henkemeyer M, Sadek HA, Kandil ES. Identification of tetracycline combinations as EphB1 tyrosine kinase inhibitors for treatment of neuropathic pain. Proc Natl Acad Sci. 2021;118(10):e2016265118.
[32] Stockstill K, Doyle TM, Yan X, Chen Z, Janes K, Little JW, Braden K, Lauro F, Giancotti LA, Harada CM. Dysregulation of sphingolipid metabolism contributes to bortezomib-induced neuropathic pain. J Exp Med. 2018;215(5):1301–1313.
[33] Sim-Selley LJ, Wilkerson JL, Burston JJ, Hauser KF, McLane V, Welch SP, Lichtman AH, Selley DE. Differential tolerance to fty720-induced antinociception in acute thermal and nerve injury mouse pain models: Role of sphingosine-1-phosphate receptor adaptation. J Pharmacol Exp Ther. 2018;366 (3):509–518.
[34] Fujita M, Tamano R, Yoneda S, Omachi S, Yogo E, Rokushima M, Shinohara S, Sakaguchi G, Hasegawa M, Asaki T. Ibudilast produces anti-allodynic effects at the persistent phase of peripheral or central neuropathic pain in rats: Different inhibitory mechanism on spinal microglia from minocycline and propentofylline. Eur J Pharmacol. 2018;833:263–274.
[35] Howard R, McShane R, Lindesay J, Ritchie C, Baldwin A, Barber R, Burns A, Dening T, Findlay D, Holmes C. Donepezil and memantine for moderate-to-severe Alzheimer's disease. N Engl J Med. 2012;366(10):893–903.
[36] Swinney DC, Anthony J. How were new medicines discovered?. Nat Rev Drug Discov. 2011;10 (7):507–519.
[37] Pandey UB, Nichols CD. Human disease models in Drosophila melanogaster and the role of the fly in therapeutic drug discovery. Pharmacol Rev. 2011;63(2):411–436.
[38] Deshmukh VA, Tardif V, Lyssiotis CA, Green CC, Kerman B, Kim HJ, Padmanabhan K, Swoboda JG, Ahmad I, Kondo T. A regenerative approach to the treatment of multiple sclerosis. Nature. 2013;502 (7471):327–332.
[39] Najm FJ, Madhavan M, Zaremba A, Shick E, Karl RT, Factor DC, Miller TE, Nevin ZS, Kantor C, Sargent A. Drug-based modulation of endogenous stem cells promotes functional remyelination in vivo. Nature. 2015;522(7555):216–220.
[40] Xiao L, Xu H, Zhang Y, Wei Z, He J, Jiang W, Li X, Dyck L, Devon R, Deng Y. Quetiapine facilitates oligodendrocyte development and prevents mice from myelin breakdown and behavioral changes. Mol Psychiatry. 2008;13(7):697–708.
[41] Sanseau P, Agarwal P, Barnes MR, Pastinen T, Richards JB, Cardon LR, Mooser V. Use of genome-wide association studies for drug repositioning. Nat Biotechnol. 2012;30(4):317–320.
[42] Rastegar-Mojarad M, Ye Z, Kolesar JM, Hebbring SJ, Lin SM. Opportunities for drug repositioning from phenome-wide association studies. Nat Biotechnol. 2015;33(4):342–345.

[43] Lamb J, Crawford ED, Peck D, Modell JW, Blat IC, Wrobel MJ, Lerner J, Brunet J-P, Subramanian A, Ross KN. The connectivity map: Using gene-expression signatures to connect small molecules, genes, and disease. Science. 2006;313(5795):1929–1935.
[44] Teachey DT, Sheen C, Hall J, Ryan T, Brown VI, Fish J, Reid GS, Seif AE, Norris R, Chang YJ. mTOR inhibitors are synergistic with methotrexate: An effective combination to treat acute lymphoblastic leukemia. Blood J Am Soc Hematol. 2008;112(5):2020–2023.

Shikha Asthana, Vaibhav Lasure, Shivam Kumar Pandey,
Avtar Singh Gautam, Rakesh Kumar Singh

4 Implication of drug repurposing in the identification of drugs for neurological disorders

Abstract: Drug repurposing or repositioning is finding alternate therapeutic use of already clinically approved drugs for the treatment of disease. Drug discovery process costs billions of dollars and requires a very long period of time, whereas drug repurposing can overcome these drawbacks and can help identify potent drugs. Neurodegenerative diseases primarily affect the central nervous system and peripheral system, with complexed pathophysiology of the disease that limits the discovery of new drugs. However, drug repurposing utilizes the known drug profile of existing drugs for the development of new therapeutic uses of the drugs. There are various drugs (e.g., Minocycline fenfluramine and propranolol) that were developed for a particular indication, but drug repurposing techniques led to the development of another therapeutic use that has been proven effective in the treatment. This book chapter focuses on the different techniques used for the drug repurposing and the repurposed drugs for the neurodegenerative disease.

Keywords: Drug repurposing, drug repositioning, neurodegenerative disease

4.1 Introduction

Neurodegenerative diseases (NDs) represent the greatest health burden globally. NDs affect millions of people, accounting for a significant induction of morbidity and mortality globally; however, the prevalence, morbidity, and mortality vary with different types of NDs [1]. There are more than hundreds of types of ND; among these some are very prevalent like, Alzheimer's disease (AD), Parkinson's disease (PD), multiple sclerosis (MS), amyotrophic lateral sclerosis (ALS) Huntington's disease (HD), etc. [2] These NDs are characterized by the loss of neurons either by deposition of unfamiliar proteins produced due to malfunctioning of cellular metabolism or by activation of neuroinflammation. In the case of AD, cognitive decline and dementia are the results of ageing and deposition of faulty protein called amyloid beta (Aβ) and hyperphosphorylation of tau proteins. Deposition of Aβ in brain causes activation of brain-resident immune cells like astrocytes and microglia that initiate neuroinflammation [3].

The development of effective therapeutic drug candidates for NDs has been the biggest challenge of medical history and is still unachieved. The complex and multifactorial pathophysiology of these NDs are the reason for slow discovery process of

https://doi.org/10.1515/9783110791150-004

finding potent therapeutic targets [4]. Although many drug candidates are developed, the last few decades have witnessed the continuous failure of efforts made in search of a potent therapeutic compound. However, many drugs have been withdrawn from the market due to less efficacy and intolerable adverse effects [5]. Therefore, reusing existing drugs with known profile has attracted the attention to the drug repurposing approach, as this consumes less time and money.

Drug repurposing or reprofiling or drug repositioning is the development of new indications or dosage forms of existing therapeutic agents. Drug repurposing is economical and saves a lot of time [6]. The discovery of an individual drug from the chemical entity to therapeutic drug requires huge money and time investment compared to drug repurposing. The drug discovery pipeline is long and tedious because it may, most of the time, result in a failure [7]. However, in drug repurposing, most of the drug candidate details are known such as its structure, targets, and type of actions it can produce [8].

The aim of this chapter is to discuss how the drug is repurposed and different techniques used in or during drug repurposing. A list of repurposed drugs for NDs is also provided for better understanding.

4.2 Therapeutic targets in NDs explored for drug repurposing

It emphasizes how the drug repositioning process is carried out before producing repurposed medications for treating neurological illnesses. In general, drug repurposing involves three steps. First, a variety of methods are used to find potential therapeutic candidates, including data mining of clinical medication interactions, in silico drug screens, cellular drug activity assays, and serendipitous clinical observation [9]. Second, preclinical studies on neurological illnesses are carried out using in vitro cell lines and in vivo animal models [10]. Finally, extensive and multicenter clinical trials are used to assess the effectiveness and safety of repurposed medications [11]. Numerous medications are currently repurposed for neurological illnesses by using the above methods. Additionally, we have discussed many therapeutic targets for neurological diseases in this chapter, including those for Alzheimer's disease, Parkinson's disease, Huntington's disease, multiple sclerosis, amyotrophic lateral sclerosis, and multiple sclerosis.

4.3 Alzheimer's disease

In the past ten years, clinical studies that were generally unsuccessful have hampered efforts to find drugs for AD [12]. Drug repositioning, the process of finding a new therapeutic application for already-approved medications or drug candidates, is a desirable and timely drug development approach, particularly for AD [13]. Time and money are saved compared to typical de novo drug development because most safety and pharmacokinetic characteristics of repositioning candidates have previously been established. Most attempts to reposition drugs for AD have been based on encouraging clinical or epidemiological findings or in vivo efficacy discovered in animal models of AD [6]. Drug repositioning for AD will be easier with more systematic, multidisciplinary methods.

4.3.1 Calcium channel blockers

These are medications for the treatment of angina and hypertension. Dihydropyridine calcium channel blockers, like nilvadipine, have good blood-brain barrier penetration and can increase brain blood flow, thanks to their vasodilatory properties [6]. In vitro, they can decrease the production, oligomerization, and accumulation of amyloid-β, improve cell survival, and reduce neurotoxicity.

4.3.2 Antidiabetics

Since Type 2 diabetes has been discovered as a risk factor for AD, many antidiabetics have also been repurposed for this condition [14]. According to studies, AD individuals' brains have a desensitized insulin signaling system [15]. Treatment with insulin has demonstrated neuroprotection, phosphorylated tau protein levels regulation, and an improvement in memory and cognition. Insulin can also cause neuronal stem cell activation, cell proliferation, and repair [15]. As a result, substances that affect insulin release may potentially be beneficial for AD. Insulin-promoting analogues of glucagon-like peptide 1 and Dapagliflozin are SGLT2 inhibitors for Type 2 Diabetes, helping increase insulin sensitivity and CNS glucose metabolism [16]. GLP-1 agonists are widely used for Type 2 diabetes; Semaglutide lowers neuroinflammation and enhances insulin transmission in the brain [17].

4.3.3 Antimicrobials

The possibility of using antimicrobials to treat AD and its symptoms has also been researched. The macrolide antibiotics, erythromycin and azithromycin have been demonstrated to block the amyloid precursor protein, resulting in lower amounts of amyloid-β in the brain tissue [18]. Tetracyclines have also been shown to enhance the breakdown of produced fibrils and to increase the resistance of amyloid- β to trypsin digestion [19]. They also reduced oxidative stress, pointing to various mechanisms of action. Doxycycline has demonstrated potential in this regard and in conjunction with rifampicin. The most popular drug for treating mycobacterium infections, rifampicin, has been shown to have effects in the reduction of amyloid-β fibrils in a dose-dependent manner [20], most likely as a result of the reduced formation and increased clearance of amyloid-β.

4.3.4 Retinoid receptor activators

They are employed to treat skin diseases like psoriasis and acne. Reduced signaling of retinoic acid, which is essential for nerve function and repair, maybe a factor in AD [21]. Studies using acitretin have revealed an elevation of antioxidant regulation and amyloid-clearing enzymes.

4.3.5 Phosphodiesterase-5 inhibitors

The erectile dysfunction medications sildenafil and tadalafil are inhibitors of phosphodiesterase-5, a cGMP controller associated with AD [22]. cGMP may be beneficial in regulating memory impairments brought on by amyloid-β, but it is harmful in regulating learning, memory, and neuroplasticity. In older mouse models, sildenafil effectively reduced amyloid-β and suppressed neuroinflammation [23]. In addition to improving cognition and promoting neuroprotection, tadalafil crossed the blood-brain barrier more efficiently than sildenafil by inhibiting phosphodiesterase-5 [24].

4.3.6 Antidepressant drugs

Interestingly, the antidepressant trazodone has demonstrated potential in suppressing signaling through the PERK/eIF2P branch of the pro-inflammatory response, which is overactivated in individuals with Alzheimer's disease and plays a negative function in controlling protein synthesis in cells [25]. In both in vitro and in vivo experiments, trazodone successfully corrected eIF2P translational attenuation [25]. Additionally, it demonstrated neuroprotection, memory restoration, and halted neurodegeneration.

4.3.7 Miscellaneous class of drugs

There have been many other classes of drugs used for treating schizophrenia, migraine, attention-deficit/hyperactivity disorder, hypertension, and asthma but they are repurposed in AD treatment with different target approaches; these drugs are, Brexpiprazole, Caffeine, Guanfacine, Hydralazine, and Montelukast.

4.4 Parkinson's disease

In addition to α-synuclein aggregation, neuroinflammation, mitochondrial dysfunction, neuronal vulnerability, iron deposition, and changes to neural networks, PD is linked to several other pathophysiological events [26]. A tailored strategy for therapy will be necessary due to the complexity of these interconnected networks and the variability of clinical presentations. The development of novel compounds and repurposed medications is likely to be facilitated by improvements in disease modelling, high-throughput small molecule screening techniques, and analytical technology, even though present treatments only alleviate symptoms. Immunotherapies may offer an innovative method for the body to improve its response to β-synuclein [27]. A better understanding of the condition has been achieved through research in the field of cell-based therapies, and some iPSC therapies may be used in individualized therapy. Together, these developments imply a favorable future for PD treatments.

Similar to AD, a medication used to treat Parkinson's disease was first created as amantadine [28]. Amantadine was initially designed to treat influenza, and it was not until much later that it was focused on treating PD. As a weak glutamate receptor antagonist, it increases dopamine while preventing its reuptake.

4.4.1 Tyrosine kinase inhibitors

A tyrosine kinase Abl inhibitor called nilotinib is used to treat chronic myeloid leukemia. It was found that an increase in alpha-synuclein expression and, consequently, its accumulation indicate that Abl is activated in neurodegeneration. Nilotinib promotes α-synuclein breakdown by inhibiting Abl phosphorylation [29].

4.4.2 Antibiotics

Doxycycline has also been investigated for its potential anti-PD effects. It was earlier discussed as a potential anti-AD option. Doxycycline concentration variations can distinguish between antibacterial and anti-inflammatory action. According to studies, lower doses than those employed for antibiotics do not alter the susceptibility of bacteria. Still, they do demonstrate anti-inflammatory activity, which is related to their neuroprotective benefits. Doxycycline's antioxidant properties and its capacity to transform early species of α-synuclein oligomers into nontoxic and nonseeding species are additional mechanisms that support neuroprotection [30]. Apart from that, the normal monomeric forms of α-synuclein are retained, but only oligomeric species of α-synuclein have been observed to bind to doxycycline [31].

4.4.3 Antiepileptic drugs

A sulfonamide antiepileptic medication with a mixed mechanism of action, zonisamide is suitable for use in a variety of illnesses [32]. The blocking of sodium and calcium channels, the regulation of the GABAA receptor, the inhibition of carbonic anhydrase, and the inhibition of glutamate release are some of these modes of action. When therapeutic doses were applied, studies with rats revealed a rise in dopamine in the striatum. On the other hand, a reduction in intracellular dopamine was seen when greater doses were used. This medication has shown good results in treating both motor and nonmotor symptoms of Parkinson's disease [33], although its exact mode of action is yet unknown. Another monoamine oxidase-B inhibitor is zonisamide [34]. This enzyme, mostly found in astrocytes, breaks down dopamine in neural and glial cells, eventually producing free radicals, which can play a significant part in the pathogenesis of PD. Its inhibition stabilizes the amount of dopamine in the synaptic cleft and enhances the effects of dopamine.

4.4.4 GLP-1 analogues

Exenatide, a glucagon-like peptide 1 used to treat Type 2 diabetes, is similar to liraglutide, which was previously addressed. It has been researched as a PD treatment and has demonstrated neuroprotection and advantageous neuroplastic change that can slow or stop the progression of the disease. It can pass the blood-brain barrier and works by activating GLP-1 receptors to protect neurons. Exenatide has also received favorable reports for treating AD. Phase II clinical studies for the previously described drug liraglutide are now in place, with results anticipated in 2019 (clinicaltrials.gov, ID: NCT02953665).

4.4.5 Anti-asthma drugs

Notably, the anti-PD activity of agonists of the β2-adrenoreceptor has been investigated. Recent research has connected the β2-adrenoreceptor to the control of the SNCA alpha-synuclein gene [35]. More precisely, it was demonstrated that stimulation of the 2-adrenoreceptors exhibited neuroprotection. Three anti-asthmatic medications that were examined showed the most promise, with salbutamol having the best blood-brain barrier penetration and the current approval for use [35]. The study revealed that all three medications might decrease the abundance of SNCA-mRNA and α-synuclein.

4.4.6 Methylphenidate

It is a stimulant of the central nervous system that reduces dopamine and noradrenaline reuptake in the striatum and prefrontal cortex by blocking the presynaptic dopamine transporter and the noradrenaline transporter. Attention-deficit hyperactivity disorder has been treated with it. Methylphenidate has been proven in numerous studies to help lower PD-related gait problems as well as nonmotor symptoms [36].

4.5 Huntington's disease

The most common monogenic neurological disease in the developed world is HD, an autosomal dominant condition. It is characterized by dementia, behavioral and mental abnormalities, and involuntary choreatic movements [37]. The multifunctional protein huntingtin develops a mutant form due to a genetic mutation, which causes toxicity as also neuronal death and malfunction. The symptoms of HD begin to appear in adults, and they worsen with time until they eventually result in death within years. The only alternative is to control the symptoms since there is no known cure for this illness.

4.6 Antipsychotic class of drugs

4.6.1 Tetrabenazine

It was initially developed as a result of research aimed at designing simple compounds with reserpine-like antipsychotic activity. These compounds function as mild blockers of the D2 dopamine postsynaptic neurons as well as high-affinity, reversible inhibitors of monoamine absorption by presynaptic neurons. Studies on this substance as an antipsychotic were conflicting. Thus, this medication was repurposed for conditions like HD that are characterized by abnormal, involuntary hyperkinetic movements [37]. Tetrabenazine has never been shown to elicit signs of dyskinesia, making it a safer drug to use in HD than dopamine receptor blockers [38]. This has led to the testing of additional medications with dopamine-antagonistic action for the treatment of HD.

4.6.2 Clozapine

It is a neuroleptic medication that is used to treat schizophrenia with little antagonistic activity toward the D2 dopaminergic receptors; it exhibits a high affinity for the D1 and D4 dopamine receptors. Although clinical trials had mixed outcomes, it was recommended as an excellent symptomatic medication for chorea due to its low prevalence of extrapyramidal side effects [39].

4.6.3 Olanzapine

This antipsychotic medication, which is yet another one, is frequently recommended to treat behavioral and motor symptoms in HD patients. While antagonizing dopamine D2 receptors this medication has a high affinity for serotonin receptors. It might be advised when chorea is present, along with irritability, sleep disorders, and weight loss, because it is safe and well-tolerated [40].

4.6.4 Risperidone and quetiapine

This antipsychotic medication, also used to treat bipolar disorder and schizophrenia, works as a serotonin agonist and D2 receptor antagonist to treat HD chorea. It demonstrated positive effects in stabilizing mental symptoms and motor deterioration [41]. Quetiapine, an atypical antipsychotic, has a strong affinity for dopamine and serotonin receptors. Although there have not been many instances of quetiapine to treat HD symptoms, those have to emphasize the drug's effectiveness in treating chorea, particularly when mental symptoms are present [42].

4.6.5 Memantine

It is an AD therapy that is a derivative of adamantane. It is an inhibitor of N-methyl-D-aspartate (NMDA) that is noncompetitive. A large influx of calcium enters the cell due to excessive NMDA receptor stimulation, ultimately resulting in cell death. Memantine can therefore stop this calcium influx in neuronal cells and prevent the death of brain cells. When memantine's effectiveness in treating HD was investigated, it was shown that it could lessen the susceptibility of neurons to glutamate-mediated excitotoxicity [41].

4.7 Multiple sclerosis

An autoimmune condition affecting the central nervous system is MS [43]. It is a protracted, inflammatory disorder in which the myelin and axons are partially or completely damaged. Unpredictable neurological abnormalities that can be reversed at first precede their increasing progression throughout time. Although there is currently no cure for MS, there are already approved medicines that lessen its symptoms and course [44]. However, numerous anticancer medications have been repurposed for treating MS and its symptoms. In addition, the following therapeutic classes and individual medications are repurposed in MS: monoclonal antibodies, alkylating agents, antimetabolites, diuretics, and some of the individual medications, mitoxantrone, and Ibudilast.

4.7.1 Monoclonal antibodies

Ofatumumab and Teriflunomide are immunomodulatory drugs that specifically and irreversibly inhibit dihydro-orotate dehydrogenase. The former is a human monoclonal antibody to CD20 that blocks early-stage B cells. Interestingly, this medication has been repurposed to treat multiple sclerosis [45].

4.7.2 Alkylating agent

The alkylating drug known as cyclophosphamide is licensed for the treatment of leukemia, lymphomas, and breast carcinoma, among other solid tumors [46]. The majority of the cells it affects are those that divide quickly. It is related to nitrogen mustards and binds to DNA, interfering with mitosis and cell replication. Cyclophosphamide is used in MS because it can have an immunosuppressive and immunomodulatory effect [47]. It explicitly affects T- and B-cells, suppressing humoral and cell-mediated immunity. Additionally, it has been demonstrated that cyclophosphamide can increase the release of anti-inflammatory cytokines in the blood and brain, while decreasing the secretion of pro-inflammatory T helper 1 cytokines, such as interleukin-12 and interferon [48]. The phenotype of T-lymphocytes is also changed, becoming less inflammatory.

Additionally, cyclophosphamide has strong absorption in the central nervous system and can cross the blood-brain barrier. By doing so, it can exercise its immunomodulation and immunosuppressive effects, stabilizing and stopping the disease's progression [48].

4.7.3 Antimetabolites

Hairy cell leukemia and other hematopoietic tumors are treated with cladribine, an antimetabolite [49]. It is an analogue of deoxyadenosine that must be phosphorylated intracellularly into a triphosphate to become active, which results in cell death. Despite being repurposed for the treatment of MS, this medication was turned down in 2013. The European Medicines Agency recently, in 2017, approved cladribine for marketing as a therapy for this illness [50]. The reduction in circulating B- and T-lymphocytes is connected to its mode of action. Cladribine can also have neuroprotective characteristics, as demonstrated by the activation of myeloid dendritic cells that produce interferon and the interference with the synaptic effects of interleukin 1, among other potential routes [51].

4.7.4 Diuretics

A diuretic medication called amiloride treats hypertension and swelling brought on by liver or high blood pressure [52]. Its potential to protect neurons in MS has been researched. Amiloride can exert its neuroprotective and myeloprotective effects by blocking the neuronal proton-gated acid-sensing ion channel 1 (ASIC1), which is overexpressed in axons and oligodendrocytes in MS lesions [53]. A further benefit of amiloride's protective action occurring after inflammation starts is that it makes it active even before inflammation develops.

4.8 Amyotrophic lateral sclerosis

Upper and lower motor neurons, which regulate voluntary muscles, die as a result of ALS, a disease [54]. It causes muscular atrophy in which muscles gradually weaken and shrink. Other signs include difficulty breathing, swallowing, speaking, and twitching, or rigid muscles [54]. Most ALS patients have unclear etiologies, with 10% having genetic inheritance as a factor. Only two medications, riluzole and edaravone are now available to slow the progression of the disease, despite the fact that several medicines are actively being studied for their potential use in the treatment of ALS [55]. Once the symptoms start to appear, they cannot be reversed. Below is a list of some of the drug classes that have been described as repurposed.

4.8.1 Tyrosine kinase inhibitors

Tyrosine kinase inhibitor masitinib is used to treat cancer in dogs [56]. Tyrosine kinase inhibitors may be effective against the aberrant glial cells that grow in ALS, explaining their usage in the disease [57]. It has been demonstrated that masitinib improved survival and reduced glial cell activation in the relevant rat model. Ibudilast's neuroprotective effects are being investigated for use in the treatment of ALS.

4.8.2 Antiretroviral drugs

Due to the fact that ALS patients have serum concentrations of reverse transcriptase comparable to those of HIV-infected patients and that a human endogenous retrovirus is expressed in ALS patients' brains, an antiretroviral drug called Triumeq®, which is typically used as an anti-HIV therapy, has been studied for the treatment of ALS [58]. Given this, anti-HIV medications may be beneficial for treating ALS. Dolutegravir, an integrase inhibitor, abacavir and lamivudine, antiretrovirals, are combined as Triumeq®, demonstrating safety and tolerability in ALS patients.

4.8.3 Miscellaneous class of drugs

4.8.3.1 Retigabine

It is a recognized treatment for epilepsy that works by attaching to voltage-gated potassium channels to boost M-current and cause membrane hyperpolarization [59]. Retigabine has the ability to extend motor neuron survival and reduce excitability, which is helpful in the treatment of ALS [60] because it is thought that neurons in this condition are hyperexcited, firing more frequently than normal, and ultimately die. The clinical trial for this medication to treat ALS is still ongoing.

4.8.3.2 Tamoxifen

It is an antiestrogen medication that has been authorized for breast cancer chemotherapy and chemoprevention [61]. The discovery of a neurological improvement in patients and a stabilization of the condition in ALS patients who had breast cancer and were taking tamoxifen led to the drug's accidental repurposing for the treatment of ALS [62]. Its neuroprotective qualities have been previously identified, and they seem to be connected to the suppression of protein kinase C, which is overexpressed in ALS patients' spinal cords [63]. Furthermore, it was shown that tamoxifen, as an autophagy regulator, could alter a proteinopathy observed in ALS.

All the compounds presented in this chapter have been under clinical trial and some of them are stopped for the treatment of ALS, as this disease has been getting a lot of attention in the last years.

4.9 Approaches used in drug repurposing for NDs

4.9.1 Computational and in silico molecular docking approach

For computational drug repurposing, a variety of biomedical informatics databases, web servers, software modules, and cheminformatics toolkits are available. There are databases of drug-induced transcriptional signatures as well as higher-level drug annotations such as drug-drug interactions, side effects, small molecule libraries, bioactivities, and compound-target binding kinetics datasets [64]. Databases that could be beneficial for drug repurposing in neurodegenerative disorders are as follows:

i. ChEMBL: It is an open database that includes binding, functional, and ADMET (absorption, distribution, metabolism, excretion, and toxicity) details for numerous bioactive molecules that resemble drugs. These data are often manually extracted from the primary published literature, then filtered, and standardized to increase their quality and applicability across a wide range of chemical biology and drug discovery research concerns [65].

ii. DRAR-CPI: It is a web server for drug repurposing and adverse reaction prediction. A single compound may be submitted to this web server, which will then provide candidate off-target information and association scores between the selected molecule and a drug library using chemical-protein interactions [66].

iii. Drug versus Disease (DvD): DvD offers a pipeline for comparing medication and disease gene expression patterns from open microarray sources, which is accessible using Cytoscape. Positively correlated profiles can be used to infer pharmacological side effects, whilst negatively correlated profiles can be used to develop hypotheses about drug repurposing. Array Express, Gene Expression Omnibus, and Connectivity Map data are dynamically accessed by DvD, enabling users to compare medication and disease signatures [67].

iv. DrugBank 5.0: It is a database of drug and drug target information. It provides in-depth information on the nomenclature, ontology, chemistry, structure, function, and action of drugs as well as information on their pharmacology, pharmacokinetics, pharmaco-metabolomics (metabolite levels), pharmaco-transcriptomics (gene expression levels), and pharmaco-proteomics (protein expression levels) [68].

v. DrugCentral: The database provides drugs in a categorized manner based on their market availability and intellectual property rights. It divides drugs into three categories that are: OFP, (off-patent) that is, the drugs included are out of patent but in the market; ONP (on-patent), and thus the manufacturer still has rights to sell drug in the market; and OFM (off-market), that is, the drugs are not available in the market or they are withdrawn from the market. Based on the current intellectual property landscape, this classification scheme enables researchers to decide on drug repositioning [69].

vi. e-Drug3D: It is a database of 3D chemical structures for pharmaceuticals and offers various sets of ready-to-screen SD files for pharmaceuticals and commercial drug fragments. They serve as natural inputs for research on fragment-based drug design (FBDD) and drug repurposing [70].

vii. IUPHAR-DB: This database is a well-known online reference source for numerous significant classes of human therapeutic targets and associated proteins. The database incorporates information on the chemical, genetic, functional, and pathophysiological aspects of receptors and ion channels, information that has been selected and peer-reviewed from the biomedical literature by a network of experts [71].

viii. K-Map: It is a unique and user-friendly web-based database that matches a group of query kinases to kinase inhibitors systematically, using quantitative profiles of activities of kinase inhibitors. In eukaryotic cells, protein kinases are crucial regulators of signal transduction. Due to evolutionary conserved binding sites in the catalytic domain of the kinases, the majority of inhibitors that target these sites could inhibit a variety of kinases. Quantitative analysis can expose the complicated and unexpected relationship between protein kinases and their inhibitors, offering opportunities for finding multi-targeted inhibitors of different kinases for drug repurposing and development [72].

ix. NCGC pharmaceutical collection (NPC): National Institute of Health (NIH) Chemical Genomic Centre (NCGC) has built both a freely available electronic database and screening resource. It is a comprehensive compendium of all small-molecule medications approved for human use, which would be beneficial for systematic repurposing across human disorders, particularly for uncommon and neglected diseases, where the cost and time required to produce a new chemical entity are typically prohibitive [73].

x. Promiscuous: It is a comprehensive network-focused resource of protein-protein and protein-drug interactions enhanced with side effects and structural data to supply a uniform data set for additional analysis while integrating fundamental graph theoretical analysis techniques. It helps investigate and comprehend off-target effects and analyze how different drugs interact with different targets. Furthermore, this tool provides a unique starting point to find an indication and a tab for drug-repurposing [74, 75].

xi. PubChem: It provides thorough chemical information that is helpful for drug discovery, making it a valuable tool for computer-aided drug discovery. In order to create automated virtual screening processes that utilize PubChem data, it provides a variety of programmatic access mechanisms. Resources from PubChem have been employed in numerous studies for the development of bioactivity and toxicity prediction models, the discovery of polypharmacological (multi-target) ligands, and the identification of new macromolecule targets for compounds, such as for the repurposing of drugs or the prediction of off-target side effects [76].

xii. Therapeutic target database (TTD): It supports research on drug-ability, systems pharmacology, new trends, and molecular landscapes and creates drug discovery tools by focusing on target regulators and patented therapeutic agents [77].

xiii. RE-fine Drugs: It is an interactive website that is freely available for integrated medication repurposing candidate discovery and searches across databases such as PheWas (Phenome-wide association studies) and GWAS (Gene-wide association studies). Based on the transitive Drug-Gene-Disease triad theory, it may detect and rank the uniqueness of possibilities for drug repurposing [78].

4.10 Artificial intelligence and machine learning approach

Artificial intelligence is a branch of computer science that creates computer systems that mimic human problem-solving and learning abilities, whereas machine learning is a branch of artificial intelligence that aims to build a mathematical model from training data that includes input features like gene expression, laboratory results, etc. and outcome states like disease versus control [13].

4.10.1 Artificial intelligence (AI)

Artificial intelligence (AI) is a system based on technology that can replicate human intelligence by using a variety of advanced tools and networks. AI makes use of hardware and software that can analyze and learn from input data to make independent judgments for achieving predetermined goals. The goal may include reasoning, knowledge representation, and solution research and it could be possible with the use of AI and machine learning models [79].

4.10.2 Machine learning (ML)

It is a revolutionary computational approach based on state-of-the-art machine learning (ML) algorithms that predicts drug repositioning. There are various machine learning algorithms widely available such as linear regression, logistic regression, decision tree, SVM algorithm, Naive Bayes algorithm, KNN algorithm, K-means, Random-forest algorithm, Dimensionality reduction algorithms, Gradient boosting algorithm, AdaBoosting algorithm, etc.

ML is a drug-centered approach that predicts the therapeutic class of FDA-approved compounds without taking into account disease data. Information needed for drug repurposing can be categorized based on how similar the drugs in their chemical structures and their targets within the protein-protein interaction network are and the degree to which the patterns of gene expression after treatment are correlated. ML classifiers, after getting higher accuracy levels, could predict the therapeutic usefulness of a drug and its repurposing beneficence in the target disease [80].

Recent ML applications have primarily concentrated on using supervised learning algorithms to train the classifier to anticipate compounds that are active against the target protein. Over the recent few decades, various methodologies for data preparation, feature selection, model training, and prediction have been established, which include ANN (Artificial Neural Networks)-, SVM (Support vector machine)-, and DNN (Deep neural network)-based models [81]. (ANN) is one group of ML algorithms. Despite being around for more than 60 years, ANNs lost popularity in the 1990s and 2000s. ANNs have experienced a comeback in the last five years under a new name: deep artificial networks also known as Deep Learning [82]. SVM is primarily used for classification tasks, such as finding novel inhibitors against a target by distinguishing chemical classes into active and inactive subgroups based on molecule descriptors. SVMs are typically used for binary classification, but they can also produce actual values for regression analysis. These regression analyses assist in establishing the connection between the drug and the target, making it easier to anticipate the activity [83]. DNN-based models benefited significantly for their activity and toxicity prediction. Models based on DNN have made a remarkable breakthrough by winning the Merck Kaggle challenge for activity prediction and the NIH Tox21 toxicity prediction challenge [84]. Since then, DNN models are preferably used by researchers for different kinds of predictions like ADMET, toxicity, reactivity, docking, solubility, and activity studies [85].

Neurodegenerative diseases need specialized kind of frameworks as they are complex and yet not completely understood. AD framework, also well-known as DRIAD (Drug repurposing in Alzheimer's disease) finds agents that could be repurposed in AD. DRIAD can find the drugs that target proteins in signaling networks controlling innate immunity, autophagy, and microtubule dynamics and were revealed to be the best predictors of Alzheimer's disease risk. This represents previously unresearched avenues for prospective Alzheimer's treatment. It requires two types of inputs that are mRNA gene expression profiles from the human brain at various stages of AD progression based on the proposed theory of Braak and a dataset comprising DGL's list of genes expressed in the neuronal environment including microglia and astrocytes [86].

4.11 Polypharmacology approach

Polypharmacology suggests that the drug should have multiple targets, so that it can elicit multiple pharmacological actions and thus could be helpful in complex diseases such as cancer and central nervous system disorders [87]. Polypharmacology deals with the identification of off-targets of small chemical moieties or marketed drugs by the use of computational approaches such as in silico docking studies, network pharmacology, and ML techniques, and hence the identification of active compounds are beneficial for drug repurposing [88].

Several platforms have been formed to ease polypharmacology including CANDO (Computational Analysis of Novel Drug Opportunities) and Binding MOAD (Mother of all databases), which can be used for drug repurposing. Binding MOAD with the use of bound biologically relevant ligands offers users a data collection that is concentrated on high-quality x-ray crystal structures; it comprises 32,747 structures made up of 16,044 different ligands and 9,117 protein families. Information about a sequence, binding site, similar ligand, and similar binding site, protein-ligand information, etc. have been included in this database to identify potential polypharmacology compounds [89], whereas the CANDO platform predicts interactions between all drugs (a library of 3,733 approved drugs for human use) with all proteins (a library of 48,278 proteins) to assess its efficacy, allowing it to simultaneously predict the presumed effectivity of drugs for selective indications [90].

4.12 Experimental and preclinical approaches

Experimental repurposing methodologies include binding assays and phenotypic screening approaches, which can be used to investigate the binding interaction of drug-molecule to assay small molecules as well as drugs and to identify lead compounds from a large number of chemical libraries [91]. Phenotypic screening can involve screening a compound library against cell lines to evaluate the cellular response; once the compounds that affect the phenotype have been found, the disease state and the mechanism of action can then be determined [92]. The conditions required for effective drug repurposing are illustrated through the evaluation of a series of compounds in a variety of separate models to detect efficacy in one or more of the examined models [93]. In vitro phenotypic screening is required to discover and confirm candidates for repositioning by taking into account known medications or drug-like compounds that were initially used in disease-relevant phenotypic tests [94]. The path to discovering new target-based drugs frequently begins with phenotypic screens such as HDAC inhibitors like vorinostat, mTOR inhibitors like rapamycin, BET bromodomain inhibitors, and Niemann-Pick C1-Like 1 inhibitor, which were all screened using phenotypic assay and molecular target-based approach was then applied [95].

There are mainly three phenotypic approaches to screen drug repurposing for neurodegenerative diseases, which includes toxin removal, neuroprotection, and neuronal regeneration [96]. Toxin removal studies conducted to remove toxic substances, which may responsible for disease generation or progression like amyloid plaque deposition and tau hyperphosphorylation are toxicants in AD, Lewy bodies deposition in PD, etc. A study of such toxic substance removal conducted on M17 neuroblastoma cells, which endogenously have high levels of tau and were treated with several natural products found bayberry extract to reduce tau levels, and which was then screened in several other cell lines and forebrain slices of mice [97]. Neuroprotection studies are important in neurodegenerative diseases as neurons get degenerated due to toxin deposition; thus neuroprotection is a prominent strategy to slow down the progression of the disease. Among approaches of neuroprotection, one is targeting neuroinflammation as several studies have shown that neuroinflammation can lead to the degeneration of neurons. In a study to assess neuroprotection by inhibition of neuroinflammation conducted using iPSC technique, culturing mouse stem cells to differentiate into motor neurons co-cultured along with microglial BV2 cells, treated with varying doses of LPS and IFNγ resultedinBV2 cells activation and thus nitric oxide (NO) production causing degeneration of neurons. Several drugs were then treated on this culture and the levels of NO production were assessed, and then the drugs were screened out, which lowers NO levels; a few of them were also shown to activate the Nrf2 pathway, which is helpful for neuroprotection [98]. Neuronal regeneration is mainly emphasized in stem cell therapy, which either to focuses on the transplantation of stem cells to repair damaged neural networks or promotes neurogenesis by stimulating resident neural stem cells to differentiate into new neurons and form a new neural network [99, 100]. Phenotypic studies carried out in cell lines and small animals like drosophila and zebrafish are used for high-throughput screening (HTS) but do not give information about pharmacokinetic studies, the effect of a drug on the complex biological network; thus, to have a better understanding of disease pathology and varying targets of a drug as well as toxicokinetic study, drugs need to be screened in preclinical models [101, 102].

4.12.1 Preclinical approach

The preclinical approach plays a vital role in de novo drug screening as a new chemical entity or an unapproved drug is been tested to know its safety as well as efficacy profile. But this approach is not similar to drug repurposing, as drugs to be tested on animals are already FDA-approved and have established safety profiles; thus drugs will be tested on animals to find newer therapeutic indications [103]. The computational/in silico approach identifies targets as well as drugs that could be better targeted, whereas phenotypic assays performed in cell lines can identify lead molecules. These approaches suggest prominent molecules or drug candidates for disease pathology, which further need to be tested in animal models to validate their safety and efficacy [104]. Preclinical shreds of evidence are important in the case of neurodegenerative diseases to understand the underlying mechanism of drugs shown in the pathological model. AD is widely studied in animal models for de novo synthesis of drug molecules, but AD pathology is complex and newer drugs have failed to show efficacy, so scientists find drug repurposing as a promising strategy to discover newer drugs. From this perspective, various drugs are in clinical trials to assess their effectiveness in AD patients. To reach clinical trials, repurposing hit drugs have to go through preclinical studies to check their efficacy in disease models. Various drugs, which are FDA-approved medications for certain diseases, including riluzole, nicotinamide, quercetin, rapamycin, etc. are repurposed for AD. Riluzole tested in 5X FAD transgenic mice that exhibit early Aβ accumulation have found to reverse the gene expression of NMDA receptors, which are essential in learning and memory thus manifest improved cognition and are also able to reduce Aβ load [105]. Nicotinamide has proved to inhibit Aβ accumulation as well as astrocyte activation, and thus inhibit neuroinflammation and cognitive impairment in the APP/PS1 mice model of AD [106]. Quercetin, a flavonoid known for its antioxidant activity, has shown to protect cognitive and emotional function in aged triple transgenic (3x-Tg) AD mice model [107], whereas, study conducted in transgenic AD mice, rapamycin reverses memory impairment [108]. PD is also a progressive neurodegenerative disorder and the demand to discover newer medications could be possible with a better drug repurposing strategy. For PD, various drugs are being repurposed and are currently in different phases of the clinical trial, as zonisamide has appeared to improve motor symptoms in PD mice model [109], nilotinib protects dopaminergic neurons by inhibition of c-Abl in MPTP (1-methyl-4-phenyl-1,2,3,6-tetrahydropyridine)-induced mice model [110], simvastatin showed protection against oxidative stress in PD mice model [111], etc. On the other hand, for the ALS animal model, rasagiline alone or in combination has shown antiapoptotic and neuroprotective potential, and thereby increased the lifespan of the animal [112]. Antidiabetic and antihyperlipidemic drugs are most prominent among all that are currently being repurposed for neurodegenerative diseases. Metformin, which is a widely accepted Type 2 diabetes mellitus medication, has shown a reduction in motor and neuropsychiatric symptoms in the knock-in zQ175 mouse model of Huntington's disease [113].

4.13 Current repurposed drugs for ND (clinical and preclinical status)

Approximately 600 kinds of neuropathological conditions exist like Alzheimer's disease (AD), Parkinson's disease (PD), amyotrophic lateral sclerosis (ALS), multiple sclerosis (MS), and Huntinston's disease (HD). With current treatment, severe neurological problems are typically incurable in the advanced stages of the disease. Meanwhile, because each person reacts differently to drugs, medication therapy frequently lacks effectiveness and has negative side effects. Therefore, finding new drugs is the primary reason for the researcher to develop CNS-targeted medications, which are more potent and/or less harmful. In this situation, drug repurposing contributes to identifying new uses of approved drugs. It is cost-effective and time-efficient; these are reasons why drugs are being repurposed for neurodegenerative disease. Currently there are various categories of drugs repurposed for treatment of several neurodegenerative diseases as shown in Figure 4.1.

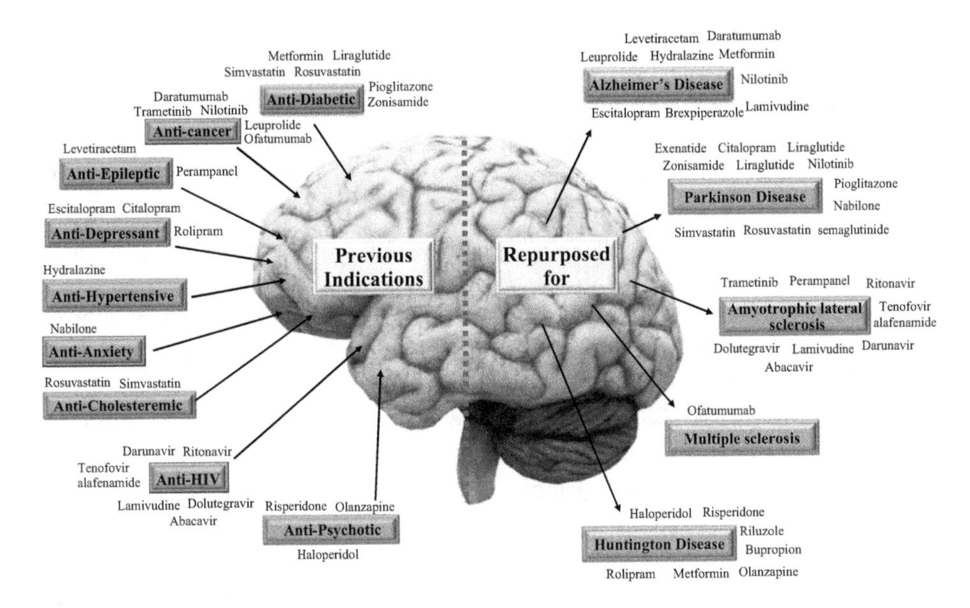

Figure 4.1: Repurposed drugs for CNS along with previous indications.

4.14 Alzheimer's disease

Alzheimer's disease (AD) is a fatal, progressing neurological condition that has a significant personal and financial impact on sufferers, their families, and society. Existing medications are being explored for AD by serendipity, side effect observations, target findings, or innovative thinking. They can function through the same mechanism of action as their conventional use or through new mechanisms. Studies have been conducted to determine whether cancer medications can be used to treat AD, for example, nilotinib, which is a protein tyrosine kinase inhibitor that is used to treat people with a blood cancer called Philadelphia chromosome-positive chronic myeloid leukemia (CML). However, now, this drug is repurposed for AD and is in phase-3 trials (NCT05143528) (Table 4.1). It was found that in AD models, nilotinib prevents the degeneration of dopamine neurons and improves memory performance [114]. Leuprolide is used as an antineoplastic and it suppresses the gonadotrophic secretion of LH and FSH hormones. This drug is also under phase 2 clinical trials (NCT00076440) (Table 4.1) for the management of AD. Dronabinol is a synthetic delta-9-THC that is used in the treatment of anorexia and weight loss in HIV patients, as well as nausea and vomiting in cancer chemotherapy. As for its intervention in AD, the drug is in phase-2 trials (NCT02792257) (Table 4.1). In a pilot study, it was found that safe and effective interventions are greatly needed for severe agitation. The safety and efficacy of dronabinol for agitation in AD are examined in this pilot trial [115].

Antidiabetics have also been investigated to see whether they could be used to treat AD and its symptoms, since Type 2 diabetes has been established as a risk factor for AD. In one of the literature studies, it was found that desensitization of insulin signaling occurs in AD patients' brains. Antidiabetic drugs that are repurposed for AD are Metformin, Semaglutinide, and dapagliflozin. These drugs are under phase-3 trials (NCT04098666) (Table 4.1), (NCT04777409) (Table 4.1) and dapagliflozin are in phase 1 and phase 2 trials (NCT03801642) (Table 4.1). It was found that antidiabetic agents may improve cognition in patients with AD [116].

Some monoclonal antibodies like canakinumab, which is a recombinant, human antihuman-IL-1β monoclonal antibody are used to treat Active Still's disease, including Adult-Onset Still's Disease (AOSD). Canakinumab was repurposed for AD, which is in phase-2 trials (NCT04795466) (Table 4.1). Another drug in this category is Daratumumab, which is a human monoclonal antibody that targets CD38, which is a cell surface protein used to treat multiple myeloma; it is being repurposed for AD. The clinical drug trials are in phase 2 (NCT04070378) (Table 4.1).

An antiepileptic drug like levetiracetam modulates neurotransmitter release by binding to the synaptic vesicle protein SV2A. This is being tested in phase 2 (NCT04004702) (Table 4.1) for AD. It was found that levetiracetam improved performance on spatial memory and executive function tasks in patients with AD and epileptiform activity [117].

4.15 Parkinson's disease

Parkinson's disease (PD) is the second most common disease in neurodegenerative disease, in which loss of dopaminergic neurons occurs in the substantia nigra, which leads to some motor (like tremors, bradycardia, rigidity) and nonmotor symptoms (sleep disorder, depression). The current treatment is effective but it is not able to cure PD. Hence, to manage these symptoms, drugs that are of a different class have been repurposed for PD. Antidiabetic drugs Exenatide, Pioglitazone, liraglutide, and semaglutinide drugs were previously approved for patients with diabetes, and now, as shown during multiple studies, these drugs are reprofiled for PD patients by showing their effect in different clinical trials. Exenatide, which is in phase 3 trials (NCT04232969) (Table 4.2) and pioglitazone in phase 2 trials are unlikely to modify progression in early PD as the dose analyzed in the studies [118].

Liraglutide, which is under phase 2 trial improves MPTP-induced motor impairments [122]. Another drug, semaglutinide is undergoing phase 2 clinical trials (NCT03659682) (Table 4.2).

Antiemetics like nabilone, a synthetic cannabinoid act as an antiemetic and analgesic used to treat nausea and vomiting by agonizing endogenous cannabinoid receptors. Moreover, this drug is repurposed in several clinical trials. As mentioned in the

Table 4.1: Repurposed drug in Alzheimer's disease.

S. no.	Drug	Class	Company	Previous indication	Phase	Status	Outcomes	References
1.	Escitalopram	Antidepressant	JHSPH Center for Clinical Trials	Depression, anxiety, and other mood disorders	Phase 3	Recruiting	–	NCT03108846
			University of Rochester		Phase 4	Completed	Data not reported.	NCT00260624
2.	Hydralazine	Antihypertensive agents	Shahid Sadoughi University of Medical Sciences and Health Services	High blood pressure (hypertension)	Phase 3	Recruiting	–	NCT04842552
3.	Metformin	Antidiabetic	Columbia university	Type 2 diabetes	Phase 3	Recruiting	–	NCT04098666
			University of Pennsylvania		Phase 2	Completed	Data not reported.	NCT01965756
4.	Nabilone	Antianxiety agents	Sunnybrook health science center	Nausea, vomiting caused by cancer chemotherapy	Phase 2	Completed	Nabilone may be an effective treatment for agitation. However, sedation and cognition should be closely monitored.	[119]
5.	Nilotinib	Anticancer	KeifeRx LLC	Chronic myeloid leukemia	Phase 3	Not yet recruiting	–	NCT05143528

(continued)

Table 4.1 (continued)

S. no.	Drug	Class	Company	Previous indication	Phase	Status	Outcomes	References
6.	Semaglutide	Antidiabetic	Novo Nordisk A/S	High blood pressure, Type 2 diabetes, or high cholesterol	Phase 3	Recruiting	–	NCT04777409
7.	Allopregnanolone	Neurosteroids; Antidepressants	University of Arizona	Postpartum depression	Phase 1	Active, not Recruiting	–	NCT03748303
			University of Southern California		Phase 1	Completed	Allopregnanolone was well-tolerated and safe across all doses in persons with early AD. Safety, MTD, and PK profiles support advancement of allopregnanolone as a regenerative therapeutic for AD to a phase 2 efficacy trial.	[120]
8.	Canakinumab	Monoclonal antibody	Novartis Pharmaceuticals	Active Still's disease, including Adult-Onset Still's Disease (AOSD)	Phase 2	Recruiting	–	NCT04795466
9.	Dapagliflozin	Antidiabetic	Jeff Burns, MD	Type 2 diabetes medicines	Phase 1	Active, not recruiting	–	NCT03801642
10.	Daratumumab	Antineoplastic agents	Marc L Gordon, MD	Multiple myeloma	Phase 2	Recruiting	–	NCT04070378

11.	Dronabinol	Analgesics	Johns Hopkins University	Nausea and vomiting caused by chemotherapy	Phase 2	Recruiting	–	NCT02792257
12.	Lamivudine	Anti-HIV agents	Bess Frost, PhD	HIV disease	Phase 2	Recruiting	–	NCT04552795
13.	Levetiracetam	Anticonvulsants	Walter Reed National Military Medical Center	Partial-onset seizures, Adjunctive therapy of myoclonic seizures	Phase 2	Not yet recruiting	–	NCT04004702
			University of Minnesota	myoclonic seizures	Phase 2	Completed	Levetiracetam improved performance on spatial memory and executive function tasks in patients with AD and epileptiform activity.	[121]
14.	Leuprolide	Antineoplastic Agents	Voyager Pharmaceutical Corporation	Central precocious puberty	Phase 2	Completed	Data not reported.	NCT00076440

study of Pinball, Seppi et al. 2022, it is in phase-2 trials and has a beneficial effect on sleep outcomes in PD Patients who are experiencing a sleeping problem [123].

Zonisamide is an antiepileptic drug that targets T-type calcium channels in neurons to treat partial seizures and mixed actions as it can also act to inhibit monoamine oxidase-B enzyme. This drug is in a clinical trial under recruiting status (NCT04182399) (Table 4.2). In PD, this drug has shown good activity in treating motor and nonmotor symptoms.

As in AD, some anticancer drugs like carmustine, which is a nitrosourea, are used as an alkylating agent in brain cancer. As mentioned in the literature, it is a lipophilic drug; due to this, it can penetrate the blood-brain barrier, and hence its effect on AD. In literature, it has been mentioned that bis-chloroethylnitrosourea (BCNU or carmustine) shows a strong reduction in amyloid beta production [124].

4.16 Amyotrophic lateral sclerosis

Amyotrophic lateral sclerosis (ALS) is also called Lou Gehrig's disease in which nerve cells are affected in the brain and spinal cord, causing loss of muscle control and muscle to become weaker and decrease in size. The only two drugs that are used in the treatment of ALS are riluzole and edaravone [125]. Still, many things are not clear about this disease. To provide effective therapeutic solutions for ALS, drugs have been repurposed.

As mentioned in AD and PD, some antiepileptic drugs are also repurposed for ALS, like Perampanel, which is a noncompetitive, selective antagonist at the AMPA glutamate receptor and is in an early phase-2 trial (NCT03793868) (Table 4.3). Perampanel compared to placebo significantly improves cortical hyperexcitability, but it has been shown that some of the adverse effects include aggression, anger, and dysarthria [126].

Vasodilatory drugs like fasudil, which is a Rho-kinase (ROCK) inhibitor are used in Raynaud's disease. This drug is in phase 2 (NCT05218668) (Table 4.3) of an ALS trial. In preclinical trials it increases, inhibits axonal degeneration motor neuron survival, and enhances axonal regeneration. In a trial, the researcher used three ALS patients to test the drug's acceptability, and they discovered that fasudil 30 mg iv twice daily for 20 days is well tolerated with no side effects [135].

Some blood coagulation factors like Albutein 5% are used in hypovolemia. This medication reprofiling for ALS has completed its phase-2 trials (NCT02872142) (Table 4.3). It was revealed in investigations that it has a tolerable safety profile in ALS patients [136].

Table 4.2: Repurposed drug in Parkinson's disease.

S. no.	Drug	Category	Company	Previously used for	Phase	Status	Outcomes	References
1.	Zonisamide	Antiepileptic	Ain Shams University	Partial seizures	NA	Recruiting	–	NCT04182399
2.	Exenatide	Antidiabetic	University College, London	Type- 2 diabetes mellitus	3	Active, not recruiting	–	NCT04232969
3.	Nabilone	Antianxiety Agents	Medical University Innsbruck	Treat nausea and vomiting caused by cancer chemotherapy	Phase 2	Completed	Beneficial effects on sleep outcomes in PD patients experiencing sleep problems at baseline.	[127]
					Phase 3	Completed	Data not reported.	NCT03773796
4.	Isradipine	Calcium channel blockers	University of Rochester	High blood pressure	Phase 3	Completed	Long-term treatment with immediate-release isradipine did not slow the clinical progression of early-stage PD.	[128]
5.	Pioglitazone	Anti diabetic	University of Rochester	Type 2 diabetes	Phase 2	Completed	Pioglitazone at the doses studied here is unlikely to modify progression in early Parkinson's disease.	[129]
6.	Nilotinib	kinase inhibitors	Northwestern University	Chronic myeloid leukemia	Phase 2	Completed	Should not be further tested in PD.	[130]

(continued)

Table 4.2 (continued)

S. no.	Drug	Category	Company	Previously used for	Phase	Status	Outcomes	References
7.	Liraglutide	Anti diabetic	Cedars-Sinai Medical Center	Type 2 diabetes mellitus	Phase 2	Active, not recruiting	DA-CH5 is superior to liraglutide and could be a therapeutic treatment for PD.	[131]
8.	Citalopram	Anti depressant	University of Michigan	Depression, anxiety, and other mood disorders	Phase 2	Recruiting	–	NCT04497168
9.	Semaglutinide	Antidiabetic	Oslo University Hospital	Type 2 diabetes	Phase 2	Not yet recruiting	–	NCT03659682
10.	Varenicline	Smoking cessation aids	Rush University Medical Center	Smoking cessation	Phase 2	Completed	Gait dysfunction and akinesia improved markedly.	[132]
11.	Modafinil	CNS stimulants	University of Arkansas	Narcolepsy, obstructive sleep apnea/hypopnea syndrome (OSAHS), or shift work sleep disorder (SWSD)	Phase 2	Completed	Data not reported.	NCT03083132
12.	Lubiprostone	Chloride channel agonist	Baylor College of Medicine	Chronic idiopathic constipation (CIC) – Adults Irritable bowel syndrome with constipation (IBS-C): adult women	Phase 4	Completed	Placebo-controlled trial of lubiprostone for constipation associated with Parkinson's disease.	[133]

13.	Simvastatin	Anticholesteremic agents	University Hospital Plymouth NHS Trust	Cholesterol-lowering	Phase 2	Completed	The primary outcome is the change in the Movement Disorder Society Unified Parkinson's Disease Rating Scale part III motor subscale score in the practically defined OFF medication state (OFF state) between baseline and 24 months.	[134]
14.	Mirabegron	Adrenergic Agents	Daniel Burdick, MD	Overactive bladder	Phase 4	Completed	Data not reported.	NCT02092181
15.	Rosuvastatin	Anticholesteremic agents	Kyowa Kirin Co., Ltd	Reduce blood levels of low-density lipoprotein	Phase 1	Completed	Data not reported.	NCT03970798

4.17 Multiple sclerosis

Multiple sclerosis (MS) is an anti-inflammatory disease that is characterized by Lhermitte's phenomenon (electric shock-like sensation in the spine or limbs on neck flexion), Uhthoff's phenomenon(worsening of symptoms with a rise in body temperature), and optic neuritis [43]. There is no proper treatment for MS; however several drugs are available to reduce the symptoms and the progression of the disease, such as glatiramer, interferons, and fingolimod, which is the first-line drug for MS. However, some drugs are repurposed for MS.

Some antitumor agents like mitoxantrone, which is used to treat breast and prostate cancer due to their immunosuppressant properties and inhibit the activation of T-cells and cyclophosphamide can permeate the blood-brain barrier and stabilize the progression of MS.

4.18 Huntington's disease

Huntington's disease (HD), also known as Huntington's chorea, is a rare autosomal-dominant, progressive neurodegenerative disease with distinct phenotypes like dystonia and chorea, cognition decline, and behavioral difficulties [137]. There is no exact medicine to cure the disease, so the only option is to decrease the symptoms.

Antipsychoactive drugs like Haloperidol, Risperidone, and olanzapine are the drugs that are used for schizophrenia with a different mechanism. These drugs work as D2 receptor antagonists. However, these are repurposed in HD in which haloperidol and risperidone are in phase-1 trials and the latter is in phase 3 trials (NCT04071639) (Table 4.5). Some investigators found that atypical antipsychotics tend to be used more in HD [138].

The psychoactive cannabinoids like delta-9-tetrahydrocannabinol and cannabidiol counteract nausea associated with cancer chemotherapy by acting as a partial agonist at both the CB1R and the CB2R. This drug has completed its phase-2 trials in which they have mentioned that Sativex (cannabis-based medicine) is safe and well-tolerated in patients with HD [140].

An antidiabetic drug like metformin is widely used as an anti-diabetic agent, which acts by activating muscle AMPK and promoting glucose uptake. Moreover, reprofiling of this drug is under phase-3 trials (NCT04826692) (Table 4.5). In different studies, it was found that metformin decreases the progression and supports the AMPK-activated protein kinase as a target against HD [113].

Drugs that are approved for treating ALS like riluzole are showing their neuroprotective action by blocking glutaminergic neurotransmission in ALS patients. But now, investigators are repositioning this drug in HD and it is in phase-3 trials (NCT00277602)

Table 4.3: Repurposed drug in Amyotrophic lateral sclerosis.

S. no	Drug	Category	Company	Previously used for	Phase	Status	Outcomes	References
1.	Biotin	Growth Substances	American University of Beirut Medical Center	Stimulates keratin production in hair	Phase 2	Completed	Data not reported.	NCT03427086
2.	Trametinib	Antineoplastic Agents	Genuv Inc.	Advanced malignant melanoma	Phase 1 Phase 2	Recruiting	–	NCT04326283
3.	Dolutegravir, Abacavir and Lamivudine	Anti-HIV agents	Macquarie University, Australia	HIV/AIDS	Phase 3	Recruiting	–	NCT05193994
5.	Colchicine 1 MG Oral Table	Anti-gout	Azienda Ospedaliero-Universitaria di Modena	Gout	Phase 2	Active, not recruiting	–	NCT03693781
6.	Darunavir, Ritonavir, Dolutegravir, Tenofovir alafenamide (TAF)	Anti-HIV agents	National Institute of Neurological Disorders and Stroke (NINDS)	HIV	Phase 1	Recruiting	–	NCT02437110
7.	*Fasudil*	Calcium Channel Blockers	Woolsey Pharmaceuticals	Raynaud's disease	Phase 2	Recruiting	–	NCT05218668
8.	Tideglusib	NSAID and neuroprotective agent	University Hospital, Geneva	Alzheimer's and progressive supranuclear palsy	Phase 2	Not yet recruiting	–	NCT05105958
9.	Enoxacin	Antibacterial Agents	McGill University	Urinary-tract infections and gonorrhea	Phase 1	Recruiting	–	NCT04840823
10	Clenbuterol	Adrenergic Agents	Dwight Koeberl, M.D., Ph.D.	Asthma	Phase 2	Completed	Data not reported.	NCT04245709

(continued)

Table 4.3 (continued)

S. no	Drug	Category	Company	Previously used for	Phase	Status	Outcomes	References
11.	Perampanel	Antiepileptic	Mayo Clinic	Partial seizures and generalized tonic-clonic seizures	Early phase 2	Completed	Data not reported.	NCT03793868
12.	Rasagiline	Central Nervous System Agents	University of Ulm	Parkinson's disease	Phase 2	Completed	Rasagiline might modify disease progression in patients with an initial slope of Amyotrophic Lateral Sclerosis Functional Rating Scale Revised greater than 0 · 5 points per month at baseline.	[139]
13.	Deferiprone	Chelating Agents	University Hospital, Lille	Thalassemia, sickle cell disease	Phase 2	Active, not Recruiting	–	NCT03293069
14.	Albutein 5%	Serum albumin; blood coagulation factors	Grifols Therapeutics LLC	Hypovolemia	Phase 2	Completed	Data not reported.	NCT02872142
15.	Pegcetacoplan (APL-2)	Complement C3 antagonist	Apellis Pharmaceuticals, Inc.	Geographic Atrophy	Phase 2	Active, not recruiting	–	NCT04579666

(Table 4.5). Researchers found that the antichoreatic and more sustained effects are shown by riluzole in HD patients [141].

4.19 Some failed drugs

Drugs may pass the in silico, in vitro, and preclinical phase, but that does not mean it passes the clinical phase; some drugs even failed when they were in the third phase of a clinical trial. Some attempts at repurposing can also fail.

Isradipine relaxes the blood vessels by blocking the calcium channel and treats hypertensive patients. At the moment, the drug completed its phase-3 trials for PD, which states that long-term treatment with immediate-release isradipine did not slow the clinical progression of early-stage PD [142].

Nilotinib is a protein tyrosine kinase inhibitor that is used to treat people with a blood cancer called Philadelphia chromosome-positive CML. On the other hand, the repurposing of the drug for PD has completed the phase-2 trials (NCT05143528) (Table 4.2). In different kinds of assessments, researchers evaluated the effect of this drug on PD disability, pharmacokinetics, cerebrospinal fluid penetration, and biomarkers after which they suggested that nilotinib is not suitable for further testing in PD [143].

Decongestant agents like clenbuterol are adrenergic agents, which are used in asthma therapy. Reprofling of the drug is in phase-2 trials (NCT04245709) (Table 4.3) and the analyses indicate that the efficacy of clenbuterol as symptomatic treatment has not been proved by investigators [144].

4.20 Opportunities and future directions of drug repurposing for neurodegenerative diseases

Drug repurposing provides opportunities to explore the drug candidates with known safety profiles that could be repurposed for ND. The known profiles of drugs help minimize the time duration and cost for the development [103].

The repurposed drugs for ND have shown promising results in the prevention of ND progression. However, there are still no potent drugs for the treatment of ND; the available drugs for ND provide symptomatic relief but not a permanent solution for the disease [58].

Drug repurposing turns to be more convenient because some of the diseases share pathological similarities like inflammation, mitochondrial dysfunction, and generation of misfolded proteins. Hence, interesting results have been seen when targeting common signaling in different diseases [6]. For example, galantamine was first developed for peripheral neuropathy, and when its ability to inhibit the acetylcholin-

Table 4.4: Repurposed drug in Multiple sclerosis.

S. no	Drug	Category	Company	Previously used for	Phase	Status	Outcomes	References
1.	Ofatumumab subcutaneous injection, Teriflunomide-matching placebo capsules, Teriflunomide capsule, Matching placebo of ofatumumab subcutaneous injections	Chronic lymphocytic leukemia	Novartis Pharmaceuticals	Anti-neoplastic Agents	Phase 3	Completed	Subcuatneous administration of ofatumumab more effective than oral terifunomide in decreasing the annualized relapse rate by 50%.	[145]

Table 4.5: Repurposed drug in Huntington's disease.

S. no.	Drug	Category	Company	Previously used for	Phase	Status	Outcomes	References
1.	Delta-9-tetrahydrocannabinol (THC) and cannabidiol (CBD)	Psychoactive cannabinoid	Fundacion para la Investigacion Biomedica del Hospital Universitario Ramon y Cajal	Counteract the nausea associated with cancer chemotherapy	Phase 2	Completed	Sativex is safe and well-tolerated in patients with HD, with no SAE or clinical worsening.	[146]
2.	Haloperidol 2 Mg Tab, Risperidone 1 Mg Tab, Zoloft 50 Mg Tablet: Idebenone Deutetrabenazine Oral Tablet [Austedo]	Anti-dyskinesia agents Antipsychotic Agents	Second Affiliated Hospital, School of Medicine, Zhejiang University	Schizophrenia	Phase 1	Recruiting	–	NCT04071639
3.	Ubiquinol	Antioxidant	University of Rochester		Phase 1	Completed	Data not reported.	NCT00980694
4.	Fenofibrate	Lipid-regulating agents	University of California, Irvine	To, reduces both cholesterol and triglycerides	Phase 2	Active, not recruiting	–	NCT03515213
5.	Minocycline	Antibacterial Agents	Merit Cudkowicz	Bacterial infection such as UTI, respiratory and skin infection	Phase 2,3	Completed	Data not reported.	NCT00277355

(continued)

Table 4.5 (continued)

S. no.	Drug	Category	Company	Previously used for	Phase	Status	Outcomes	References
6.	Dimebon (Latrepirdine)	Antihistamine drug	Medivation, Inc.	Allergy	Phase 2	Completed	Short-term administration of latrepirdine is well-tolerated in patients with HD and may have a beneficial effect on cognition. Further investigation of latrepirdine is warranted in this population with HD.	[147]
7.	Triheptanoin oil	Glycerolipids	institute National de la Santé Et de la Recherche Médicale, France	Metabolic disorders	Phase 2	Completed	Data not reported.	NCT02453061
8.	Laquinimod	Immunomodulator	Teva Branded Pharmaceutical Products R&D, Inc.	Multiple sclerosis	Phase 2	Completed	Laquinimod can directly downregulate neuronal apoptosis pathways relevant for axonal degeneration in addition to its known effects on astrocytes and microglia in the CNS. It targets a pathway that is relevant for the pathogenesis of HD, supporting the hypothesis that laquinimod may provide clinical benefit.	[148]

9.	Olanzapine Xenazine Tiapridal	Atypical antipsychotic	Assistance Publique – Hôpitaux de Paris	Schizophrenia	Phase 3	Completed	Data not reported.	NCT00632645
10.	Riluzole	Anticonvulsants	Sanofi	ALS	Phase 3	Completed	Data not reported.	NCT00277602
11.	Bupropion	Antidepressive Agents	Charite University, Berlin, Germany	Depression	Phase 2	Completed	Data not reported.	NCT01914965
12.	Metformin	Antidiabetic	Instituto de Investigacion Sanitaria La Fe	Type-2 Diabetes mellitus	Phase 3	Recruiting	–	NCT04826692
13.	Risperidone	Antipsychotic agents	University of Rochester	Schizophrenia	Phase 2	Recruiting	–	NCT04201834
14.	Rolipram	Antidepressive agents	GlaxoSmithKline	Depression	Phase 1	Completed	–	NCT01602900
15.	Amantadine	Antiviral	National Institute of Neurological Disorders and Stroke (NINDS)	Influenza A	Phase 2	Completed	–	NCT00001930

esterase was found, it was investigated for AD and was found to be effective [149]. Another drug, Imatinib, an anticancer tyrosine kinase inhibitor has shown to reduce the amyloid-beta production; however its only drawback is low blood brain barrier permeability [150].

Hence, exploring existing drugs with known profiles may provide a great opportunity to find effective treatment in a very short period of time.

4.21 Conclusion

The most common NDs are AD, PD, HD, MS, and ALS, which are still being investigated for the search of a potent therapeutic target. In light of this, drug repurposing or repositioning has been proven to be the most reliable due to very less time consumption and known drug profile. Currently, a wide range of approaches are available like computational approaches, artificial intelligence and machine learning, polypharmacology, experimental, and preclinical approaches that can be used to repurpose an existing drug.

References

[1] Erkkinen MG, Kim MO, Geschwind MD. Clinical neurology and epidemiology of the major neurodegenerative diseases. Cold Spring Harbor Perspect Biol. 2018;10. doi: 10.1101/cshperspect.a033118.

[2] Khatri DK, Kadbhane A, Patel M, Nene S, Atmakuri S, Srivastava S, Singh SB. Gauging the role and impact of drug interactions and repurposing in neurodegenerative disorders. Curr Res Pharmacol Drug Discov. 2021;2:100022. doi: https://doi.org/10.1016/j.crphar.2021.100022

[3] Jellinger KA. Basic mechanisms of neurodegeneration: A critical update. J Cell Mol Med. 2010;14:457–487. doi: 10.1111/j.1582-4934.2010.01010.x.

[4] Kakoti BB, Bezbaruah R, Ahmed N. Therapeutic drug repositioning with special emphasis on neurodegenerative diseases: Threats and issues. Front Pharmacol. 2022;13:1007315. doi: 10.3389/fphar.2022.1007315.

[5] Gribkoff VK, Kaczmarek LK. The need for new approaches in cns drug discovery: Why drugs have failed, and what can be done to improve outcomes. Neuropharmacology. 2017;120:11–19. doi: 10.1016/j.neuropharm.2016.03.021.

[6] Ballard C, Aarsland D, Cummings J, O'Brien J, Mills R, Molinuevo JL, Fladby T, Williams G, Doherty P, Corbett A, Sultana J. Drug repositioning and repurposing for Alzheimer's disease. Nat Rev Neurol. 2020;16:661–673. doi: 10.1038/s41582-020-0397-4.

[7] Low ZY, Farouk IA, Lal SK. Drug repositioning: New approaches and future prospects for life-debilitating diseases and the covid-19 pandemic outbreak. Viruses. 2020;12. doi: 10.3390/v12091058.

[8] Sahoo BM, Ravi Kumar BVV, Sruti J, Mahapatra MK, Banik BK, Borah P. Drug repurposing strategy (drs): Emerging approach to identify potential therapeutics for treatment of novel coronavirus infection. Front Mol Biosci. 2021;8:628144. doi: 10.3389/fmolb.2021.628144.

[9] O'Connor KA, Roth BL. Finding new tricks for old drugs: An efficient route for public-sector drug discovery. Nat Rev Drug Discov. 2005;4:1005–1014. doi: 10.1038/nrd1900.

[10] Turanli B, Grøtli M, Boren J, Nielsen J, Uhlen M, Arga KY, Mardinoglu A. Drug repositioning for effective prostate cancer treatment. Front Physiol. 2018;9:500. doi: 10.3389/fphys.2018.00500.

[11] Lago SG, Bahn S. Clinical trials and therapeutic rationale for drug repurposing in schizophrenia. ACS Chem Neurosci. 2019;10:58–78. doi: 10.1021/acschemneuro.8b00205.

[12] Schneider LS, Mangialasche F, Andreasen N, Feldman H, Giacobini E, Jones R, Mantua V, Mecocci P, Pani L, Winblad B, Kivipelto M. Clinical trials and late-stage drug development for Alzheimer's disease: An appraisal from 1984 to 2014. J Internal Med. 2014;275:251–283. doi: 10.1111/joim.12191.

[13] Paranjpe MD, Taubes A, Sirota M. Insights into computational drug repurposing for neurodegenerative disease. Trends Pharmacol Sci. 2019;40:565–576. doi: 10.1016/j.tips.2019.06.003.

[14] Li W, Wang T, Xiao S. Type 2 diabetes mellitus might be a risk factor for mild cognitive impairment progressing to Alzheimer's disease. Neuropsychiatr Dis Treat. 2016;12:2489–2495. doi: 10.2147/ndt.S111298.

[15] Sebastião I, Candeias E, Santos MS, De Oliveira CR, Moreira PI, Duarte AI. Insulin as a bridge between type 2 diabetes and Alzheimer's disease – How anti-diabetics could be a solution for dementia. Front Endocrinol. 2014;5:110. doi: 10.3389/fendo.2014.00110.

[16] Sa-Nguanmoo P, Tanajak P, Kerdphoo S, Jaiwongkam T, Pratchayasakul W, Chattipakorn N, Chattipakorn SC. Sglt2-inhibitor and dpp-4 inhibitor improve brain function via attenuating mitochondrial dysfunction, insulin resistance, inflammation, and apoptosis in hfd-induced obese rats. Toxicol Appl Pharmacol. 2017;333:43–50. doi: 10.1016/j.taap.2017.08.005.

[17] Nowell J, Blunt E, Edison P. Incretin and insulin signaling as novel therapeutic targets for Alzheimer's and Parkinson's disease. Mol Psychiatry. 2022. doi: 10.1038/s41380-022-01792-4.

[18] Appleby BS, Nacopoulos D, Milano N, Zhong K, Cummings JL. A review: Treatment of Alzheimer's disease discovered in repurposed agents. Dementia Geriatric Cognit Disord. 2013;35:1–22. doi: 10.1159/000345791.

[19] Stoilova T, Colombo L, Forloni G, Tagliavini F, Salmona M. A new face for old antibiotics: Tetracyclines in treatment of amyloidoses. J Med Chem. 2013;56:5987–6006. doi: 10.1021/jm400161p.

[20] Umeda T, Ono K, Sakai A, Yamashita M, Mizuguchi M, Klein WL, Yamada M, Mori H, Tomiyama T. Rifampicin is a candidate preventive medicine against amyloid-β and tau oligomers. Brain A J Neurol. 2016;139:1568–1586. doi: 10.1093/brain/aww042.

[21] Maden M. Retinoic acid in the development, regeneration and maintenance of the nervous system. Nat Rev Neurosci. 2007;8:755–765. doi: 10.1038/nrn2212.

[22] Zuccarello E, Acquarone E, Calcagno E, Argyrousi EK, Deng SX, Landry DW, Arancio O, Fiorito J. Development of novel phosphodiesterase 5 inhibitors for the therapy of Alzheimer's disease. Biochem Pharmacol. 2020;176:113818. doi: 10.1016/j.bcp.2020.113818.

[23] Sanders O. Sildenafil for the treatment of Alzheimer's disease: A systematic review. J Alzheimer's Dis Rep. 2020;4:91–106. doi: 10.3233/adr-200166.

[24] García-Barroso C, Ricobaraza A, Pascual-Lucas M, Unceta N, Rico AJ, Goicolea MA, Sallés J, Lanciego JL, Oyarzabal J, Franco R, Cuadrado-Tejedor M, García-Osta A. Tadalafil crosses the blood-brain barrier and reverses cognitive dysfunction in a mouse model of ad. Neuropharmacology. 2013;64:114–123. doi: 10.1016/j.neuropharm.2012.06.052.

[25] Halliday M, Radford H, Zents KAM, Molloy C, Moreno JA, Verity NC, Smith E, Ortori CA, Barrett DA, Bushell M, Mallucci GR. Repurposed drugs targeting eif2α-p-mediated translational repression prevent neurodegeneration in mice. Brain A J Neurol. 2017;140:1768–1783. doi: 10.1093/brain/awx074.

[26] Elkouzi A, Vedam-Mai V, Eisinger RS, Okun MS. Emerging therapies in Parkinson disease – Repurposed drugs and new approaches. Nat Rev Neurol. 2019;15:204–223. doi: 10.1038/s41582-019-0155-7.

[27] Teng JS, Ooi YY, Chye SM, Ling APK, Koh RY. Immunotherapies for Parkinson's disease: Progression of clinical development. CNS Neurol Disord Drug Targets. 2021;20:802–813. doi: 10.2174/1871527320666210526160926.
[28] Rascol O, Fabbri M, Poewe W. Amantadine in the treatment of Parkinson's disease and other movement disorders. Lancet Neurol. 2021;20:1048–1056. doi: 10.1016/s1474-4422(21)00249-0.
[29] Lonskaya I, Hebron ML, Desforges NM, Franjie A, Moussa CE. Tyrosine kinase inhibition increases functional parkin-beclin-1 interaction and enhances amyloid clearance and cognitive performance. EMBO Mol Med. 2013;5:1247–1262. doi: 10.1002/emmm.201302771.
[30] González-Lizárraga F, Socías SB, Ávila CL, Torres-Bugeau CM, Barbosa LR, Binolfi A, Sepúlveda-Díaz JE, Del-Bel E, Fernandez CO, Papy-Garcia D, Itri R, Raisman-Vozari R, Chehín RN. Repurposing doxycycline for synucleinopathies: Remodelling of α-synuclein oligomers towards non-toxic parallel beta-sheet structured species. Sci Rep. 2017;7:41755. doi: 10.1038/srep41755.
[31] Dominguez-Meijide A, Parrales V, Vasili E, González-Lizárraga F, König A, Lázaro DF, Lannuzel A, Haik S, Del Bel E, Chehín R, Raisman-Vozari R, Michel PP, Bizat N, Outeiro TF. Doxycycline inhibits α-synuclein-associated pathologies in vitro and in vivo. Neurobiol Disease. 2021;151:105256. doi: 10.1016/j.nbd.2021.105256.
[32] Kothare SV, Kaleyias J. Zonisamide:Review of pharmacology, clinical efficacy, tolerability, and safety. Expert Opin Drug Metab Toxicol. 2008;4:493–506. doi: 10.1517/17425255.4.4.493.
[33] Murata M, Hasegawa K, Kanazawa I. Zonisamide improves motor function in Parkinson's disease: A randomized, double-blind study. Neurol. 2007;68:45–50. doi: 10.1212/01.wnl.0000250236.75053.16.
[34] Özdemir Z, Alagöz MA, Bahçecioğlu Ö, Gök F. Monoamine oxidase-b (mao-b) inhibitors in the treatment of Alzheimer's and Parkinson's disease. Curr Med Chem. 2021;28:6045–6065. doi: 10.2174/0929867328666210203204710.
[35] Mittal S, Bjørnevik K, Im DS, Flierl A, Dong X, Locascio JJ, Abo KM, Long E, Jin M, Xu B, Xiang YK, Rochet JC, Engeland A, Rizzu P, Heutink P, Bartels T, Selkoe DJ, Caldarone BJ, Glicksman MA, Khurana V, Schüle B, Park DS, Riise T, Scherzer CR. B2-adrenoreceptor is a regulator of the α-synuclein gene driving risk of Parkinson's disease. Science (New York, N Y). 2017;357:891–898. doi: 10.1126/science.aaf3934.
[36] Seppi K, Ray Chaudhuri K, Coelho M, Fox SH, Katzenschlager R, Perez Lloret S, Weintraub D, Sampaio C. Update on treatments for nonmotor symptoms of Parkinson's disease-an evidence-based medicine review. Mov Disord. 2019;34:180–198. doi: 10.1002/mds.27602.
[37] Frank S. Tetrabenazine:The first approved drug for the treatment of chorea in us patients with Huntington disease. Neuropsychiatr Dis Treat. 2010;6:657–665. doi: 10.2147/ndt.S6430.
[38] Kenney C, Hunter C, Davidson A, Jankovic J. Short-term effects of tetrabenazine on chorea associated with Huntington's disease. Mov Disord. 2007;22:10–13. doi: 10.1002/mds.21161.
[39] Bampton TJ, Hack D, Galletly CA. Clozapine treatment for Huntington's disease psychosis. Aust N Z J Psychiatry. 2022;56:200. doi: 10.1177/00048674211013082.
[40] Bonelli RM, Mahnert FA, Niederwieser G. Olanzapine for Huntington's disease: An open label study. Clin Neuropharmacol. 2002;25:263–265. doi: 10.1097/00002826-200209000-00007.
[41] Cankurtaran ES, Ozalp E, Soygur H, Cakir A. Clinical experience with risperidone and memantine in the treatment of Huntington's disease. J Natl Med Assoc. 2006;98:1353–1355.
[42] Seitz DP, Millson RC. Quetiapine in the management of psychosis secondary to Huntington's disease: A case report. Can J Psychiatry Revue Canadienne de Psychiatrie. 2004;49:413. doi: 10.1177/070674370404900617.
[43] Doshi A, Chataway J. Multiple sclerosis, a treatable disease. Clin Med (London, England). 2016;16: s53–s59. doi: 10.7861/clinmedicine.16-6-s53.
[44] Dobson R, Giovannoni G. Multiple sclerosis – A review. Eur Neurol. 2019;26:27–40. doi: 10.1111/ene.13819.

[45] Hauser SL, Bar-Or A, Cohen JA, Comi G, Correale J, Coyle PK, Cross AH, De Seze J, Leppert D, Montalban X, Selmaj K, Wiendl H, Kerloeguen C, Willi R, Li B, Kakarieka A, Tomic D, Goodyear A, Pingili R, Häring DA, Ramanathan K, Merschhemke M, Kappos L. Ofatumumab versus teriflunomide in multiple sclerosis. New Engl J Med. 2020;383:546–557. doi: 10.1056/NEJMoa1917246.
[46] Vega-Stromberg T. Chemotherapy-induced secondary malignancies. J Infus Nurs. 2003;26:353–361. doi: 10.1097/00129804-200311000-00004.
[47] Perini P, Calabrese M, Rinaldi L, Gallo P. Cyclophosphamide-based combination therapies for autoimmunity. Neurol Sci. 2008;29(Suppl 2):S233–234. doi: 10.1007/s10072-008-0947-9.
[48] Neuhaus O, Archelos JJ, Hartung HP. Immunomodulation in multiple sclerosis: From immunosuppression to neuroprotection. Trends Pharmacol Sci. 2003;24:131–138. doi: 10.1016/s0165-6147(03)00028-2.
[49] Cannon T, Mobarek D, Wegge J, Tabbara IA. Hairy cell leukemia: Current concepts. Cancer Invest. 2008;26:860–865. doi: 10.1080/07357900801965034.
[50] Holmøy T, Torkildsen Ø, Myhr KM. An update on cladribine for relapsing-remitting multiple sclerosis Expert Opin Pharmacother. 2017;18:1627–1635. doi: 10.1080/14656566.2017.1372747.
[51] Balasa R, Barcutean L, Mosora O, Manu D. Reviewing the significance of blood-brain barrier disruption in multiple sclerosis pathology and treatment. Int J Mol Sci. 2021;22. doi: 10.3390/ijms22168370.
[52] Sun Q, Sever P. Amiloride: A review. J Renin Angiotensin Aldosterone Sys. 2020;21:1470320320975893. doi: 10.1177/1470320320975893.
[53] Arun T, Tomassini V, Sbardella E, De Ruiter MB, Matthews L, Leite MI, Gelineau-Morel R, Cavey A, Vergo S, Craner M, Fugger L, Rovira A, Jenkinson M, Palace J. Targeting asic1 in primary progressive multiple sclerosis: Evidence of neuroprotection with amiloride. Brain A J Neurol. 2013;136:106–115. doi: 10.1093/brain/aws325.
[54] Hulisz D. Amyotrophic lateral sclerosis: Disease state overview. Am J Manag Care AM J MANAG CARE. 2018;24:S320–s326.
[55] Chiò A, Mazzini L, Mora G. Disease-modifying therapies in amyotrophic lateral sclerosis. Neuropharmacology. 2020;167:107986. doi: 10.1016/j.neuropharm.2020.107986.
[56] Hahn KA, Ogilvie G, Rusk T, Devauchelle P, Leblanc A, Legendre A, Powers B, Leventhal PS, Kinet JP, Palmerini F, Dubreuil P, Moussy A, Hermine O. Masitinib is safe and effective for the treatment of canine mast cell tumors. J Vet Intern Med. 2008;22:1301–1309. doi: 10.1111/j.1939-1676.2008.0190.x.
[57] Trias E, Ibarburu S, Barreto-Núñez R, Babdor J, Maciel TT, Guillo M, Gros L, Dubreuil P, Díaz-Amarilla P, Cassina P, Martínez-Palma L, Moura IC, Beckman JS, Hermine O, Barbeito L. Post-paralysis tyrosine kinase inhibition with masitinib abrogates neuroinflammation and slows disease progression in inherited amyotrophic lateral sclerosis. J Neuroinflammation. 2016;13:177. doi: 10.1186/s12974-016-0620-9.
[58] Durães F, Pinto M, Sousa E. Old drugs as new treatments for neurodegenerative diseases. Pharmaceuticals (Basel, Switzerland). 2018;11. doi: 10.3390/ph11020044.
[59] Meldrum BS, Rogawski MA. Molecular targets for antiepileptic drug development. Neurotherapeutics. 2007;4:18–61. doi: 10.1016/j.nurt.2006.11.010.
[60] Wainger BJ, Kiskinis E, Mellin C, Wiskow O, Han SS, Sandoe J, Perez NP, Williams LA, Lee S, Boulting G, Berry JD, Brown RH Jr., Cudkowicz ME, Bean BP, Eggan K, Woolf CJ. Intrinsic membrane hyperexcitability of amyotrophic lateral sclerosis patient-derived motor neurons. Cell Rep. 2014;7:1–11. doi: 10.1016/j.celrep.2014.03.019.
[61] Jordan VC. Tamoxifen: A most unlikely pioneering medicine. Nat Rev Drug Discov. 2003;2:205–213. doi: 10.1038/nrd1031.
[62] Hua Y, Dai X, Xu Y, Xing G, Liu H, Lu T, Chen Y, Zhang Y. Drug repositioning: Progress and challenges in drug discovery for various diseases. Eur J Med Chem. 2022;234:114239. doi: 10.1016/j.ejmech.2022.114239.

[63] Martinez A, Palomo Ruiz MD, Perez DI, Gil C. Drugs in clinical development for the treatment of amyotrophic lateral sclerosis. Expert Opin Investig Drugs. 2017;26:403–414. doi: 10.1080/13543784.2017.1302426.

[64] Hodos RA, Kidd BA, Shameer K, Readhead BP, Dudley JT. In silico methods for drug repurposing and pharmacology. Wiley Interdiscip Rev Syst Biol Med. 2016;8:186–210. doi: 10.1002/wsbm.1337.

[65] Gaulton A, Bellis LJ, Bento AP, Chambers J, Davies M, Hersey A, Light Y, McGlinchey S, Michalovich D, Al-Lazikani B, Overington JP. Chembl: A large-scale bioactivity database for drug discovery. Nucl Acid Res. 2012;40:D1100–1107. doi: 10.1093/nar/gkr777.

[66] Luo H, Chen J, Shi L, Mikailov M, Zhu H, Wang K, He L, Yang L. Drar-cpi:A server for identifying drug repositioning potential and adverse drug reactions via the chemical–protein interactome. Nucl Acid Res. 2011;39:W492–W498. doi: 10.1093/nar/gkr299%.

[67] Pacini C, Iorio F, Gonçalves E, Iskar M, Klabunde T, Bork P, Saez-Rodriguez J. Dvd: An r/cytoscape pipeline for drug repurposing using public repositories of gene expression data. Bioinform. 2012;29:132–134. doi: 10.1093/bioinformatics/bts656.

[68] Wishart DS, Feunang YD, Guo AC, Lo EJ, Marcu A, Grant JR, Sajed T, Johnson D, Li C, Sayeeda Z, Assempour N, Iynkkaran I, Liu Y, Maciejewski A, Gale N, Wilson A, Chin L, Cummings R, Le D, Pon A, Knox C, Wilson M. Drugbank 5.0: A major update to the DrugBank database for 2018. Nucl Acid Res. 2018;46:D1074–d1082. doi: 10.1093/nar/gkx1037.

[69] Avram S, Bologa CG, Holmes J, Bocci G, Wilson TB, Nguyen D-T, Curpan R, Halip L, Bora A, Yang JJ, Knockel J, Sirimulla S, Ursu O, Oprea TI. Drugcentral 2021 supports drug discovery and repositioning. Nucl Acid Res. 2020;49:D1160–D1169. doi: 10.1093/nar/gkaa997.

[70] Pihan E, Colliandre L, Guichou J-F, Douguet D. E-drug3d: 3d structure collections dedicated to drug repurposing and fragment-based drug design. Bioinform. 2012;28:1540–1541. doi: 10.1093/bioinformatics/bts186.

[71] Sharman JL, Mpamhanga CP, Spedding M, Germain P, Staels B, Dacquet C, Laudet V, Harmar AJ. Iuphar-db: New receptors and tools for easy searching and visualization of pharmacological data. Nucl Acid Res. 2011;39:D534–538. doi: 10.1093/nar/gkq1062.

[72] Kim J, Yoo M, Kang J, Tan AC. K-map: Connecting kinases with therapeutics for drug repurposing and development. Hum Genomics. 2013;7:20. doi: 10.1186/1479-7364-7-20.

[73] Huang R, Southall N, Wang Y, Yasgar A, Shinn P, Jadhav A, Nguyen DT, Austin CP. The NCGC pharmaceutical collection: A comprehensive resource of clinically approved drugs enabling repurposing and chemical genomics. Sci Transl Med. 2011;3:80ps16. doi: 10.1126/scitranslmed.3001862.

[74] Von Eichborn J, Murgueitio MS, Dunkel M, Koerner S, Bourne PE, Preissner R. Promiscuous: A database for network-based drug-repositioning. Nucl Acid Res. 2011;39:D1060–1066. doi: 10.1093/nar/gkq1037.

[75] Gallo K, Goede A, Eckert A, Moahamed B, Preissner R, Gohlke B-O. Promiscuous 2.0: A resource for drug-repositioning. Nucl Acid Res. 2020;49:D1373–D1380. doi: 10.1093/nar/gkaa1061.

[76] Kim S. Getting the most out of pubchem for virtual screening. Expert Opin Drug Discov. 2016;11:843–855. doi: 10.1080/17460441.2016.1216967.

[77] Wang Y, Zhang S, Li F, Zhou Y, Zhang Y, Wang Z, Zhang R, Zhu J, Ren Y, Tan Y, Qin C, Li Y, Li X, Chen Y, Zhu F. Therapeutic target database 2020: Enriched resource for facilitating research and early development of targeted therapeutics. Nucl Acid Res. 2019;48:D1031–D1041. doi: 10.1093/nar/gkz981.

[78] Moosavinasab S, Patterson J, Strouse R, Rastegar-Mojarad M, Regan K, Payne PR, Huang Y, Lin SM. 'Re:Finedrugs': An interactive dashboard to access drug repurposing opportunities. Database. 2016. doi: 10.1093/database/baw083.

[79] Paul D, Sanap G, Shenoy S, Kalyane D, Kalia K, Tekade RK. Artificial intelligence in drug discovery and development. Drug Discov Today. 2021;26:80–93. doi: 10.1016/j.drudis.2020.10.010.

[80] Napolitano F, Zhao Y, Moreira VM, Tagliaferri R, Kere J, D'Amato M, Greco D. Drug repositioning: A machine-learning approach through data integration. J Cheminf. 2013;5:30. doi: 10.1186/1758-2946-5-30.
[81] Kashyap K, Siddiqi MI. Recent trends in artificial intelligence-driven identification and development of anti-neurodegenerative therapeutic agents. Mol Divers. 2021;25:1517–1539. doi: 10.1007/s11030-021-10274-8.
[82] Jones LD, Golan D, Hanna SA, Ramachandran M. Artificial intelligence, machine learning and the evolution of healthcare: A bright future or cause for concern?. Bone Jt Res. 2018;7:223–225. doi: 10.1302/2046-3758.73.Bjr-2017-0147.R1.
[83] Heikamp K, Bajorath J. Support vector machines for drug discovery. Expert Opin Drug Discov. 2014;9:93–104. doi: 10.1517/17460441.2014.866943.
[84] Ma J, Sheridan RP, Liaw A, Dahl GE, Svetnik V. Deep neural nets as a method for quantitative structure–activity relationships. J Chem Inf Model. 2015;55:263–274. doi: 10.1021/ci500747n.
[85] Hu S, Chen P, Gu P, Wang B. A deep learning-based chemical system for qsar prediction. IEEE J Biomed Health Inform. 2020;24:3020–3028. doi: 10.1109/JBHI.2020.2977009.
[86] Rodriguez S, Hug C, Todorov P, Moret N, Boswell SA, Evans K, Zhou G, Johnson NT, Hyman BT, Sorger PK, Albers MW, Sokolov A. Machine learning identifies candidates for drug repurposing in Alzheimer's disease. Nat Commun. 2021;12:1033. doi: 10.1038/s41467-021-21330-0.
[87] Anighoro A, Bajorath J, Polypharmacology: RG. Challenges and opportunities in drug discovery. J Med Chem. 2014;57:7874–7887. doi: 10.1021/jm5006463.
[88] Achenbach J, Tiikkainen P, Franke L, Proschak E. Computational Tools for Polypharmacology and Repurposing, Vol. 3, 2011, 961–968. doi: 10.4155/fmc.11.62.
[89] Smith RD, Clark JJ, Ahmed A, Orban ZJ, Dunbar JB, Carlson HA. Updates to binding moad (mother of all databases): Polypharmacology tools and their utility in drug repurposing. J Mol Biol. 2019;431:2423–2433. doi: https://doi.org/10.1016/j.jmb.2019.05.024
[90] Chopra G, Samudrala R. Exploring polypharmacology in drug discovery and repurposing using the cando platform. Curr Pharm Des. 2016;22:3109–3123. doi: 10.2174/1381612822666160325121943.
[91] Pushpakom S, Iorio F, Eyers PA, Escott KJ, Hopper S, Wells A, Doig A, Guilliams T, Latimer J, McNamee C, Norris A, Sanseau P, Cavalla D, Pirmohamed M. Drug repurposing: Progress, challenges and recommendations. Nat Rev Drug Discov. 2019;18:41–58. doi: 10.1038/nrd.2018.168.
[92] Lage OM, Ramos MC, Calisto R, Almeida E, Vasconcelos V, Vicente F. Current screening methodologies in drug discovery for selected human diseases. 2018;16:279.
[93] Reaume AG. Drug repurposing through nonhypothesis driven phenotypic screening. Drug Discov Today Ther Strateg. 2011;8:85–88. doi: https://doi.org/10.1016/j.ddstr.2011.09.007
[94] Wilkinson GF, Pritchard KJJOBS. In Vitro Screening for Drug Repositioning, Vol. 20, 2015, 167–179.
[95] Brown DG, Wobst HJ. Opportunities and challenges in phenotypic screening for neurodegenerative disease research. J Med Chem. 2020;63:1823–1840. doi: 10.1021/acs.jmedchem.9b00797.
[96] Zhang M, Luo G, Zhou Y, Wang S, Zhong Z. Phenotypic screens targeting neurodegenerative diseases. SLAS Discovery. 2014;19:1–16. doi: https://doi.org/10.1177/1087057113499777
[97] Jones JR, Lebar MD, Jinwal UK, Abisambra JF, Koren J, Blair L, O'Leary JC, Davey Z, Trotter J, Johnson AG, Weeber E, Eckman CB, Baker BJ, Dickey CA. The diarylheptanoid (+)-ar,11s-myricanol and two flavones from bayberry (myrica cerifera) destabilize the microtubule-associated protein tau. J Nat Prod. 2011;74:38–44. doi: 10.1021/np100572z.
[98] Höing S, Rudhard Y, Reinhardt P, Glatza M, Stehling M, Wu G, Peiker C, Böcker A, Parga JA, Bunk E, Schwamborn JC, Slack M, Sterneckert J, Schöler HR. Discovery of inhibitors of microglial neurotoxicity acting through multiple mechanisms using a stem-cell-based phenotypic assay. Cell Stem Cell. 2012;11:620–632. doi: https://doi.org/10.1016/j.stem.2012.07.005
[99] Aboody K, Capela A, Niazi N, Stern JH, Temple SJN. Translating stem cell studies to the clinic for cns repair: Current state of the art and the need for a rosetta stone. 2011;70:597–613.

[100] Ming G-L, Song HJARN. Adult Neurogenesis in the Mammalian Central Nervous System, Vol. 28, 2005, 223–250.
[101] Hellerstein MK. Exploiting complexity and the robustness of network architecture for drug discovery. J Pharmacol Exp Ther. 2008;325:1–9. doi: 10.1124/jpet.107.131276.
[102] Singh VK, Seed TM. How necessary are animal models for modern drug discovery?. Expert Opin Drug Discov. 2021;16:1391–1397. doi: 10.1080/17460441.2021.1972255.
[103] Mithun R, Shubham JK, Anil GJ. Drug Repurposing (Dr): An Emerging Approach in Drug Discovery, 2020. doi: 10.5772/intechopen.93193.
[104] Agamah FE, Mazandu GK, Hassan R, Bope CD, Thomford NE, Ghansah A, Chimusa ER. Computational/in silico methods in drug target and lead prediction. Brief Bioinform. 2020;21:1663–1675. doi: 10.1093/bib/bbz103.
[105] Okamoto M, Gray JD, Larson CS, Kazim SF, Soya H, McEwen BS, Pereira AC. Riluzole reduces amyloid beta pathology, improves memory, and restores gene expression changes in a transgenic mouse model of early-onset Alzheimer's disease. Transl Psychiatry. 2018;8:153. doi: 10.1038/s41398-018-0201-z.
[106] Xie X, Gao Y, Zeng M, Wang Y, Wei TF, Lu YB, Zhang WP. Nicotinamide ribose ameliorates cognitive impairment of aged and Alzheimer's disease model mice. Metab Brain Dis. 2019;34:353–366. doi: 10.1007/s11011-018-0346-8.
[107] Sabogal-Guáqueta AM, Muñoz-Manco JI, Ramírez-Pineda JR, Lamprea-Rodriguez M, Osorio E, Cardona-Gómez GP. The flavonoid quercetin ameliorates Alzheimer's disease pathology and protects cognitive and emotional function in aged triple transgenic Alzheimer's disease model mice. Neuropharmacology. 2015;93:134–145. doi: https://doi.org/10.1016/j.neuropharm.2015.01.027
[108] Van Skike CE, Hussong SA, Hernandez SF, Banh AQ, DeRosa N, Galvan V. Mtor attenuation with rapamycin reverses neurovascular uncoupling and memory deficits in mice modeling Alzheimer's disease. 2021;41:4305–4320. doi: 10.1523/JNEUROSCI.2144-20.2021.
[109] Uemura MT, Asano T, Hikawa R, Yamakado H, Takahashi R. Zonisamide inhibits monoamine oxidase and enhances motor performance and social activity. Neurosci Res. 2017;124:25–32. doi: https://doi.org/10.1016/j.neures.2017.05.008
[110] Karuppagounder SS, Brahmachari S, Lee Y, Dawson VL, Dawson TM, Ko HS. The c-abl inhibitor, nilotinib, protects dopaminergic neurons in a preclinical animal model of Parkinson's disease. Sci Rep. 2014;4:4874. doi: 10.1038/srep04874.
[111] Tong H, Zhang X, Meng X, Lu L, Mai D, Qu S. Simvastatin Inhibits Activation of Nadph Oxidase/p38 Mapk Pathway and Enhances Expression of Antioxidant Protein in Parkinson Disease Models, Vol. 11, 2018. doi: 10.3389/fnmol.2018.00165.
[112] Waibel S, Reuter A, Malessa S, Blaugrund E, Ludolph AC. Rasagiline alone and in combination with riluzole prolongs survival in an ALS mouse model. J Neurol. 2004;251:1080–1084. doi: 10.1007/s00415-004-0481-5.
[113] Sanchis A, García-Gimeno MA, Cañada-Martínez AJ, Sequedo MD, Millán JM, Sanz P, Vázquez-Manrique RP. Metformin treatment reduces motor and neuropsychiatric phenotypes in the zq175 mouse model of Huntington's disease. Exp Mol Med. 2019;51:1–16. doi: 10.1038/s12276-019-0264-9.
[114] La Barbera L, Vedele F, Nobili A, Krashia P, Spoleti E, Latagliata EC, Cutuli D, Cauzzi E, Marino R, Viscomi MT, Petrosini L, Puglisi-Allegra S, Melone M, Keller F, Mercuri NB, Conti F, D'Amelio M. Nilotinib restores memory function by preventing dopaminergic neuron degeneration in a mouse model of Alzheimer's disease. Prog Neurobiol. 2021;202:102031. doi: 10.1016/j.pneurobio.2021.102031.
[115] Outen J, Rosenberg P, Vandrey R, Amjad H, Burhanullah H, Agronin M, Castaneda R, Isesalaya M, Walsh P, Ash E. Pilot trial of dronabinol adjunctive treatment of agitation in Alzheimer's disease (thc-ad). Am J Geriatric Psychiatry. 2021;29:S115–S117.

[116] Michailidis M, Tata DA, Moraitou D, Kavvadas D, Karachrysafi S, Papamitsou T, Vareltzis P, Papaliagkas V. Antidiabetic drugs in the treatment of Alzheimer's disease. Int J Mol Sci. 2022;23. doi: 10.3390/ijms23094641.

[117] Vossel K, Ranasinghe KG, Beagle AJ, La A, Ah Pook K, Castro M, Mizuiri D, Honma SM, Venkateswaran N, Koestler M, Zhang W, Mucke L, Howell MJ, Possin KL, Kramer JH, Boxer AL, Miller BL, Nagarajan SS, Kirsch HE. Effect of levetiracetam on cognition in patients with Alzheimer's disease with and without epileptiform activity: A randomized clinical trial. JAMA Neurol. 2021;78:1345–1354. doi: 10.1001/jamaneurol.2021.3310.

[118] Investigators. NETI-PDN-PF-Z. Pioglitazone in early Parkinson's disease: A phase 2, multicentre, double-blind, randomised trial. Lancet Neurol. 2015;14:795–803. doi: 10.1016/s1474-4422(15)00144-1.

[119] Herrmann N, Ruthirakuhan M, Gallagher D, Verhoeff NPL, Kiss A, Black SE, Lanctôt KLJTAJOGP. Randomized Placebo-controlled Trial of Nabilone for Agitation in Alzheimer's Disease, Vol. 27, 2019, 1161–1173.

[120] Hernandez GD, Solinsky CM, Mack WJ, Kono N, Rodgers KE, Wu CY, Mollo AR, Lopez CM, Pawluczyk S, Bauer GJAS, Research DT. Interventions, C. Safety, Tolerability, and Pharmacokinetics of Allopregnanolone as A Regenerative Therapeutic for Alzheimer's Disease: A Single and Multiple Ascending Dose Phase 1b/2a Clinical Trial, Vol. 6, 2020, e12107.

[121] Vossel K, Ranasinghe KG, Beagle AJ, La A, Pook KA, Castro M, Mizuiri D, Honma SM, Venkateswaran N, Koestler MJJN. Effect of Levetiracetam on Cognition in Patients with Alzheimer's Disease with and without Epileptiform Activity: A Randomized Clinical Trial, Vol. 78, 2021, 1345–1354.

[122] Zhang L, Zhang L, Li Y, Li L, Melchiorsen JU, Rosenkilde M, Hölscher C. The novel dual glp-1/gip receptor agonist da-ch5 is superior to single glp-1 receptor agonists in the mptp model of Parkinson's disease. J Parkinsons Dis. 2020;10:523–542. doi: 10.3233/jpd-191768.

[123] Peball M, Seppi K, Krismer F, Knaus HG, Spielberger S, Heim B, Ellmerer P, Werkmann M, Poewe W, Djamshidian A. Effects of nabilone on sleep outcomes in patients with Parkinson's disease: A post-hoc analysis of nms-nab study. Mov Disord Clin Pract. 2022;9:751–758. doi: 10.1002/mdc3.13471.

[124] Hayes CD, Dey D, Palavicini JP, Wang H, Patkar KA, Minond D, Nefzi A, Lakshmana MK. Striking reduction of amyloid plaque burden in an Alzheimer's mouse model after chronic administration of carmustine. BMC Med. 2013;11:81. doi: 10.1186/1741-7015-11-81.

[125] Meyer T. [amyotrophic lateral sclerosis (als) – Diagnosis, course of disease and treatment options]. Dtsch Med Wochenschr (1946). 2021;146:1613–1618. doi: 10.1055/a-1562-7882.

[126] Turalde CWR, Moalong KMC, Espiritu AI, Prado MB Jr. Perampanel for amyotrophic lateral sclerosis: A systematic review and meta-analysis. Neurol Sci. 2022;43:889–897. doi: 10.1007/s10072-022-05867-6.

[127] Peball M, Seppi K, Krismer F, Knaus HG, Spielberger S, Heim B, Ellmerer P, Werkmann M, Poewe W, Djamshidian AJMDCP. Effects of Nabilone on Sleep Outcomes in Patients with Parkinson's Disease: A Post-hoc Analysis of Nms-nab Study, Vol. 9, 2022, 751–758.

[128] medicine -PSGS-PIIJAOI. Isradipine versus Placebo in Early Parkinson Disease: A Randomized Trial, Vol. 172, 2020, 591–598.

[129] Simuni T, Kieburtz K, Tilley B, Elm JJ, Ravina B, Babcock D, Emborg M, Feigin A, Zweig R. Ninds Exploratory Trials, P. J. L. N. Pioglitazone in Early Parkinson's Disease: A Phase 2, Multicentre, Double-blind, Randomised Trial, 2015, 14.

[130] Simuni T, Fiske B, Merchant K, Coffey CS, Klingner E, Caspell-Garcia C, Lafontant D-E, Matthews H, Wyse RK, Brundin PJJN. Efficacy of Nilotinib in Patients with Moderately Advanced Parkinson Disease: A Randomized Clinical Trial, Vol. 78, 2021, 312–320.

[131] Zhang L, Zhang L, Li Y, Li L, Melchiorsen JU, Rosenkilde M, Hölscher CJJOPSD. The novel dual glp-1/gip receptor agonist da-ch5 is superior to single glp-1 receptor agonists in the mptp model of Parkinson's disease. 2020;10:523–542.

[132] Zesiewicz TA, Sullivan KLJCN. Treatment of ataxia and imbalance with varenicline (chantix): Report of 2 patients with spinocerebellar ataxia (types 3 and 14). 2008;31:363–365.

[133] Ondo WG, Kenney C, Sullivan K, Davidson A, Hunter C, Jahan I, McCombs A, Miller A, Zesiewicz TJN. Placebo-controlled trial of lubiprostone for constipation associated with Parkinson disease. 2012;78:1650–1654.

[134] Carroll CB, Webb D, Stevens KN, Vickery J, Eyre V, Ball S, Wyse R, Webber M, Foggo A, Zajicek JJBO. Simvastatin as a neuroprotective treatment for Parkinson's disease (pd stat): Protocol for a double-blind, randomised, placebo-controlled futility study. 2019;9:e029740.

[135] Koch JC, Kuttler J, Maass F, Lengenfeld T, Zielke E, Bähr M, Lingor P. Compassionate use of the rock inhibitor fasudil in three patients with amyotrophic lateral sclerosis. Front Neurol. 2020;11:173. doi: 10.3389/fneur.2020.00173.

[136] Povedano M, Paipa A, Barceló M, Woodward MK, Ortega S, Domínguez R, Aragonés ME, Horrillo R, Costa M, Páez A. Plasma exchange with albumin replacement and disease progression in amyotrophic lateral sclerosis: A pilot study. Neurol Sci. 2022;43:3211–3221. doi: 10.1007/s10072-021-05723-z.

[137] Walker FO. Huntington's disease. Lancet (London, England). 2007;369:218–228. doi: 10.1016/s0140-6736(07)60111-1.

[138] Unti E, Mazzucchi S, Palermo G, Bonuccelli U, Ceravolo R. Antipsychotic drugs in Huntington's disease. Expert Rev Neurother. 2017;17:227–237.

[139] Ludolph AC, Schuster J, Dorst J, Dupuis L, Dreyhaupt J, Weishaupt JH, Kassubek J, Weiland U, Petri S, Meyer TJTLN. Safety and efficacy of rasagiline as an add-on therapy to riluzole in patients with amyotrophic lateral sclerosis: A randomised, double-blind, parallel-group, placebo-controlled, phase 2 trial. 2018;17:681–688.

[140] López-Sendón Moreno JL, García Caldentey J, Trigo Cubillo P, Ruiz Romero C, García Ribas G, Alonso Arias MA, García de Yébenes MJ, Tolón RM, Galve-Roperh I, Sagredo O, Valdeolivas S, Resel E, Ortega-Gutierrez S, García-Bermejo ML, Fernández Ruiz J, Guzmán M, García de Yébenes Prous J. A double-blind, randomized, cross-over, placebo-controlled, pilot trial with sativex in Huntington's disease. J Neurol. 2016;263:1390–1400. doi: 10.1007/s00415-016-8145-9.

[141] Seppi K, Mueller J, Bodner T, Brandauer E, Benke T, Weirich-Schwaiger H, Poewe W, Wenning GK. Riluzole in Huntington's disease (hd): An open label study with one year follow up. J Neurol. 2001;248:866–869. doi: 10.1007/s004150170071.

[142] Investigators. -PSGS-PI. Isradipine versus placebo in early parkinson disease: A randomized trial. Ann Internal Med. 2020;172:591–598. doi: 10.7326/m19-2534.

[143] Simuni T, Fiske B, Merchant K, Coffey CS, Klingner E, Caspell-Garcia C, Lafontant DE, Matthews H, Wyse RK, Brundin P, Simon DK, Schwarzschild M, Weiner D, Adams J, Venuto C, Dawson TM, Baker L, Kostrzebski M, Ward T, Rafaloff G. Efficacy of nilotinib in patients with moderately advanced parkinson disease: A randomized clinical trial. JAMA Neurol. 2021;78:312–320. doi: 10.1001/jamaneurol.2020.4725.

[144] Puls I, Beck M, Giess R, Magnus T, Ochs G, Toyka KV. clenbuterol in amyotrophic lateral sclerosis. No indication for a positive effect. Der Nervenarzt. 1999;70:1112–1115. doi: 10.1007/s001150050548.

[145] Gärtner J, Hauser SL, Bar-Or A, Montalban X, Cohen JA, Cross AH, Deiva K, Ganjgahi H, Häring DA, Li B, Pingili R, Ramanathan K, Su W, Willi R, Kieseier B, Kappos L. Efficacy and safety of ofatumumab in recently diagnosed, treatment-naive patients with multiple sclerosis: Results from asclepios i and ii. Mult Scler. 2022;28:1562–1575. doi: 10.1177/13524585221078825.

[146] López-Sendón Moreno JL, García Caldentey J, Trigo Cubillo P, Ruiz Romero C, García Ribas G, Arias A, Alonso M, García de Yébenes MJ, Tolón RM, Galve-Roperh IJJON. A Double-blind, Randomized, Cross-over, Placebo-controlled, Pilot Trial with Sativex in Huntington's Disease, Vol. 263, 2016, 1390–1400.

[147] Kieburtz K, McDermott MP, Voss TS, Corey-Bloom J, Deuel LM, Dorsey ER, Factor S, Geschwind MD, Hodgeman K, Kayson EJAON. A Randomized, Placebo-controlled Trial of Latrepirdine in Huntington Disease, Vol. 67, 2010, 154–160.

[148] Ehrnhoefer DE, Caron NS, Deng Y, Qiu X, Tsang M, Hayden MRJEN. Laquinimod Decreases Bax Expression and Reduces Caspase-6 Activation in Neurons, Vol. 283, 2016, 121–128.
[149] Mucke HA. The case of galantamine: Repurposing and late blooming of a cholinergic drug. Future Sci OA. 2015;1:Fso73. doi: 10.4155/fso.15.73.
[150] Weintraub MK, Bisson CM, Nouri JN, Vinson BT, Eimerbrink MJ, Kranjac D, Boehm GW, Chumley MJ. Imatinib methanesulfonate reduces hippocampal amyloid-β and restores cognitive function following repeated endotoxin exposure. Brain Behav Immun. 2013;33:24–28. doi: 10.1016/j.bbi.2013.05.002.

Priyank Purohit, Ravi K. Mittal, Akanksha Bhatt

5 Implication of drug repurposing in the identification of drugs as antiviral agents

Abstract: The drug repurposing approach has several advantages over designing a whole new medication for a specific indication. The repurposing of approved drugs for curing several disorders becomes one of the widely used approaches because its lower risk chances and previously known data such as preclinical, pharmacokinetic, and pharmacodynamic data that helps in bypassing the phase 1 clinical trial and getting direct entry into phase 3 or 4 clinical trials minimizes the period of drug development and reduces cost as well. The demand for new antiviral drugs in the treatment of chronic infectious diseases as well as the emergence of more efficient new viruses drives research into new targets and processes for antiviral development. The medication repurposing technique, which involves identifying new uses for present FDA-approved pharmaceuticals, is a potential way to speed up the production of infectious disease therapies.

Keywords: Preclinical, pharmacokinetic, pharmacodynamic, clinical trials, antiviral drugs, DA-approved pharmaceuticals

5.1 Introduction

Drug repurposing is a technique for discovering innovative uses for licensed medications that are not covered by the original medical indications and is also called drug repositioning, reprofiling, and retasking (Figure 5.1). The most scientifically gratifying research goal is to identify novel medicines for unmet clinical needs, a process that appears to be more achievable through drug repurposing [1]. This approach has several advantages over designing a whole new medication for a specific indication. Approximately 75% of currently accessible medication may be repurposed to cure numerous disorders. This is thought to be a cost-reductive and capable procedure [2]. The repurposing avoids some of the critical steps of drug discovery, such as preclinical and initial clinical study, which directly impacts on the cost and time of the drug disovery [3]. This has resulted in WHO and other agencies turning to reassess the efficacy of licensed and experimental cures for treating developing health issues [4–7].

https://doi.org/10.1515/9783110791150-005

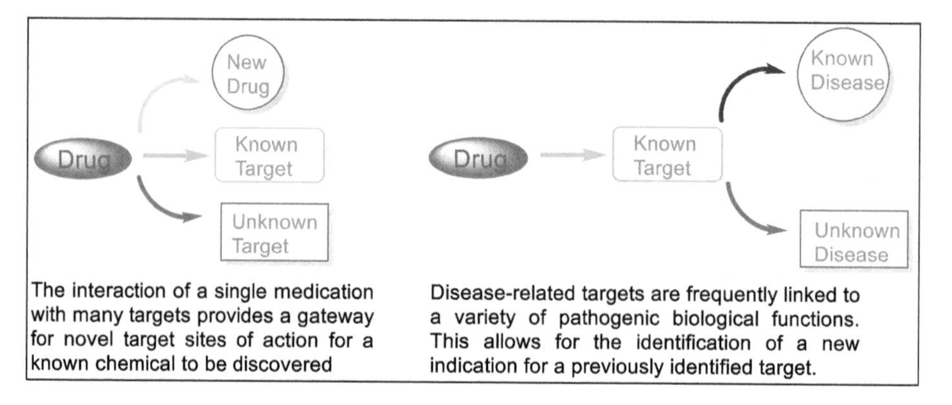

Figure 5.1: Concepts of drug repurposing.

The discipline of drug repositioning is quickly expanding, thanks to the promise of lower costs and faster approval times [8]. Medicine repositioning, or revealing a drug's new roles, is gaining traction in the pharmaceutical industry as a way to combat high failure rates and long-term development in drug development, while medication combinations or drug cocktails, which involve combining different drugs to treat diseases, are primarily intended to address the problem of recurrent drug resistance and to disclose their synergistic effects [9]. The timeline for drug repurposing is illustrated in Figure 5.2.

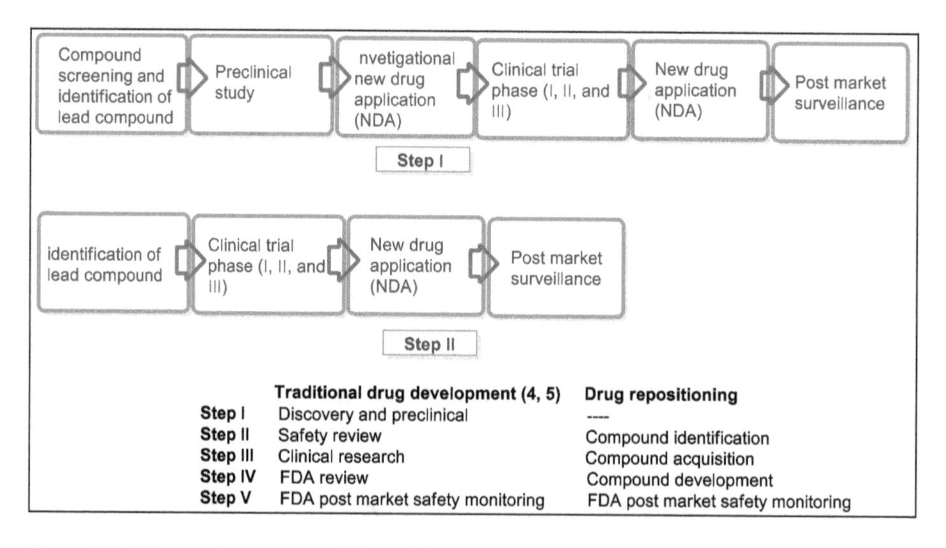

Figure 5.2: Difference between steps followed for traditional and repositioning drug development strategies (I: steps in traditional drug development; II: steps in drug repurposing) [10].

Drug repositioning reduces the time cost of the drug development process greatly due to the rapid increase of bioinformatics expertise and biology big data. When medicine is repurposed, it goes straight into phases 2 and 3 of a clinical trial. Traditional drug development takes about 15–20 years, while drug repurposing takes about 10–12 years. On average, it takes 1–2 years for researchers to uncover novel pharmacological targets and 8 years to produce a repositioned medicine. Currently, around 30% of newly FDA-approved medications in the United States are repurposed solely [11]. Viagra is a well-known example, which was developed by Pfizer to treat hypertension and angina. Clinical testing revealed the compound's potential for treating erectile dysfunction [12].

5.2 Advantages of drug repurposing [13–15]

- The majority of the repositioned medications are authorized components with safety, toxicity, and bioavailability profiles that have been studied. They also have a well-established formulating and manufacturing process.
- There is a lower risk of failure; because the repurposed medicine has previously exhibited sufficient and secure results in preclinical models and humans if initial level trials have been done, there is negligible chance of failure in later success trials, at least from the point of view of safety.
- Relaunching a repositioned medicine rather than a new drug saves a corporation millions of dollars.
- Drug development time can be cut in half because most preclinical research, safety evaluations, and a few circumstances in developing the formulation will have previously been accomplished.
- Low cost is required, albeit this could differ substantially based upon the repurposing candidate's phase and development method. Companies may be motivated to out-license some of their clinical treatments that have been shelved or abandoned for various reasons if their repositioned drugs are successful.

The benefits of drug repurposing are highlighted in Figure 5.3.

Although regulatory and phase III costs for a repurposed medicinal drug can be much like the ones for brand new drug remedies with equal indication, there are probably a few widespread financial savings in preclinical and phase I and II expenses.

Together, those blessings can bring about decreased danger and quickly go back on funding the improvement of repurposed drugs, in addition to decreasing common associated costs as soon as disasters are factored in (in general, the expenses of bringing a repurposed drug to marketplace had been predicted to be $300 million, as compared to a predicted $2–3 billion for a brand-new chemical entity). Moreover, the repurposed medicinal drugs can also additionally find new goals and pathways that may be investigated further.

Figure 5.3: Benefits of drug repurposing.

5.3 Drug repurposing in antiviral research

Viruses are a broad class of pathogens that can cause serious infectious diseases. Many antiviral medicines that target viral proteins or host factors have been developed effectively over the last 30 years. Chronic viral infectious disorders like HIV, influenza, and hepatitis C virus (HCV), as well as the arrival of many novel infections such as picornaviruses and coronaviruses, as well as resistance to currently accessible antiviral medications, are driving the rising need for antiviral agents. The demand for new antiviral drugs in the treatment of chronic infectious diseases, as well as the emergence of more efficient new viruses, drives research into new targets and processes for antiviral development [16].

The Food and Drug Administration (FDA) in the United States has approved four medications, two of which are a combination of anti-human immunodeficiency virus (HIV) therapies, for treating the HCV-based diseases. Simultaneously, governments and the World Health Organization (WHO) face a global danger from several emerging and reemerging viruses, which have caused worrying outbreaks in the past few years.

In the area of antiviral medication development, there has been a notable increase in attention towards drug repositioning in the last 10 years, fueled by the undeniable fact that many identified viral illnesses still need specialized treatment. This enthusiasm is inversely related to the relatively small number of traditional antiviral compounds approved in the last five years, largely for curing the HCV- or HIV-associated diseases. Rising viruses like Ebola, Zika, and MERS CoV are the ideal examples of diseases driving antiviral drug repurposing efforts since they have an urgent and cost-effective need for medicines. One interesting method that quickly offers a cure in the term of a viral outbreak is looking at existing pharmacopeia used for curing the diseases. For example, chloroquine, a common antimalarial medicine, has been advocated for treating filoviral infections, as well as other rising pathogens because it focused on endosomal acidification, a critical phase in the reproduction cycle of many viruses [17–19]. Favipiravir is another notable example, which has been proven to have repurposing therapeutic potential for curing Zika and Ebola virus infections [20, 21].

5.4 Drug repurposing for viral infection

The medication repurposing technique, which involves identifying new uses for present FDA-approved pharmaceuticals, is a potential way to speed up the production of infectious disease therapies. The brief methodology for drug repurposing for antiviral drugs is suggested pictorially in Figure 5.4.

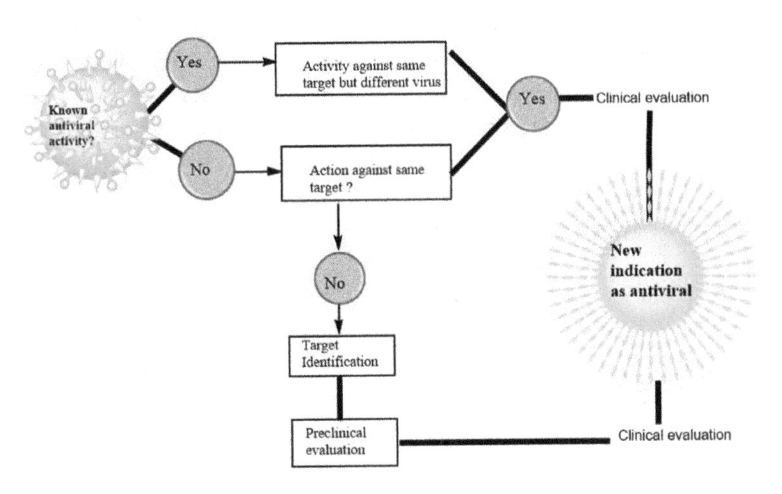

Figure 5.4: Drug repurposing for antiviral drugs.

This displays a list of drugs that have been repurposed for a particular illness, along with the original indication for which it was developed [22]. The following are the advances made in repurposing existing and candidate drugs to treat viral infections:

5.5 Repurposing of drugs in Zika virus infection

Zika virus (ZIKV) is an arbovirus that is spread by mosquitos. In most situations, ZIKV infection is self-limiting; but, when it happens during pregnancy, it has been related to neurologic illnesses (such as Guillain–Barré syndrome and others) as well as significant congenital malformations in babies (microcephaly and ophthalmological alterations). Potential ZIKV pandemics are of special concern for global public health since the virus can transmit both vertically (through transplacental transmission) and sexually. There are no specific antiviral drugs or vaccines available to treat ZIKV infection [23]. In distinct brain cell lines targeted by ZIKV fetal infection, the FDA-approved medication niclosamide (antihelminthic) and the macrolide azithromycin (antibacterial) were discovered to be strong inhibitors of ZIKV replication [24]. Niclosamide and azithromycin are commonly prescribed medicines with clinically attainable active concentrations opposite to ZIKV that can also be used during pregnancy (category B). The immunosuppressant medication mycophenolic acid and the antibiotic daptomycin have been found to be potential inhibitors of ZIKA virus replication due to their widespread use during pregnancy to treat other diseases and their ability to cross the placenta. Xu et al. used a high-throughput screening approach to screen around 6,000 compounds, which included FDA-approved medicines, molecules in clinical trials, and pharmacological active components. More than 100 components were found to inhibit ZIKV-induced caspase-3 action in SNB-19 cells [25].

The brief methodology for drug repurposing against Zika virus is suggested pictorially in Figure 5.5.

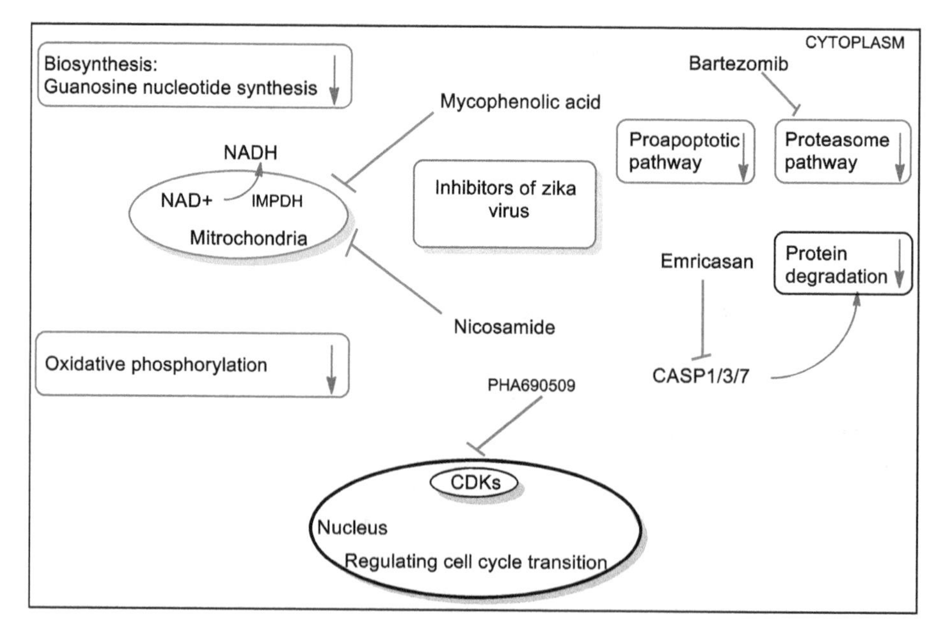

Figure 5.5: Drug repurposing in ZIKA virus infection.

5.6 Influenza viral infection-based drug repurposing

Notably, repurposing is thought to account for 30% of newly marketed drugs that are permitted by the FDA. The term “drug repurposing” refers to a wide range of different, but not jointly limited, experimental approaches to identifying possible new uses outer the extent of the novel medical indication [26, 27]. The unique evolving nature of the virus, high infectivity, host dependency, suboptimal vaccine efficacy, limited antiviral weapons, makes as an attractive drug discovery target. Despite various fascinating omics-based techniques and high-throughput screening of particular pharmacological libraries, like kinase inhibitors, no anti-influenza treatment created from drug repurposing has yet been approved by the FDA [28, 29]. Some drugs repurposed for influenza infection are mentioned in Table 5.1.

Table 5.1: Drug used for anti-influenza activity that show potential in repurposing.

Drug	Initial		Anti-influenza action	References
	Indication	**Action**		
Nitazoxanide	Antiparasitic; chronic hepatitis	Inhibition of the pyruvate: ferredoxin/flavodoxin	HA maturation and transport inhibition	[30, 31]
Celecoxib	Anti-inflammatory	COX-2 inhibitor	Immunomodulator	[32, 33]
Metformin	Type 2 diabetes drug	Hepatic glucogenesis inhibitor	Immunomodulator Autophagy induction	[34, 35]
Gemcitabine	Anticancer drug	Ribonucleotide reductase inhibitor	Immunomodulator	[36]
Lisinopril	Antihypertensive Drug	Peptidyl dipeptidase inhibitor	NA inhibitor	[37]
Trametinib	Anticancer drug	MEK1/2 inhibitor	vRNP transport inhibition	[38]

5.7 Coronavirus

Coronaviruses (CoVs) are RNA viruses that cause gastrointestinal, respiratory, and neurological disorders and zoonotic infections in both humans and animals. The established possibility of cross-species spread with domestic animals that behave like intermediate hosts and are accountable for human infection is especially significant. In two separate screenings of licensed medications against MERS- and SARS-CoV, two compounds, chlorpromazine (dopamine receptor antagonist) and chloroquine (antimalarial) were found. Both MERS- and SARS-CoV were reported to also have anti-flavivirus action. The above-mentioned drugs had antiviral action opposite to both viruses and were clinically feasible in human beings. These are not, however, grown more in preclinical studies. The drug repurposing method also permits the determination of the ABL tyrosine kinase oncogene pathway as being required for the entry of CoV. The ABL kinase inhibitor imatinib, which is permitted as an anticancer drug for the oral route, is responsible for inhibition of MERS- and SARS-CoV replication by preventing virus fusion with the endosomal membrane. This shows how drug repurposing could be used to identify host factors needed for pathogen replication and to identify new antiviral targets. Other anti-CoV drugs that target the host include cyclophilin A inhibitors such as cyclosporine A and alispovir. In cell culture, this later drug had antiviral action opposite to both MERS and SARS-CoV but was unsuccessful in protecting SARS-CoV infected mouse model. Despite this, cyclophilin A inhibitors should be investigated further for their anti-CoV efficacy. Drug repurposing searches also led to the discovery of HIV protease inhibitors, which were shown to be effective in MERS-CoV infected cells and a non-human model. For curing the MERS-CoV syndrome, lopinavir/ritonavir and interferon b-1b combination is now being investigated in clinical studies [39–42].

Several studies have recently demonstrated the anti-COVID effectiveness of repurposed medicines in in vitro, preclinical, and clinical trials. Statistical drug repositioning methods used in COVID-19 are classified as: (i) network-based models, (ii) structure-based techniques, or (iii) machine/deep learning methods. The need for medications to combat the COVID-19 epidemic has propelled this type of research in recent months. Computational techniques have performed and continue to play an important part in the hunt for effective weapons against the SARS-CoV-2 virus among the medications, but the findings do not appear to meet expectations [43, 44].

– **Computational study**

Drug	Target	Method used	References
Remdesivir	Main protease (PDB: 6LU7)	Covalent docking screening + MD on the top three hits	[45]

– **A randomized clinical trial study**

Drug	Clinical studies	Dose given	Result	References
Remdesivir	Patient with positive PCR test for SARS-CoV-2 infection and O_2 saturation ≤ 88%, and other resembled symptoms	200 mg IV for 1 day; should be followed by 100 mg in a single daily infusion	Randomized patients up to 5-days – received remedesivir – an observed difference in clinical status statically compared to control group	[46]

(continued)

Drug	Clinical studies	Dose given	Result	References
Ivermectin	164 patients with COVID-19 positive result were included in the study	12 mg taken daily for 3 days	It was noted that there is a trend in reduction of hospital stay in the ivermectin-treated group compared with the control	[47]

5.8 DNA virus infections

Virus-targeting drugs were also shown to have a repositioning perspective. This is true for HIV integrase inhibitors, which inhibit the viral termination and have shown broad-spectrum action resistance to numerous herpes viruses, as well as the HIV protease inhibitors lopinavir/ritonavir, which have been clinically examined as potential cure for HPV-related preinvasive cervical malignancies [48–50]. Drug repurposing is also being examined in the lookout for potential antiviral methods focused opposed to DNA viruses, which are not termed "emerging" viral pathogens but induce latent, lifelong infections, which can cause major morbidity and death in at-risk populations. Human cytomegalovirus (HCMV) is a β-herpes virus that induces chronic infection and can have reactivation in immunity-compromised people. HCMV is one of the well-studied cases of virus–host adaptation and the ability of viruses to fully undermine cellular physiological functions in infected cells. Numerous medication repurposing campaigns have identified several recognized or developing pharmaceuticals that have anti-HCMV mechanisms that are different from those of existing drugs currently present [51–53].

5.9 The scientific basis of drug repurposing

As a result, any structural alterations to the medicine are excluded from the concept of drug repositioning. Instead, repurposing relies on biological qualities of the drug that are approved previously (in the form of a new formulation, at a new dose, or through a new method of administration); the other is the drug's negative effects for a positive purpose. The use of in silico techniques such as data mining, machine learning, ligand-based, and structure-based approaches that is done to describe the factors connected to the intricate interplay between disorders, medications, and targets is one of the most significant parts of drug repurposing [54]. Now, it is possible to specify disorders related to their molecular characteristics (e.g., genes, biomarkers, signaling pathways, environmental variables, etc.) and to compare diseases that share a number of these molecular aspects using computational tools, specifically data mining. For example, 48 genes and four signaling pathways are shared by Parkinson's disease and Alzheimer's disease. The availability of protein targets that are shared by numerous disorders indicates that a single medication could treat both. Many medications today have well-defined phenotypes in terms of their core therapeutic effects as well as their (generally unwanted) side effects. Pleiotropic interactions between the drug and several (primary and secondary) biological targets are responsible for the wide range of impacts. When one of a medicine's secondary targets is implicated in a condition other than the one for which it was initially created, the drug can successfully work against it. Additionally, due to these pleiotropic interactions, medications having many all-intentional actions that function in tandem to provide better therapeutic efficacy, such as pan-kinase inhibitors used in oncology can be designed. Regardless of the therapeutic reason, drugs and disorders can be compared for phenotypic similarities. A high similarity score between two medications with distinct applications indicates that they may be beneficial in both [55–60]. Table 5.2 lists some of the empirical methodologies that have been used to find potential additional use outside the initial clinical basis:

Table 5.2: Empirical methodologies that have been used to find potential leads outside the initial clinical basis.

Serendipity	Target based	Phenotypic
– Examination and validation for novel potential therapy. – Like Thalidomide, first for nausea in pregnant women repositining for the curing leprosy and numerous type of myeloma.	– research demonstrates that the objective is significant in a condition or disease different than the primary purpose. – Favipiravir inhibits viral RNA polymerase, as a target of the anticancer medication imatinib.	– Test cell-based or in vivo disease models sets of small chemical libraries could be made publicly available to speed therapeutic repurposing through hypothesis biassed or unbiased phenotypic testing

There are numerous drug repurposing examples along with some old drug inclusion that can be found throughout medical history. The majority of them happened by chance. New methods, based primarily on data mining, have now been developed for the identification of novel candidates for drug repositioning. Drug repositioning has since been broadened to encompass active chemicals that fail in clinical trials because of toxicity or low effectiveness, and also medications that were pulled from the market due to safety concerns. However, it should not include substances that have not yet been clinically tested. This particularly prohibits chemicals stored in chemical libraries by academic and industry research groups from being screened for new biological qualities other than those for which they were developed and manufactured. These types of components often need hit-to-lead chemistry to achieve the required new therapeutic effect. Likewise, Professor Wermuth's proposal for selective optimization of side action (SOSA) does not fall under the repositioning umbrella [61].

Latest clinical trial data [62].

Status	Title	Conditions	Interventions
Not yet recruiting	"Drug repurposing using Metformin for improving the therapeutic outcome in multiple sclerosis patients"	Multiple sclerosis	Drug: Metformin 1000 mg oral tablet Drug: Interferon-beta-1a

5.10 Conclusion

Determining a new purpose for an old drug with few modifications has numerous benefits including improvement in efficacy, cost reduction, and bypass of phase 1 trial. Also having a low risk of failure minimizes the time to reach the market level, i.e., post-marketing. Inhibitors are targeted to identify and validate repurposed drugs using various methods. Furthermore, data depositing including negative outcomes into the public database should be initiated because it enhances efforts to reposition licensed or orphaned drugs and also enhances the chance of finding new action of antiviral drugs. In respect of the globalized world confronted with complexities like population, climatic conditions, and numerous other emergences of zoonotic viruses, the potency of the convectional de novo drug development system is a challenge. The concept of drug repositioning offers several possibilities for rapid and capable finding of new antiviral drugs. Theoretically, the drugs that are identified via drug repurposing are capable of bypassing the phase 1 clinical trial and moving further directly into phase 2. There are very few victorious examples that exist in the case of antiviral drug discovery against diseases like influenza, EBOV. Hence, there is need for new development in this field by finding new antiviral target drugs for therapeutic purposes with the help of drug repurposing.

References

[1] Ashburn TT, Thor KB. Drug repositioning: Identifying and developing new uses for existing drugs. Nat Rev Drug Discov. 2004 Aug;3(8):673–683.

[2] Huang F, Zhang C, Liu Q, Zhao Y, Zhang Y, Qin Y, Li X, Li C, Zhou C, Jin N, Jiang C. Identification of amitriptyline HCl, flavin adenine dinucleotide, azacitidine and calcitriol as repurposing drugs for influenza A H5N1 virus-induced lung injury. PLoS Pathog. 2020 Mar 16;16(3):e1008341.

[3] Pushpakom S, Iorio F, Eyers PA, Escott KJ, Hopper S, Wells A, Doig A, Guilliams T, Latimer J, McNamee C, Norris A. Drug repurposing: Progress, challenges, and recommendations. Nat Rev Drug Discov. 2019 Jan;18(1):41–58.

[4] Koch U, Hamacher M, Nussbaumer P. Cheminformatics at the interface of medicinal chemistry and proteomics. Biochim Biophys Acta Proteins Proteom. 2014 Jan 1;1844(1):156–161.

[5] Paolini GV, Shapland RH, Van Hoorn WP, Mason JS, Hopkins AL. Global mapping of pharmacological space. Nat Biotechnol. 2006 Jul;24(7):805–815.

[6] Hodos RA, Kidd BA, Shameer K, Readhead BP, Dudley JT. In silico methods for drug repurposing and pharmacology. Wiley Interdiscip Rev Syst Biol Med. 2016 May;8(3):186–210.

[7] Piro RM. Network medicine: Linking disorders. Hum Genet. 2012 Dec;131(12):1811–1820.

[8] Brown AS, Patel CJ. A standard database for drug repositioning. Scientific Data. 2017 Mar 14;4(1):1–7.

[9] Wu Z, Wang Y, Chen L. Network-based drug repositioning. Mol Biosyst. 2013;9(6):1268–1281.

[10] Dhir N, Jain A, Mahendru D, Prakash A, Medhi B. Drug repurposing and orphan disease therapeutics, drug repurposing-hypothesis. Mol Aspects Therapeut Appl. 2020. Apr 23. IntechOpen.

[11] Novac N. Challenges and opportunities of drug repositioning. Trends Pharmacol Sci. 2013 May 1;34 (5):267–272.

[12] Osterloh IH. The discovery and development of Viagra®(sildenafil Citrate). In Sildenafil. Birkhäuser, Basel, 2004, 1–13.

[13] Drucker DJ. Advances in oral peptide therapeutics. Nat Rev Drug Discov. 2020 Apr;19(4):277–289.

[14] Nosengo N. Can you teach old drugs new tricks? Nature. 2016;534(7607):314–316.

[15] Naylor DM. Therapeutic drug repurposing, repositioning and rescue. Drug Discovery. 2015;57.

[16] Lou Z, Sun Y, Rao Z. Current progress in antiviral strategies. Trends Pharmacol Sci. 2014 Feb 1;35 (2):86–102.

[17] Mercorelli B, Pal ÙG, Loregian A. Drug repurposing for viral infectious diseases: How far are we. 2018.

[18] Al-Bari MA. Targeting endosomal acidification by chloroquine analogs as a promising strategy for the treatment of emerging viral diseases. Pharmacol Res Perspect. 2017 Feb;5(1):e00293.

[19] Akpovwa H. Chloroquine could be used for the treatment of filoviral infections and other viral infections that emerge or emerged from viruses requiring an acidic pH for infectivity. Cell Biochem Funct. 2016 Jun;34(4):191–196.

[20] Devillers J. Repurposing drugs for use against Zika virus infection. SAR QSAR Environ Res. 2018 Feb 1;29(2):103–115.

[21] Pires de Mello CP, Tao X, Kim TH, Bulitta JB, Rodriquez JL, Pomeroy JJ, Brown AN. Zika virus replication is substantially inhibited by novel favipiravir and interferon alpha combination regimens. Antimicrob Agents Chemother. 2018;62(1):e01983–17.

[22] Mercorelli B, Palù G, Loregian A. Drug repurposing for viral infectious diseases: How far are we? Trends Microbiol. 2018 Oct 1;26(10):865–876.

[23] Wikan N, Smith DR. Zika virus: History of a newly emerging arbovirus. Lancet Infect Dis. 2016 Jul 1;16 (7):e119–26.

[24] Retallack H, Di Lullo E, Arias C, Knopp KA, Laurie MT, Sandoval-Espinosa C, Leon WR, Krencik R, Ullian EM, Spatazza J, Pollen AA. Zika virus cell tropism in the developing human brain and inhibition by azithromycin. Proc Natl Acad Sci. 2016 Dec 13;113(50):14408–14413.

[25] Xu M, Lee EM, Wen Z, Cheng Y, Huang WK, Qian X, Julia TC, Kouznetsova J, Ogden SC, Hammack C, Jacob F. Identification of small-molecule inhibitors of Zika virus infection and induced neural cell death via a drug repurposing screen. Nat Med. 2016 Oct;22(10):1101–1107.

[26] Hernandez JJ, Pryszlak M, Smith L, Yanchus C, Kurji N, Shahani VM, Molinski SV. Giving drugs a second chance: Overcoming regulatory and financial hurdles in repurposing approved drugs as cancer therapeutics. Front Oncol. 2017 Nov;14(7):273.

[27] Takahashi K, Wang F, Kantarjian H, Doss D, Khanna K, Thompson E, Zhao L, Patel K, Neelapu S, Gumbs C, Bueso-Ramos C. Preleukaemic clonal haemopoiesis and risk of therapy-related myeloid neoplasms: A case-control study. Lancet Oncol. 2017 Jan 1;18(1):100–111.

[28] Ludwig S. Will omics help to cure the flu? Trends Microbiol. 2014 May 1;22(5):232–233.

[29] Perwitasari O, Yan X, O'Donnell J, Johnson S, Tripp RA. Repurposing kinase inhibitors as antiviral agents to control influenza A virus replication. Assay Drug Dev Technol. 2015 Dec 1;13(10):638–649.

[30] Rossignol JF, La Frazia S, Chiappa L, Ciucci A, Santoro MG. Thiazolides, a new class of anti-influenza molecules targeting viral hemagglutinin at the post-translational level. J Biol Chem. 2009 Oct 23;284 (43):29798–29808.
[31] Tilmanis D, Van Baalen C, Oh DY, Rossignol JF, Hurt AC. The susceptibility of circulating human influenza viruses to tizoxanide, the active metabolite of nitazoxanide. Antiviral Res. 2017 Nov;1 (147):142–148.
[32] Carey MA, Bradbury JA, Rebolloso YD, Graves JP, Zeldin DC, Germolec DR. Pharmacologic inhibition of COX-1 and COX-2 in influenza A viral infection in mice. PloS One. 2010 Jul 15;5(7):e11610.
[33] Davidson S. Treating influenza infection, from now and into the future. Front Immunol. 2018:1946.
[34] Fedson DS. Treating influenza with statins and other immunomodulatory agents. Antiviral Res. 2013 Sep 1;99(3):417–435.
[35] Hosseini E, Grootaert C, Verstraete W, Van de Wiele T. Propionate as a health-promoting microbial metabolite in the human gut. Nutr Rev. 2011 May 1;69(5):245–258.
[36] Denisova OV, Kakkola L, Feng L, Stenman J, Nagaraj A, Lampe J, Yadav B, Aittokallio T, Kaukinen P, Ahola T, Kuivanen S. Obatoclax, saliphenylhalamide, and gemcitabine inhibit influenza a virus infection. J Biol Chem. 2012 Oct 12;287(42):35324–35332.
[37] Welch SR, Scholte FE, Flint M, Chatterjee P, Nichol ST, Bergeron É, Spiropoulou CF. Identification of 2′-deoxy-2′-fluorocytidine as a potent inhibitor of Crimean-Congo hemorrhagic fever virus replication using a recombinant fluorescent reporter virus. Antiviral Res. 2017 Nov 1;147:91–99.
[38] Rohini K, Shanthi V. Hyphenated 3D-QSAR statistical model-drug repurposing analysis for the identification of potent neuraminidase inhibitor. Cell Biochem Biophys. 2018 Sep;76(3):357–376.
[39] Coleman CM, Sisk JM, Mingo RM, Nelson EA, White JM, Frieman MB. Abelson kinase inhibitors are potent inhibitors of severe acute respiratory syndrome coronavirus and Middle East respiratory syndrome coronavirus fusion. Virol J. 2016 Sep 12;90(19):8924–8933.
[40] Dyall J, Coleman CM, Venkataraman T, Holbrook MR, Kindrachuk J, Johnson RF, Olinger JGG, Jahrling PB, Laidlaw M, Johansen LM, Lear-Rooney CM. Repurposing of clinically developed drugs for treatment of Middle East respiratory syndrome coronavirus infection. Antimicrob Agents Chemother. 2014 Aug;58(8):4885–4893.
[41] Arabi YM, Alothman A, Balkhy HH, Al-Dawood A, AlJohani S, Al Harbi S, Kojan S, Al Jeraisy M, Deeb AM, Assiri AM, Al-Hameed F. Treatment of Middle East respiratory syndrome with a combination of lopinavir-ritonavir and interferon-β1b (MIRACLE trial): Study protocol for a randomized controlled trial. Trials. 2018 Dec;19(1):1–3.
[42] Chan JF, Yao Y, Yeung ML, Deng W, Bao L, Jia L, Li F, Xiao C, Gao H, Yu P, Cai JP. Treatment with lopinavir/ritonavir or interferon-β1b improves outcome of MERS-CoV infection in a nonhuman primate model of common marmoset. J Infect Dis. 2015 Dec 15;212(12):1904–1913.
[43] Xue H, Li J, Xie H, Wang Y. Review of drug repositioning approaches and resources. Int J Bio Sci. 2018;14(10):1232.
[44] Ancy I, Sivanandam M, Kumaradhas P. Possibility of HIV-1 protease inhibitors-clinical trial drugs as repurposed drugs for SARS-CoV-2 main protease: a molecular docking, molecular dynamics and binding 75.
[45] Al-Khafaji K, Al-Duhaidahawi D, Taskin Tok T. Using integrated computational approaches to identify safe and rapid treatment for SARS-CoV-2. J Biomol Struct Dyn. 2021 Jun 13;39(9):3387–3395.
[46] Maxwell D, Sanders KC, Sabot O, Hachem A, Llanos-Cuentas A, Olotu A, Gosling R, Cutrell J, Hsiang MS. COVID-19 therapeutics for low-and middle-income countries: A review of re-purposed candidate agents with potential for near-term use and impact. MedRxiv. 2021. Jan 1.
[47] Abd-Elsalam S, Noor RA, Badawi R, Khalaf M, Esmail ES, Soliman S, Abd El Ghafar MS, Elbahnasawy M, Moustafa EF, Hassany SM, Medhat MA. Clinical study evaluating the efficacy of ivermectin in COVID-19 treatment: A randomized controlled study. J Med Virol. 2021 Oct;93(10):5833–5838.

[48] Yan Z, Bryant KF, Gregory SM, Angelova M, Dreyfus DH, Zhao XZ, Coen DM, Burke JTR, Knipe DM. HIV integrase inhibitors block replication of alpha-, beta-, and gamma-herpes viruses. MBio;5: e01318–14.
[49] Nadal M, Mas PJ, Blanco AG, Arnan C, Solà M, Hart DJ, Coll M. Structure and inhibition of herpesvirus DNA packaging terminase nuclease domain. Proc Natl Acad Sci. 2010 Sep 14;107(37):16078–16083.
[50] Hampson L, Maranga IO, Masinde MS, Oliver AW, Batman G, He X, Desai M, Okemwa PM, Stringfellow H, Martin-Hirsch P, Mwaniki AM. A single-arm, proof-of-concept trial of lopimune (lopinavir/ritonavir) as a treatment for HPV-related pre-invasive cervical disease. PLoS One. 2016 Jan 29;11(1):e0147917.
[51] Mercorelli B, Lembo D, Palù G, Loregian A. Early inhibitors of human cytomegalovirus: State-of-art and therapeutic perspectives. Pharmacol Ther. 2011 Sep 1;131(3):309–329.
[52] Mercorelli B, Luganini A, Nannetti G, Tabarrini O, Palù G, Gribaudo G, Loregian A. Drug repurposing approach identifies inhibitors of the prototypic viral transcription factor IE2 that block human cytomegalovirus replication. Cell Chem Biol. 2016 Mar 17;23(3):340–351.
[53] Gardner TJ, Cohen T, Redmann V, Lau Z, Felsenfeld D, Tortorella D. Development of a high-content screen for the identification of inhibitors directed against the early steps of the cytomegalovirus infectious cycle. Antiviral Res. 2015 Jan 1;113:49–61.
[54] March-Vila E, Pinzi L, Sturm N, Tinivella A, Engkvist O, Chen H, Rastelli G. On the integration of in silico drug design methods for drug repurposing. Front Pharmacol. 2017:298.
[55] Iwata H, Sawada R, Mizutani S, Yamanishi Y. Systematic drug repositioning for a wide range of diseases with integrative analyses of phenotypic and molecular data. J Chem Inf Model. 2015 Feb 23;55(2):446–459.
[56] Dovrolis N, Kolios G, Spyrou G, Maroulakou I. Laying in silico pipelines for drug repositioning: A paradigm in ensemble analysis for neurodegenerative diseases. Drug Discov Today. 2017 May 1;22(5):805–813.
[57] Rehman W, Arfons LM, Lazarus HM. The rise, fall and subsequent triumph of thalidomide: Lessons learned in drug development. Ther Adv Hematol. 2011 Oct;2(5):291–308.
[58] Ghofrani HA, Osterloh IH, Grimminger F. Sildenafil: From angina to erectile dysfunction to pulmonary hypertension and beyond. Nat Rev Drug Discov. 2006 Aug;5(8):689–702.
[59] Moffat JG, Vincent F, Lee JA, Eder J, Prunotto M. Nat Rev Drug Discovery. 2017;16:531–543.
[60] Melo A, Monteiro L, Lima RM, De Oliveira DM, De Cerqueira MD, El-Bachá RS. Oxidative stress in neurodegenerative diseases: Mechanisms and therapeutic perspectives. Oxid Med Cell Longev. 2011 Oct;2011.
[61] Wermuth CG. Selective optimization of side activities: The SOSA approach. Drug Discov Today. 2006 Feb 1;11(3–4):160–164.
[62] https://clinicaltrials.gov/ct2/results?cond=drug+repurposing&term=antiviral&cntry=&state=&city=&dist=.

Vibhu Jha, Leif A. Eriksson

6 Implication of drug repurposing in the identification of antibacterial agents

Abstract: Significant efforts have been made over the last decade with regards to the repurposing of existing drugs for new and unexplored therapeutic indications. In comparison to the conventional drug discovery process, which is time-consuming and expensive, drug repurposing is proven to be a cost-effective strategy due to pre-established pharmacokinetic profile of the existing candidates. With particular reference to antibacterial drug discovery, enormous challenges are nowadays seen due to the development of severe drug resistance with the use of antibiotics against both gram-positive and gram-negative bacteria such as multidrug-resistant tuberculosis (MDR-TB) with isoniazid and rifampicin, methicillin-resistant *S. aureus* (MRSA), and *S. pseudintermedius* (MRSP), vancomycin-resistant Enterococcus (VRE), etc. New and more effective ways are therefore needed to overcome the drug resistance associated with antibiotics. Drug repurposing, which is characterized by reduced risk, timeline, and cost to a drug discovery process, can be thus considered as a promising alternative strategy to combat multidrug-resistant bacterial infections (MDRBIs). Furthermore, drug repurposing opens an affordable opportunity for the growth of small-scale and medium-scale companies rather than just limiting it to the big pharmaceutical companies. In this review, we have discussed a broad category of existing drugs: anti-hyperlipidemia, anticancer, antiparasitic, antifungal, antimalarial and anti-inflammatory drugs that have been repurposed to treat bacterial infections. The authors believe that the notable outcomes from the case studies discussed in this work will further guide researchers working in the development of new antibacterial agents by drug repurposing approach.

Keywords: Drug repurposing, drug resistance, antibacterial agents, anti-hyperlipidemia drugs, anticancer drugs, antiparasitic drugs

6.1 Introduction

Drug repurposing is an important field in the drug discovery process and involves identification of new therapeutic opportunities from the existing drugs. The synonymous terms "drug repurposing, drug reprofiling and drug repositioning" outline the process involving discovery of new applications for an existing drug that were not

Acknowledgments: The authors sincerely thank the Wenner-Gren Foundations (V.J.), the Swedish Science Research Council (VR; grant number 2019-3,684; L.A.E), and the Swedish Cancer Foundation (grant number 211447-Pj; L.A.E) for funding.

https://doi.org/10.1515/9783110791150-006

previously reported and are not currently prescribed. Pharmaceutical companies usually adopt this strategy to strengthen their productivity by introducing new drugs to market. One of the biggest advantages with drug repurposing is reducing the drug development timeline and subsequently circumventing the high cost and risk associated with the drug discovery processes [1]. Since the pharmacokinetic profiles (absorption, distribution, metabolism, excretion, and toxicology) and the clinical trials of the repositioned drug candidates are already established and validated, this potentiates lowering the overall budget of drug discovery processes and ultimately introducing drugs with new therapeutic indications to the market [2]. During emergency situations such as the Covid-19 pandemic and epidemics caused by Ebola and Zika viruses, drug repurposing has been significantly implemented and has produced beneficial outcomes to the society [2, 3].

In the past, scientists/researchers have already identified the repositioned drug candidates by serendipity [4], target searching, phenotype screening, high-throughput screening, and in silico screening [1]. However, drug repurposing can be broadly classified into two categories: (a) drug-based repurposing and (b) disease/target-based repurposing. Drug-based repurposing approach is characterized by the availability of information on drugs/drug-like candidates, whereas disease/target-based repurposing is considered when there is no information on the pharmacology of drug/drug-like candidates with respect to a particular target [5]. Both the aforementioned approaches have their own advantages and limitations, depending on the particular disease that has been targeted for drug repurposing. It is therefore recommended to adopt a combinatorial approach using the two drug repurposing strategies for the identification of new molecules against specific target/disease [6–8]. It has been estimated that approximately 75% of the existing drug molecules can be repurposed to treat several pathophysiological conditions [8].

A number of case studies on drug repurposing have already been reported in scientific literature and magazines (Figure 6.1). One such famous case study presents repurposing of sildenafil (also known as Viagra). Sildenafil is a phosphodiesterase (PDE)-type inhibitor that was originally developed to treat angina. Due to the therapeutically relevant side effect, sildenafil was repurposed to treat male erectile dysfunction [9]. Another well-known example is thalidomide, which was originally developed as a sedative and received much attention in the early 1960s due to its teratogenic effects [10]. Despite this, thalidomide was repositioned to treat erythema nodosum leprosum (ENL) and multiple myeloma, and was approved by the Food and Drug Administration (FDA) [11]. Similarly, zidovudine was repurposed for the management of human immunodeficiency virus (HIV), which causes acquired immunodeficiency syndrome (AIDS). Zidovudine was initially developed to treat cancer; however, due to successful repurposing campaign, zidovudine was enlisted as the first FDA-approved drug to treat AIDS [12]. Other significant examples of drug repurposing include the clinically approved drugs such as favipiravir (originally approved for the treatment of the influenza virus) and

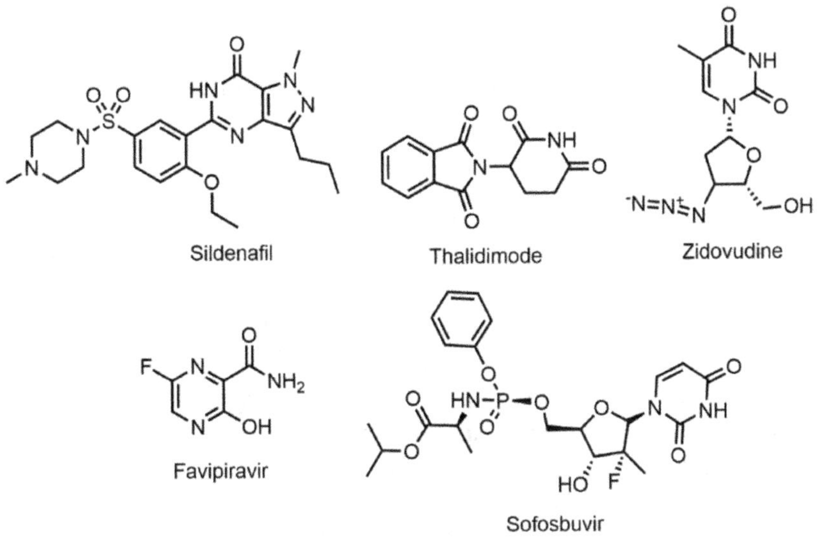

Sildenafil Thalidimode Zidovudine

Favipiravir Sofosbuvir

Figure 6.1: Significant examples of some repositioned drugs.

sofosbuvir (originally approved for the treatment of hepatitis C) for the treatment of epidemics caused by Ebola and Zika viruses [13–15].

In the context of antibacterial drug discovery, bacterial infections pose significant threats to human health, causing mortality worldwide at faster rates due to the development of resistance to the available drugs. About 700,000 deaths have been estimated to occur every year due to antibiotic-resistant infections, and the number is expected to increase to 10 million per year by 2050 [16]. Despite ongoing drug discovery efforts to identify new drugs or alternatives to antibiotics, no new classes of antibiotics or their alternatives have been clinically approved in the last three decades [17]. Only a few new classes of antibiotics such as daptomycin [18] have been approved by the FDA in recent decades. In order to combat this situation, repurposing nonantibiotic drugs (also known as antibiotic adjuvants) and the existing antibiotics (by combination) that already have surpassed extensive pharmacokinetic screening, is a promising approach to reduce the time, cost, and risks associated with conventional antibiotic drug discovery [19, 20]. In this review, we point out some recent advancement in antibacterial drug discovery by the implication of drug repurposing. The following section describes some of the notable drug repurposing case studies that have shown promising results and the potential to be developed as new antibacterial agents. A broad category of existing drugs: anti-hyperlipidemia, anticancer, antiparasitic, antifungal, antimalarial and anti-inflammatory drugs that were repositioned to treat bacterial infections are discussed in this work.

6.2 Case studies

6.2.1 Anti-hyperlipidemia drugs repositioned as antibacterial agents

Anti-hyperlipidemia drugs, also known lipid-lowering drugs, are used to reduce serum cholesterol. Statins (atorvastatin, fluvastatin, simvastatin, rosuvastatin, etc.) are the most widely used anti-hyperlipidemia medications (Figure 6.2). The mode of action of statins involves inhibition of HMG-CoA reductase (a key rate-limiting enzyme in the mevalonate pathway), which leads to the reduction of cholesterol levels in serum. Interestingly, statins possess strong potential antibacterial activity against gram-positive bacteria including oral microbiota (*Streptococcus spp.* and *Staphylococcus spp.*), gut microbiota (*Enterococcus spp.*, *Lactobacillus casei*, and *S. aureus* [Methicillin-resistant *Staphylococcus aureus* – MRSA and Methicillin-sensitive *Staphylococcus aureus* – MSSA]), drug-resistant bacteria (Vancomycin-intermediate *S. aureus* – VISA, Vancomycin-resistant *Enterococci* – VRE, and Vancomycin-resistant *S. aureus* – VRSA), and environmental bacteria (*Listeria monocytogenes* and *Bacillus anthracis*). Moreover, statins have also shown notable antibacterial activity against gram-negative bacteria including nasopharyngeal microbiota (*M. catarrhalis* and *H. influenzae*), oral microbiota (*P. gingivalis* and *A. inomycetemcomitans*), gut microbiota (*E. aerogenes, C. freundii, E. coli, E. cloacae, P. mirabilis,* and *K. pneumoniae*), and environmental bacteria (*S. Typhimurium*, and *A. baumannii, P. aeruginosa*) [21, 22].

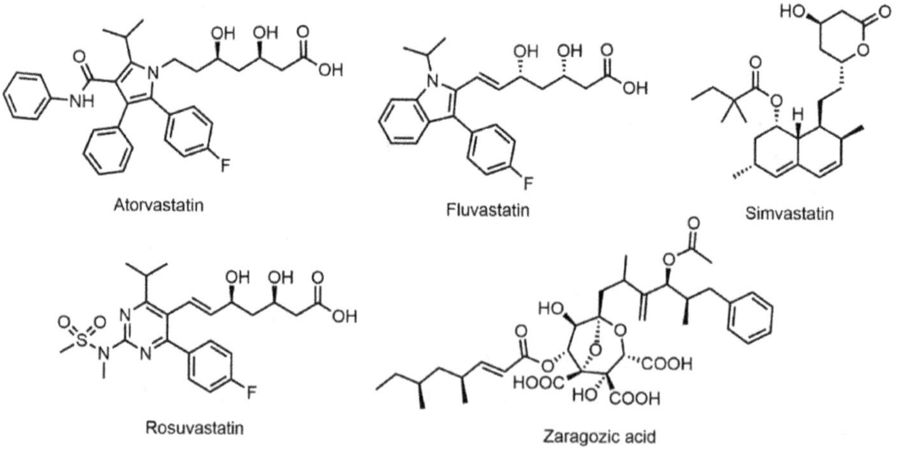

Atorvastatin

Fluvastatin

Simvastatin

Rosuvastatin

Zaragozic acid

Figure 6.2: Anti-hyperlipidemia drugs repurposed as antibacterial agents.

Simvastatin was found to be the most potent antibacterial agent against gram-positive bacteria among all statins. It demonstrated antibacterial activity against Enterococci at a minimum inhibitory concentration (MIC) of 32 µg/mL, and inhibits the biofilms of

S. aureus, reducing their formation and viability. Atorvastatin exhibited more potent antibacterial activity than simvastatin but against gram-negative bacteria *A. baumannii*, with an MIC of 16 µg/mL [21–23]. Additionally, statins have shown inflammation-reducing and immunomodulatory effects [24], and thus are used for the management of early stages of cardiovascular disease [25]. Some studies on statins reported a reduction in mortality rate of *S. aureus* bacteremia-infected patients during clinical trials [26, 27].

Apart from statins and with respect to MRSA infection, another lipid-lowering drug, zaragozic acid (Figure 6.2), has shown promising antibacterial effect. Esther et al. [28] reported that membrane carotenoid interaction with the scaffold protein flotillin led to the formation of functional membrane microdomains (FMMs) in MRSA [28]. FMMs facilitate oligomerization of multimeric protein complexes involving peptidoglycan transpeptidase (PBP2A). PBP2A is a key element in developing penicillin resistance with MRSA. Strikingly, zaragozic acid was found to disrupt FMM assembly, further interfering with oligomerization of PBP2A and eventually reversing MRSA penicillin resistance both in vitro and in vivo [28]. Taken together, repurposing of anti-hyperlipidemia drugs such as statins and zaragozic acid have demonstrated encouraging results and potential for the development of new antibiotics to treat multidrug-resistant bacterial infections (MDRBIs).

6.2.2 Anticancer drugs repositioned as antibacterial agents

Repurposing of anticancer drugs to treat antibacterial infections has drawn great attention due to several similarities between growing tumors and bacterial infections such as fast rate of replications, tendency to rapidly disseminate, virulence, and increased resistance to the immune system [29, 30]. Anticancer drugs that are originally prescribed for different kinds of cancers such as breast cancer (mitomycin C, thiotepa, 5-fluorouracil, gemcitabine, methotrexate, raloxifene, toremifene, tamoxifen, doxorubicin, epirubicin, and zoledronic acid) [31–38], brain tumors (carmustine and lomustine) [39, 40], pancreatic cancer (streptozotocin) [41], etc., have shown notable antibacterial activities against both gram-positive and gram-negative bacteria [30]. In particular, mitomycin C, tamoxifen, toremifene, and gallium nitrate (Figure 6.3) were found to be the most remarkable candidates, discussed further below.

6.2.2.1 Mitomycin C

Mitomycin C, abbreviated as MMC, is a DNA alkylating agent and FDA-approved anticancer drug for the management of various cancers such as bladder cancer, breast cancer, pancreatic cancer, and esophageal carcinoma [42]. Mitomycin C is an amphipathic molecule, which upon reduction diffuses freely [30, 43] through cellular membranes to form a spontaneously reacting methide structure [44], further reacting with

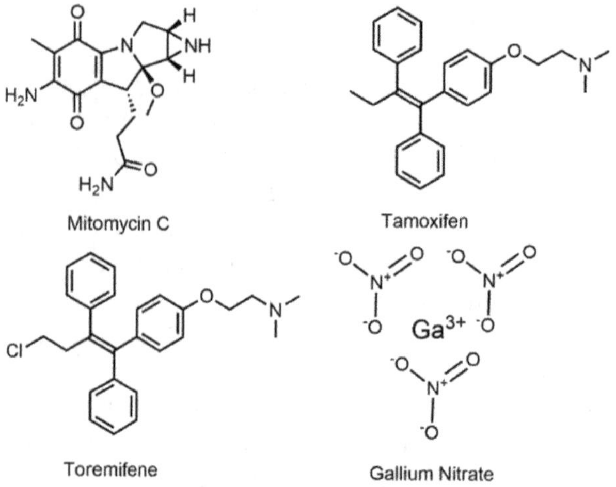

Figure 6.3: Anticancer drugs repurposed as antibacterial agents.

two adjacent guanine residues in 5′-CG sequences to form an inter-strand DNA cross-link [45]. Likewise, bacteria are also susceptible to mitomycin C due to the reductive nature of the bacterial cytoplasm [44]. Mitomycin C has shown prominent antibacterial activities against various bacterial pathogens such as *P. aeruginosa*, *S. aureus*, and *E. coli* [46]. As reported by Muniz et al. [47], mitomycin C was found to be an outstanding drug candidate against *A. baumannii* with MIC of 7 µg/mL (50% growth reduction) and completely inhibited the growth at 25 µg/mL concentration when tested along with other anticancer drugs (melphalan, cisplatin, and 5-fluorouracil). The aforementioned outcomes highlight the inhibition potential of mitomycin C against bacteria, in particular, *A. baumannii*, by a drug repurposing approach, further providing great scope for new drug development.

6.2.2.2 Gallium nitrate

Gallium nitrate (GaN) is an FDA-approved drug for the treatment of hypercalcemia and hepatocellular carcinoma and contains the gallium salt of nitric acid [48]. As reported by Runci et al. [49], gallium nitrate has shown to efficiently control the growth and biofilm formation of *A. baumannii* at a concentration of 16 µM in human serum. Furthermore, increasing the concentration to 64 µM, gallium nitrate resulted in substantial disruption of preexisting *A. baumannii* biofilm. These findings open for further development of gallium nitrate as an effective antibacterial agent.

6.2.2.3 Tamoxifen

Tamoxifen is a selective estrogen receptor modulator (SERM), chemically a triphenylethylene derivative and a well-known FDA-approved drug prescribed for the management of breast cancer [50]. It has been reported by Flores et al. [51] that tamoxifen potentiates the pro-inflammatory pathways of human neutrophils, which involve phagocytosis, chemotaxis, and neutrophil extracellular trap (NET) formation. Neutrophils play an important role against *Methicillin-resistant S. aureus* (MRSA) via a spontaneous immune response. The treatment of Neutrophils with tamoxifen boosts the bactericidal activity against MRSA, *E. coli*, and *P. aeruginosa* in vitro, which opens up new opportunities of drug development against MRSA.

6.2.2.4 Toremifene

Toremifene is another SERM and an FDA-approved drug for the treatment of metastatic breast cancer [52, 53] and prostate cancer [54]. Toremifene possesses structural resemblance to tamoxifen in that it consists of a triphenylethylene-derived structure, however, the chlorinated one (Figure 6.3). Toremifene is a mixed agonist–antagonist of the estrogen receptor (ER), with estrogenic actions in some tissues (bone, uterus and liver) and antiestrogenic actions in other tissues (breasts). Toremifene reportedly inhibits the growth of bacterial pathogens such as *S. mutans* and *P. gingivalis* at MIC ranging between 12.5 and 25 μM, as well as the prevention of bacterial biofilm formation in the MIC range of 25 to 50 μM. Furthermore, toremifene inhibits the biofilm formation of *S. mutans* and *P. gingivalis* on titanium surfaces [55].

6.3 Antiparasitic drugs repositioned as antibacterial agents

Antiparasitic drugs are a group of medications used in the management and treatment of infections by parasites, including protozoa, helminths, and ectoparasites. Antiparasitic drugs include several classes of drugs that cover a broad range of diseases caused by parasites such as malaria, pneumocystis, trypanosomiasis, and scabies [56]. In this section of the review, we discuss some of the important antiparasitic drugs such as auranofin [57], niclosamide [58], and nitazoxanide [59] (Figure 6.4) that have been successfully repositioned to exhibit potential antibacterial activities against gram-positive bacteria such as *S. aureus, E. faecalis, S. pneumoniae,* and *S. agalactiae* (antimalarial drugs are discussed separately in section E of the case studies).

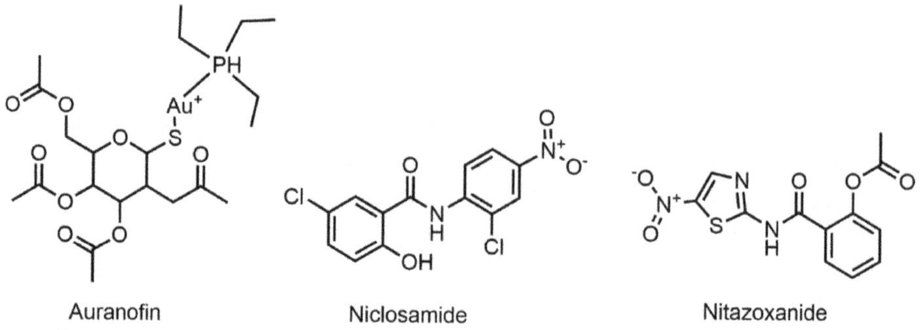

Figure 6.4: Antiparasitic drugs repurposed as antibacterial agents.

6.3.1 Auranofin

Auranofin is an FDA-approved drug for the treatment of rheumatoid arthritis. It has also been approved as an orphan drug by the FDA for treatment of human amebiasis. With the application of drug repurposing approach, auranofin emerges as a promising antimicrobial agent [57, 60]. As reported by Thangamani et al., auranofin has shown potent inhibition of gram-positive bacteria such as *E. faecium*, *S. aureus*, *E. faecalis*, *S. pneumoniae*, and *S. agalactiae* with an average MIC value of 0.125 μg/mL [61]. Furthermore, auranofin was tested both in vitro and in vivo against *S. aureus* including MRSA, VISA, and VRSA, showing MIC values in the range of 0.0625–0.125 μg/mL [62]. In vivo results highlight that auranofin demonstrated significantly improved antibacterial activity in comparison with the antibiotics mupirocin and fusidic acid, eventually lowering the bacterial load in the skin infections caused by MRSA. Likewise, auranofin resulted in a reduction of bacterial load and the production of inflammatory cytokines interleukin-1 beta (IL-1β), interleukin-6 (IL-6), tumor necrosis factor-α (TNFα), and monocyte chemoattractant protein-1 (MCP-1) when tested topically. Interestingly, auranofin in combination with linezolid or fosfomycin (already existing antibiotics) has demonstrated synergistic antimicrobial activities against MRSA and MSSA, during in vitro and in vivo tests [63]. Similar to the repositioning of other drug classes such as anti-hyperlipidemia and antifungal drugs, auranofin has shown a marked reduction in the biofilm formation of *S. epidermidis* and *S. aureus*, relative to antibiotics such as vancomycin and linezolid. The repurposed antibacterial properties of auranofin indicate that it can be considered as a potential drug candidate for the development of a topical antibiotic for skin infections caused by bacteria [62].

6.3.2 Niclosamide

Niclosamide is an anthelmintic drug used for the treatment of tapeworm infections such as diphyllobothriasis, hymenolepiasis, taeniasis, etc. [58] Niclosamide has also been prescribed recently in the management of SARs-CoV (severe acute respiratory syndrome coronavirus) [64]. As reported by Gwisai et al. [65], niclosamide inhibits biofilm formation of the gram-positive bacteria *S. aureus* and *S. epidermidis*, showing antibacterial activity at a remarkable 0.01 µg/mL concentration. In addition, niclosamide is also used as an antibacterial coating agent in medical devices [66].

6.3.3 Nitazoxanide

Nitazoxanide (NTZ) is a broad-spectrum antiparasitic and antiviral drug, used in treatment of various protozoal, helminthic, viral infections [59, 67, 68], and nosocomial infections [69]. Nitazoxanide was found to possess antibacterial activity against *Mycobacterium tuberculosis* (Mtb) as per the study conducted by Bailey et al. [70]. Furthermore, in in vitro test, radiorespirometry, and live-dead staining, nitazoxanide was found to possess bactericidal activity against *M. leprae*. Nitazoxanide displayed antibacterial activity in *M. leprae*-infected mice at a concentration of 25 mg/kg as compared to rifampicin at 10 mg/kg concentration [69].

6.4 Antifungal drugs repositioned as antibacterial agents

A broad range of studies have been carried out on the antifungal drugs that were repurposed for the bacterial inhibitory activities against *C. aquaticum, M. luteus, P. acnes*, MRSA, *P. aeruginosa, Microsporum canis, T. mentagrophytes, Candida albicans, Malassezia furfur, E. floccosum*, and *T. rubrum* [71]. In particular, the antifungal drugs clotrimazole, miconazole, and naftifine, which are discussed below (Figure 6.5), have demonstrated notable antibacterial activities against both gram-positive and gram-negative bacteria.

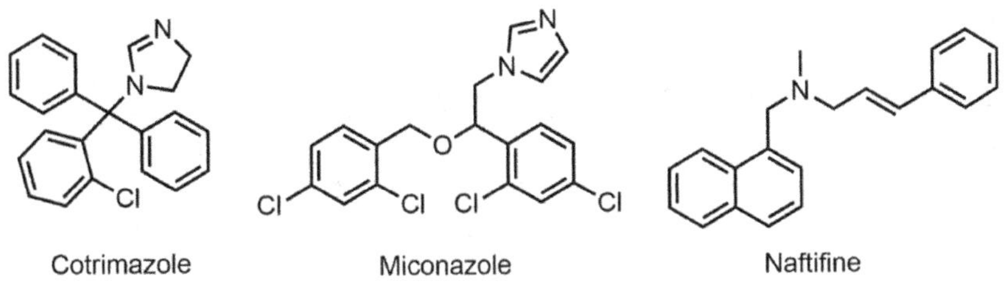

Cotrimazole Miconazole Naftifine

Figure 6.5: Antifungal drugs repurposed as antibacterial agents.

6.4.1 Clotrimazole

It is an FDA-approved antifungal drug, chemically an imidazole derivative, used in the treatment of vaginal and skin infections caused by yeasts and dermatophytes [72]. Clotrimazole and similar imidazole derivatives have shown significant antibacterial activity against *S. aureus*, *Streptomyces spp.*, *M. smegmatis*, and *S. epidermidis*, as reported by Sawyer et al. [72]. Frosini et al. [73] tested clotrimazole against 25 Methicillin-resistant *Staphylococcus pseudintermedius* (MRSP) and 25 Methicillin-susceptible *Staphylococcus pseudintermedius* (MSSP) and found comparable results with respect to MIC50 and MIC90 values against both the organisms.

6.4.2 Miconazole

Miconazole is an antifungal prescription medicine approved by the FDA for the treatment of mucocutaneous candidiasis, which includes oropharyngeal candidiasis (fungal infection of the part of the throat at the back of the mouth) and vulvovaginal candidiasis (vaginal fungal infection) [74]. Nenoff et al. [75] inspected the inhibition potential of miconazole against 80 wild-type strains of gram-positive and gram-negative bacteria and with 14 ATCC (American type culture collection) reference strains as controls. Miconazole demonstrated positive results against gram-positive aerobic bacteria (62 species), with the MICs ranging between 0.78 and 6.25 µg/mL against *Streptococcus spp.*, *Staphylococcus spp.*, *Enterococcus spp.*, and *Corynebacterium spp*; however, it did not show substantial results against gram-negative bacteria even when tested at MIC > 200 µg/mL. Despite this, the gram-positive bacterial inhibition efficacy of miconazole seems to be notable, and can thus be further used for the development of new antibacterial agents as an application of a drug repurposing strategy [75].

6.4.3 Naftifine

This is an FDA-approved drug, chemically an allyl amine derivative, used for the treatment of topical fungal infections caused by *tinea pedis* (Athlete's foot), *tinea corporis*, and *tinea cruris* [76]. 4,4'-diapophytoene desaturase (CrtN) is a key enzyme in carotenoid pigment synthesis in *S. aureus* that protects the bacteria from the antioxidants of the host and thus considered an essential virulence factor of *S. aureus*. Naftifine prevents the biosynthesis of carotenoid pigment by inhibiting the enzyme CrtN, eventually showing antibacterial action against MRSA [77].

6.5 Antimalarial drugs repositioned as antibacterial agents

Antimalarial drugs such as artemisinin [78] and artesunate [79] (Figure 6.6) are some of the most widely used medications for the treatment of malaria caused by protozoan parasite *Plasmodium falciparum*. Both artemisinin and artesunate have shown significant antibacterial activities, as discussed below.

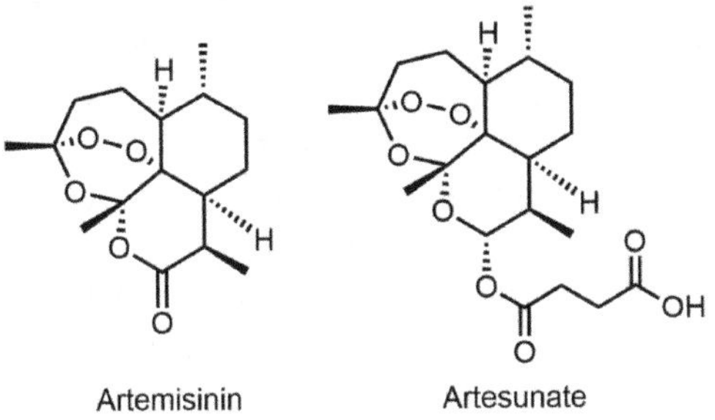

Artemisinin Artesunate

Figure 6.6: Antimalarial drugs repurposed as antibacterial agents.

6.5.1 Artemisinin

Artemisinin and its semisynthetic derivatives are the group of drugs used against malaria caused by *Plasmodium falciparum* [78]. Kim W. et al. [80] described that the antibacterial activity of artemisinin against the periodontopathic microorganisms *Aggregatibacter actinomycetemcomitans, Fusobacterium nucleatum subsp. animalis, Fusobacterium nucleatum subsp. polymorphum*, and *Prevotella intermedia*. Different concentrations of artemisinin extracts were prepared in methanol, ethanol, acetone and water, and termed Artemisinin-containing solutions (ACS). Antibacterial activity of these ACS preparations were evaluated by disc diffusion assay and MIC assay. The ACS concentrations used for the disc diffusion assay were 14 mg/mL. Antibacterial activities of these extracts were noted against *A. actinomycetemcomitans, F. nucleatum subsp. animalis, F. nucleatum subsp. polymorphum*, and *P. intermedia*. All three extracts (methanol, ethanol, and acetone preparations) of artemisinin demonstrated antibacterial activity against *P. intermedia* while the water and acetone ACS preparations showed inhibition of *A. actinomycetemcomitans* and *F. nucleatum subsp. animalis*. The MICs of these ACS preparations ranged between 7 to 14 mg/mL, underlining the bacterial inhibition potential of artemisinin [80].

6.5.2 Artesunate

Artesunate is used as first-line treatment for severe malaria, as recommended by the World Health Organization (WHO) [81]. Artesunate was developed as a more hydrophilic derivative of artemisinin [79]. Apart from the antimalarial activity, artesunate has shown promising antibacterial activity against Mycobacterium tuberculosis (Mtb). Artesunate was tested by Choi et al. [82] using different antitubercular indicator assays, such as the resazurin microtiter assay, the Mycobacteria Growth Indicator Tube (MGIT) 960 system assay, and the Ogawa slant medium assay, as well as in vivo tests. Artesunate demonstrated antitubercular activity by effectively inhibiting the growth of Mtb, as observed in the MGIT 960 system and in Ogawa slant medium for 21 days with a single dose. The MIC of artemisinin was reported to be 300 µg/mL when tested in vitro. Moreover, artesunate exhibited a consistent inhibition of Mtb for 4 weeks with a daily dose of 3.5 mg/kg when tested in vivo, without showing any toxicity or adverse effects. The overall results indicate that artesunate inhibits the growth and proliferation of Mtb by a new and unexplored pharmacological intervention, further highlighting its potential as a promising candidate for drug development and optimization by repositioning, to provide alternatives against the multidrug-resistant tuberculosis (MDR-TB) [82].

6.6 Anti-inflammatory drugs repositioned as antibacterial agents

Nonsteroidal anti-inflammatory drugs (NSAIDs) are medications that are widely used in pain, inflammation, fever, and arthritis [83]. In the present day scenario, NSAIDs have been largely used in the management of Covid-19 as one of the important drugs along with other medications. Apart from the aforementioned broad therapeutic uses, NSAIDs also exhibit noteworthy antibacterial activities against various gram-positive and gram-negative bacteria [84]. Some of the NSAIDs such as diflunisal, ebselen, diclofenac sodium, aspirin, ibuprofen, and indomethacin (Figure 6.7) have demonstrated important activities against different bacterial strains and are discussed in the following sections.

6.6.1 Diflunisal

Hendrix et al. [85] assessed the antibacterial action of diflunisal, highlighting its higher potential relative to other NSAIDs in inhibiting skeletal cell death in vitro and bone destruction caused by *S. aureus* osteomyelitis infections. The inhibition of *S. aureus* occurs via accessory gene regulator (agr) locus, which plays a crucial role in

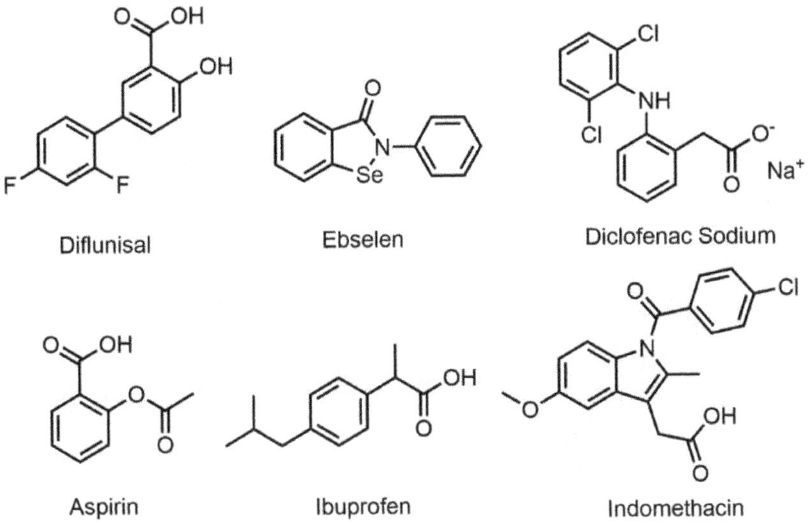

Diflunisal Ebselen Diclofenac Sodium

Aspirin Ibuprofen Indomethacin

Figure 6.7: Anti-inflammatory drugs repurposed as antibacterial agents.

remodeling of bone structures infected with staphylococcal osteomyelitis. In another study carried out by Pandey et al. [86], the antibacterial activity of diflunisal inhibiting the growth of *Helicobacter pylori* (*H. pylori* – causing peptic ulcer, gastric cancer) was evaluated. The β-clamp is the processivity-promoting factor for most of the enzymes in prokaryotic DNA replication, thus considered as a pivotal drug target. When tested in vitro, diflunisal was found to inhibit the β-clamp of *H. pylori* in the micromolar range. An X-ray structure of diflunisal in complex with the β-clamp of *H. pylori* was published (PDB code: 5G48) to reveal the binding site of diflunisal [86]. As apparent from Figure 6.8, diflunisal is accommodated into the β-clamp binding site of *H. pylori*, surrounded by residues Leu178, Thr175, Ile248, and Leu368, thus capable of forming lipophilic contacts, which contribute to the binding affinity for the target. Taking together these results, diflunisal is considered as a promising candidate by the drug repurposing approach, which can be further subjected to drug development and optimization as a potent antibacterial agent.

6.6.2 Ebselen

This is a synthetic organoselenium compound with antioxidant, anti-inflammatory, and cytoprotective activity, used for reperfusion stroke, injury, tinnitus, hearing loss, and bipolar disorder [87]. In addition, ebselen has shown antifungal activity against *Aspergillus fumigatus* [88], which causes fungal infection in the individuals with immunodeficiency. About 33 derivatives were synthesized by Ngo et al. [89], replacing the selenium in ebselen with a sulfur group and investigated against the drug-resistant and drug-

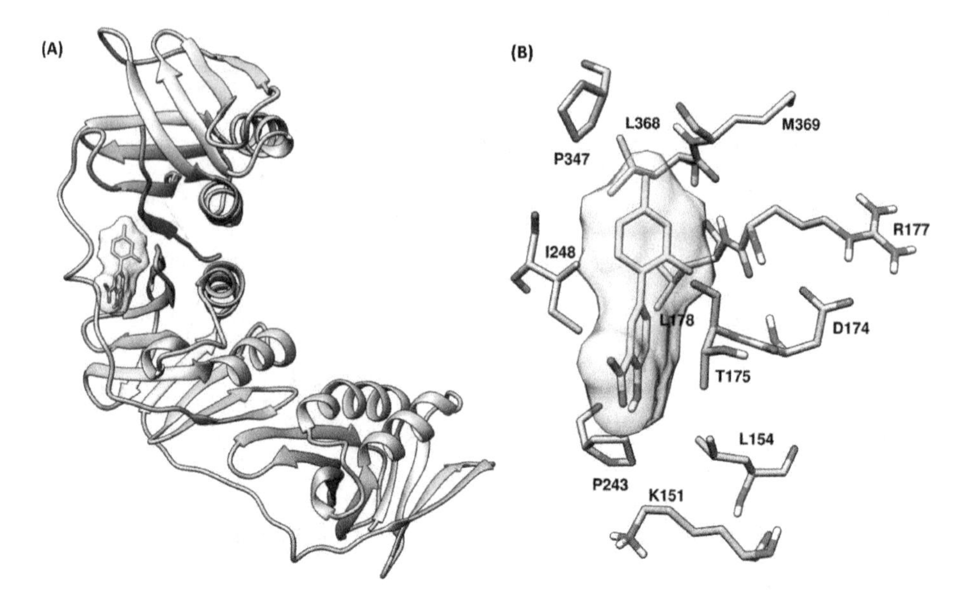

Figure 6.8: Diflunisal in the binding site of β-clamp of *H. pylori*.: (A) ribbon view. (B) binding site view. Diflunisal is colored in yellow; β-clamp of *H. pylori* is colored in pink ribbons/residues.

sensitive *S. aureus* and non-*S. aureus* bacteria. Three derivatives showed antibacterial activity with MIC values of 2 μg/mL against *S. aureus*, while the MIC values of other derivatives ranged from 1–7.8 μg/mL [89]. In another study carried out by Gustafsson et al. [90], a new library of ebselen analogues was evaluated against *Bacillus anthracis* thioredoxin reductase (involved in building blocks and protection against oxidative stress) and showed their antibacterial activity on *Bacillus subtilis*, *Staphylococcus aureus*, *Bacillus cereus*, and Mycobacterium tuberculosis. The most potent compounds of the series possessed IC_{50} values within 70 nM for the pure enzyme (*B. anthracis* thioredoxin reductase), further showing MIC values of 0.4 μM (0.12 μg/mL), 1.5 μM (0.64 μg/mL), 2 μM (0.86 μg/mL) and 10 μg/mL for *B. subtilis*, *S. aureus*, *B. cereus*, and *M. tuberculosis*, respectively. Furthermore, minimal bactericidal concentrations (MBCs) were also calculated and were found to be 1–1.5 times the MICs, highlighting a bactericidal mode of action [90]. The overall results indicate that ebselen and its analogues have strong potential to be further developed as a new antibiotic class to treat infections caused by *B. anthracis*, *S. aureus*, *M. tuberculosis*, etc. Furthermore, Zou et al. [91] investigated the effects of combination therapy using ebselen and silver against MDR gram-negative bacterial infections and revealed that five bacteria (*K. pneumonia*, *A. baumannii*, *P. aeruginosa*, *E. cloacae*, and *E. coli*) were notably sensitive to the combination therapy, further showing no toxicity or side effects on mammalian cells. Mechanistically, the combination therapy proves to be an integral element in lowering the glutathione and the thioredoxin system of bacteria [91].

6.6.3 Diclofenac sodium, aspirin, ibuprofen, and indomethacin

Ahmed et al. [92] investigated the antibacterial properties of NSAIDs – diclofenac sodium, aspirin, ibuprofen, and indomethacin – against the pathogens *E. coli, Coagulase-negative Staphylococci* (CoNS), *S.aureus, Klebsiella spp., Enterococcus faecalis, Pseudomonas spp. aeruginosa, Streptococci spp., Proteus spp.*, and *Bacillus spp.*, which cause urinary tract infection (UTI). The effects of NSAIDs in combination with β-lactam antibiotics were furthermore evaluated against the standard *K. pneumoniae* (ATCC10031) and *P. aeruginosa* (ATCC 10,145) strains by the checkerboard dilution technique. The outcomes of this study and the similar studies conducted to evaluate the antibacterial efficacy of NSAIDs are discussed as following [92].

Diclofenac Sodium demonstrated the lowest MIC90 values of 1 μg/mL and 0.5 μg/mL against *Streptococci spp.* and *Bacilli*, respectively. The highest MIC90 values were found against CoNS and *S. aureus* at 512 μg/mL and 1,024 μg/mL concentrations, respectively [92]. Aspirin exhibited lowest MIC90 value against *bacilli* at 4 μg/mL and the highest MIC90 value against *E. coli*, CoNS, *Klebsiella, S. aureus*, and *Streptococci* at1,024 μg/mL [92]. The lowest MIC90 value was noted for Ibuprofen at 2 μg/mL against *Bacilli*, whereas the MIC90 values are significantly higher (1,024 μg/mL) against *S. aureus, E. coli*, CoNS, *Pseudomonas spp.*, and *Streptococci* [92]. Indomethacin showed the lowest MIC90 value at 1 μg/mL against *Bacilli*, and the highest MIC90 values at 1024 μg/mL against *Pseudomonas spp., E. coli*, and *S. aureus* [92]. The aforementioned NSAIDs seem to be most promising against bacilli, showing lower MIC90 values, whereas MIC90 values increase drastically against other bacterial strains that indicate relatively poor inhibition. Taking into account the selectivity toward bacilli, NSAIDs exhibit great potential for further development and optimization as new antibiotics specifically targeting bacilli infections. In another research work conducted by Chan et al. [93], the antibacterial activities of NSAIDs (aspirin, diclofenac, ibuprofen, and mefenamic acid) in combination therapy with the existing antibiotics cefuroxime and chloramphenicol against MRSA were investigated [93]. Cefuroxime demonstrated improvement in potency in combination with ibuprofen/aspirin when treated against MRSA infections [93].

The results and outcomes from the aforementioned case studies by the application of drug repurposing approach are summarized in Table 6.1, highlighting two keys points – activity of the FDA-approved drugs against different bacterial species and the selected drug candidates showing promising minimum inhibitory concentration (MIC).

Table 6.1: Antibacterial activity of the existing FDA-approved drugs by drug repurposing approach.

Drug class	Existing FDA-approved drugs	Antibacterial activity	MIC	References
Anti-hyperlipidaemia	Atorvastatin Fluvastatin Simvastatin Rosuvastatin	*Streptococcus spp.*, *Staphylococcus spp.*, MRSA, MSSA, *Bacillus anthracis*, *Enterococcus spp.*, *Lactobacillus casei*, *Listeria monocytogenes*, and *A. baumannii*	32 µg/mL (Simvastatin) against Enterococci and 16 µg/mL (Atorvastatin) against *A. baumannii*	[21–23]
	Zaragozic acid	MRSA	–	[28]
Anticancer	Mitomycin C	*P. aeruginosa*, *S. aureus*, *E. coli*, and *A. baumannii*	MIC50 at 7 µg/mL against *A. baumannii*	[46, 47]
	Gallium nitrate	*A. baumannii*	16 µM against *A. baumannii* (inhibits biofilm formation)	[49]
	Tamoxifen	MRSA, *E. coli* and *P. aeruginosa*	–	[51]
	Toremifene	*S. mutans* and *P. gingivalis*	12.5 to 25 µM against *S. mutans* and *P. gingivalis*	[55]
Antiparasitic	Auranofin	*E. faecium*, *S. aureus*, *E. faecalis*, *S. pneumoniae*, *S. agalactiae*, *S. aureus*, *MRSA*, *MSSA*, *VISA*, *VRSA*, and *S. epidermidis*	0.125 µg/mL (average) against *E. faecium*, *S. aureus*, *E. faecalis*, *S. pneumoniae*, *S. agalactiae*, 0.0625–0.125 µg/mL against *S. aureus*, MRSA, VISA, and VRSA	[61, 62]
	Niclosamide	*S. aureus* and *S. epidermidis*	0.01 µg/mL against *S. aureus* and *S. epidermidis*	[65]
	Nitazoxanide	*M. tuberculosis*, *M. leprae*	25 mg/kg against *M. leprae*-infected mice	[69, 70]

Table 6.1 (continued)

Drug class	Existing FDA-approved drugs	Antibacterial activity	MIC	References
Antifungal	Clotrimazole	*S. aureus*, *Streptomyces spp.*, *M. smegmatis*, *S. epidermidis*, *MRSP*, *MSSP*	–	[72, 73]
	Miconazole	62 Gram-positive aerobic bacteria including *Streptococcus spp.*, *Staphylococcus spp.*, *Enterococcus spp.*, and *Corynebacterium spp*	0.78–6.25 μg/mL against *Streptococcus spp.*, *Staphylococcus spp.*, *Enterococcus spp.*, and *Corynebacterium spp*	[75]
	Naftifine	*S. aureus*, MRSA	–	[77]
Antimalarial	Artemisinin	*A. actinomycetemcomitans*, *F. nucleatum subsp. animalis*, *F. nucleatum subsp. polymorphum*, and *P. intermedia*	7 to 14 mg/mL against *P. intermedia* and *F. nucleatum subsp. animalis*	[80]
	Artesunate	*M. tuberculosis*	300 μg/mL against *M. tuberculosis*	[82]
Anti-inflammatory	Diflunisal	*S. aureus* and *H. pylori*	–	[85, 86]
	Ebselen	*B. subtilis*, *S. aureus*, *B. cereus*, and *M. tuberculosis*	0.12 μg/mL (*B. subtilis*), 0.64 μg/mL (*S. aureus*), 0.86 μg/mL (*B. cereus*), and (10 μg/mL) *M. tuberculosis*, 2 μg/mL (ebselen derivative against *S. aureus*)	[89, 90]
	Diclofenac sodium Aspirin Ibuprofen Indomethacin	*Streptococci spp.*, *Bacilli CoNS*, *S. aureus*, *E. coli*, *CoNS*, *Klebsiella*, and *Pseudomonas spp*	MIC90 at 1 μg/mL (diclofenac sodium against *Streptococci spp*), 0.5 μg/mL (diclofenac sodium against bacilli), 4μg/mL (aspirin against bacilli), 2 μg/mL (ibuprofen against bacilli), 1 μg/mL (indomethacin against bacilli)	[92]

6.7 Conclusions

Over the last few decades, antibiotic drug discovery has faced enormous challenges due to the emergence and development of severe drug resistance with the existing antibiotics. Drug resistance develops rapidly and can be observed with both gram-positive and gram-negative bacteria such as multidrug-resistant tuberculosis with antitubercular drugs isoniazid and rifampicin (MDR-TB), methicillin-resistant *S. aureus* (MRSA)g and *S. pseudintermedius* (MRSP), vancomycin-resistant *Enterococcus* (VRE) etc. Hence, the development of new and effective antibiotics is urgently required to overcome the current situation. However, conventional de novo drug discovery is a highly expensive and time-taking process, and thus alternative routes need to be explored for the development of new and effective antibiotics. One of the approaches for the identification of new antibacterial agents (or as drugs for other diseases) is "drug repurposing", which has demonstrated promising results within a short time frame relative to conventional drug discovery. The biggest advantages with drug repurposing approaches include reduction in the high cost and risk, and decrease in the timeline associated with a drug discovery process [1]. The availability of pre-established pharmacokinetic and clinical trials data of the drug molecules substantially help in circumventing the expensive budget and boosting the drug discovery process, aimed at exploring new therapeutic interventions by drug repurposing [2]. The existing FDA-approved drugs serve as a promising platform and have drawn significant attention to be employed for developing new antibiotics and combating MDRBIs. Furthermore, new drug development by the drug repurposing approach is not only limited to the big pharmaceutical companies but also paves an affordable way for small- and medium-scale companies. In the current study, we have outlined the antibacterial activities and inhibition potential of some of the existing FDA-approved drugs that have been repurposed as antibacterial agents targeting both gram-positive and gram-negative bacterial infections. Based on an extensive literature survey, six different case studies have been discussed for the drugs belonging to six important pharmacological categories: anti-hyperlipidaemia, anticancer, antiparasitic, antifungal, antimalarial, and anti-inflammatory drugs. The results and the outcomes of the case study discussed herein shed light on the antibacterial activity, efficacy, and potential of the selected drug molecules and their derivatives to be further developed as new and promising antibiotics against gram-positive and gram-negative bacterial infections. Moreover, the present study also provides a validation to the use of drug repurposing in the identification of new compounds, highlighting its reliability and applicability as a cost-effective and less time-consuming drug discovery process.

6.8 Future perspective

With the advent of computer-based and artificial intelligence/machine learning (AI/ML)-driven approaches, the classical drug repurposing approaches are nowadays highly influenced and led by their computational drug counterparts. Widely used computational drug repurposing strategies include data mining, machine learning, and network analysis. Data mining involves detecting potential indications from the information available in the biomedical and pharmaceutical literature (genes, drugs, diseases, and related information stored in databases). Text mining and semantic inference are the two subcategories of data mining. Text mining is a process to investigate data related to a specified gene, disease, and drug, classifying the relevant entries from the collected data and with the application of natural language processing, trying to find the commonality between the entries in a drug repurposing campaign. Semantic inference allows technologies such as topic modeling where data from multiple sources are easily combined, integrated, and predicted for new indications with promising therapeutic potential. Similar to any other science domain, machine learning approaches have gained significant attention in drug repurposing due to the open-source availability of huge biomedical and pharmaceutical databases. ML models are first trained with an extensive amount of data, followed by prediction of a test dataset, leading to vigorous decision- making and finally revealing key interrelations between the pharmaceutical entries and the biological data. The most widely used machine learning methods include logistic regression, support vector machine (SVM), random forest (RF), neural network (NN), deep learning (DL), k-nearest neighbors (kNN), etc. A very recent example is the discovery of the narrow-spectrum antibiotic Abaucin against Acinetobacter baumannii, by AI [94]. A. baumannii is recognized by the WHO as one of the main superbugs posing a critical threat to humanity. The discovery involved training the DL model with ~7,500 molecules that inhibited the growth of A. baumannii *in vitro*, followed by identifying antibacterial molecules from the Drug Repurposing Hub [95]. Network analysis is an integrated technique, which uses multiple data sources to establish relationships such drug-drug interactions, drug-protein interactions, protein-protein interactions, transcriptional and signaling networks, etc.; for instance, drug molecules exhibiting similar transcriptional response can be predicted to have similar mode of action. In a network model analysis, the nodes are used to represent genes, proteins, molecules, phenotypes, chemotypes, drugs, etc., whereas the edges are used to represent functional similarities, mode of actions, key mechanisms. Etc. [96, 97]. Taken together, the aforementioned computational drug repurposing strategies provide a great, cost-effective, and time-saving platform to the scientists/researchers working on the novel drug discovery. The conventional drug repurposing approaches seem to be taken over by the state-of-the-art of computational methods. In totality, the implication of drug repurposing (via computational or conventional means) can be considered as a promising tool and an alternative solution to circumvent the challenges associated with the long-established drug discovery processes.

References

[1] Doan TL, Pollastri M, Walters MA, et al. Chapter 23 – The future of drug repositioning: Old drugs, new opportunities. In Annual Reports in Medicinal Chemistry. Academic Press Inc, 2011. 385–401. Annual Reports in Medicinal Chemistry.

[2] Ciliberto G, Cardone L. Boosting the arsenal against COVID-19 through computational drug repurposing. Drug Discov Today. 2020;25:946–948.

[3] Jiménez-Alberto A, Ribas-Aparicio RM, Aparicio-Ozores G, et al. Virtual screening of approved drugs as potential SARS-CoV-2 main protease inhibitors. Comput Biol Chem. 2020;88:107325.

[4] Ashburn TT, Thor KB. Drug repositioning: Identifying and developing new uses for existing drugs. Nat Rev Drug Discov. 2004;3:673–683.

[5] Alam S, Kamal TB, Sarker MMR, et al. Therapeutic effectiveness and safety of repurposing drugs for the treatment of COVID-19: Position standing in 2021. Front Pharmacol. 2021;12:659577.

[6] Peyvandipour A, Saberian N, Shafi A, et al. A novel computational approach for drug repurposing using systems biology. Bioinform. 2018;34:2817–2825.

[7] Pushpakom S, Iorio F, Eyers PA, et al. Drug repurposing: Progress, challenges and recommendations. Nat Rev Drug Discov. 2019;18:41–58.

[8] Singh TU, Parida S, Lingaraju MC, et al. Drug repurposing approach to fight COVID-19. Pharmacol Rep. 2020;72:1479–1508.

[9] Goldstein I, Lue TF, Padma-Nathan H, et al. Oral Sildenafil in the treatment of erectile dysfunction. N Engl J Med. 1998;338:1397–1404.

[10] Richardson P, Hideshima T, Anderson K. Thalidomide in multiple myeloma. Biomed Pharmacother. 2002;56:115–128.

[11] Okafor MC. Thalidomide for erythema nodosum leprosum and other applications. Pharmacother J Hum Pharmacol Drug Ther. 2003;23:481–493.

[12] Trivedi J, Mohan M, Byrareddy SN. Drug repurposing approaches to combating viral infections. J Clin Med. 2020;9:3777.

[13] Dobson J, Whitley RJ, Pocock S, et al. Oseltamivir treatment for influenza in adults: A meta-analysis of randomised controlled trials. Lancet. 2015;385:1729–1737.

[14] Lv Z, Chu Y, Wang Y. HIV protease inhibitors: A review of molecular selectivity and toxicity. HIV AIDS (Auckl). 2015;7:95–104.

[15] Muralidharan N, Sakthivel R, Velmurugan D, et al. Computational studies of drug repurposing and synergism of lopinavir, oseltamivir, and ritonavir binding with SARS-CoV-2 protease against COVID-19. J Biomol Struct Dyn. 2021;39:2673–2678.

[16] De Kraker MEA, Stewardson AJ, Harbarth S. Will 10 million people die a year due to antimicrobial resistance by 2050?. PLOS Med. 2016;13:e1002184.

[17] Liu Y, Tong Z, Shi J, et al. Drug repurposing for next-generation combination therapies against multidrug-resistant bacteria. Theranostics. 2021;11:4910–4928.

[18] Gonzalez-Ruiz A, Seaton RA, Hamed K. Daptomycin: An evidence-based review of its role in the treatment of Gram-positive infections. Infect Drug Resist. 2016;9:47–58.

[19] Dutescu IA, Hillier SA. Encouraging the development of new antibiotics: Are financial incentives the right way forward? A systematic review and case study. Infect Drug Resist. 2021;14:415–434.

[20] SL L. Challenges of antibacterial discovery. Clin Microbiol Rev. 2011;24:71–109.

[21] Ko HHT, Lareu RR, Dix BR, et al. Statins: Antimicrobial resistance breakers or makers? Peer J. 2017;5: e3952.

[22] Lee -C-C, Lee MG, Hsu T-C, et al. A population-based cohort study on the drug-specific effect of statins on sepsis outcome. Chest. 2018;153:805–815.

[23] Thangamani S, Mohammad H, Abushahba MFN, et al. Exploring simvastatin, an antihyperlipidemic drug, as a potential topical antibacterial agent. Sci Rep. 2015;5:16407.

[24] Emma H, Claire A, Jerry RF, et al. Is there potential for repurposing statins as novel antimicrobials?. Antimicrob Agents Chemother. 2016;60:5111–5121.
[25] Taylor FC, Huffman M, Ebrahim S. Statin therapy for primary prevention of cardiovascular disease. Jama. 2013;310:2451–2452.
[26] CA R, TT T, Eunsun N, et al. Evidence to support continuation of statin therapy in patients with staphylococcus aureus bacteremia. Antimicrob Agents Chemother. 2017;61:e02228–16.
[27] Shrestha P, Poudel D, Pathak R, et al. Effect of statins on the mortality of bacteremic patients: A systematic review and meta-analysis of clinical trials. N Am J Med Sci. 2016;8:250–251.
[28] García-Fernández E, Koch G, Wagner RM, et al. membrane microdomain disassembly inhibits MRSA antibiotic resistance. Cell. 2017;171:1354–1367. e20.
[29] Benharroch D, Osyntsov L. Infectious diseases are analogous with cancer. Hypothesis and implications. J Cancer. 2012;3:117–121.
[30] Soo WCV, Kwan WB, Quezada H, et al. Repurposing of anticancer drugs for the treatment of bacterial infections. Curr Topics Med Chem. 2017;17:1157–1176.
[31] Kwan BW, Chowdhury N, Wood TK. Combatting bacterial infections by killing persister cells with mitomycin C. Environ Microbiol. 2015;17:4406–4414.
[32] Quinto I, Radman M. Carcinogenic potency in rodents versus genotoxic potency in E. coli: A correlation analysis for bifunctional alkylating agents. Mutat Res Mol Mech Mutagen. 1987;181:235–242.
[33] Walz JM, Avelar RL, Longtine KJ, et al. Anti-infective external coating of central venous catheters: A randomized, noninferiority trial comparing 5-fluorouracil with chlorhexidine/silver sulfadiazine in preventing catheter colonization. Crit Care Med. 2010;38:2095–2102.
[34] Ho Sui SJ, Lo R, Fernandes AR, et al. Raloxifene attenuates Pseudomonas aeruginosa pyocyanin production and virulence. Int J Antimicrob Agents. 2012;40:246–251.
[35] Kaat DC, Nicolas D, Katrijn DB, et al. Oral Administration of the Broad-Spectrum Antibiofilm Compound Toremifene Inhibits Candida albicans and Staphylococcus aureus Biofilm Formation In Vivo. Antimicrob Agents Chemother. 2014;58:7606–7610.
[36] Corriden R, Hollands A, Olson J, et al. Tamoxifen augments the innate immune function of neutrophils through modulation of intracellular ceramide. Nat Commun. 2015;6:8369.
[37] Gajadeera C, Willby MJ, Green KD, et al. Antimycobacterial activity of DNA intercalator inhibitors of Mycobacterium tuberculosis primase DnaG. J Antibiot (Tokyo). 2015;68:153–157.
[38] Chopra S, Matsuyama K, Hutson C, et al. Identification of antimicrobial activity among FDA-approved drugs for combating Mycobacterium abscessus and Mycobacterium chelonae. J Antimicrob Chemother. 2011;66:1533–1536.
[39] Ene CI, Nerva JD, Morton RP, et al. Safety and efficacy of carmustine (BCNU) wafers for metastatic brain tumors. Surg Neurol Int. 2016;7:S295–9.
[40] Van den Bent MJ, Brandes AA, Taphoorn MJB, et al. Adjuvant procarbazine, lomustine, and vincristine chemotherapy in newly diagnosed anaplastic oligodendroglioma: long-term follow-up of EORTC brain tumor group study 26951. J Clin Oncol. 2012;31:344–350.
[41] White HL, White JR. Lethal action and metabolic effects of streptonigrin on escherichia coli. Mol Pharmacol. 1968;4:549–565.
[42] Bradner WT, Mitomycin C. A clinical update. Cancer Treat Rev. 2001;27:35–50.
[43] Byfield JE, Calabro-Jones PM. Carrier-dependent and carrier-independent transport of anti-cancer alkylating agents. Nature. 1981;294:281–283.
[44] Szybalski W, Iyer VN. Crosslinking of DNA by enzymatically or chemically activated mitomycins and porfiromycins, bifunctionally 'alkylating' antibiotics. Fed Proc. 1964;23:946–957.
[45] Tomasz M. Mitomycin C: Small, fast and deadly (but very selective). Chem Biol. 1995;2:575–579.
[46] Reich E, Shatkin AJ, Tatum EL. Bacteriocidal action of mitomycin C. Biochim Biophys Acta. 1961;53:132–149.

[47] Cruz-Muñiz MY, López-Jacome LE, Hernández-Durán M, et al. Repurposing the anticancer drug mitomycin C for the treatment of persistent Acinetobacter baumannii infections. Int J Antimicrob Agents. 2017;49:88–92.

[48] Chua MS, Bernstein LR, Li R, et al. Gallium maltolate is a promising chemotherapeutic agent for the treatment of hepatocellular carcinoma. Anticancer Res. 2006;26:1739–1743.

[49] Federica R, Carlo B, Emanuela F, et al. Acinetobacter baumannii Biofilm Formation in Human Serum and Disruption by Gallium. Antimicrob Agents Chemother. 2016;61:e01563–16.

[50] Jordan VC. Tamoxifen (ICI46,474) as a targeted therapy to treat and prevent breast cancer. Br J Pharmacol. 2006;147:S269–S276.

[51] Flores R, Insel PA, Nizet V, et al. Enhancement of neutrophil antimicrobial activity by the breast cancer drug tamoxifen. FASEB J. 2016;30. 969.14.

[52] Miller WR, Ingle JN. Endocrine Therapy in Breast Cancer. CRC Press, 2002. https://books.google.se/books?id=00_LBQAAQBAJ.

[53] Burney I. Cancer chemotherapy and biotherapy: Principles and practice. Sultan Qaboos Univ Med J. 2011;11:424–425.

[54] Price N, Sartor O, Hutson T, et al. Role of 5αReductase inhibitors and selective estrogen receptor modulators as potential chemopreventive agents for prostate cancer. Clin Prostate Cancer. 2005;3:211–214.

[55] Evelien G, Valerie D, Katleen V, et al. Repurposing toremifene for treatment of oral bacterial infections. Antimicrob Agents Chemother. 2017;61:e01846–16.

[56] Pink R, Hudson A, Mouriès M-A, et al. Opportunities and challenges in antiparasitic drug discovery. Nat Rev Drug Discov. 2005;4:727–740.

[57] Harbut MB, Vilchèze C, Luo X, et al. Auranofin exerts broad-spectrum bactericidal activities by targeting thiol-redox homeostasis. Proc Natl Acad Sci. 2015;112:4453–4458.

[58] RD P, EL H. Niclosamide Therapy for Tapeworm Infections. Ann Intern Med. 1985;102:550–551.

[59] White JAC. Nitazoxanide: A new broad spectrum antiparasitic agent. Expert Rev Anti Infect Ther. 2004;2:43–49.

[60] Debnath A, Parsonage D, Andrade RM, et al. A high-throughput drug screen for Entamoeba histolytica identifies a new lead and target. Nat Med. 2012;18:956–960.

[61] Thangamani S, Mohammad H, Abushahba MFN, et al. Antibacterial activity and mechanism of action of auranofin against multi-drug resistant bacterial pathogens. Sci Rep. 2016;6:22571.

[62] Thangamani S, Mohammad H, Abushahba MFN, et al. Repurposing auranofin for the treatment of cutaneous staphylococcal infections. Int J Antimicrob Agents. 2016;47:195–201.

[63] She P, Zhou L, Li S, et al. Synergistic microbicidal effect of auranofin and antibiotics against planktonic and biofilm-encased S. aureus and E. faecalis. Front Microbiol. 2019;10:2453.

[64] Chang-Jer W, Jia-Tsrong J, Chi-Min C, et al. Inhibition of severe acute respiratory syndrome coronavirus replication by niclosamide. Antimicrob Agents Chemother. 2004;48:2693–2696.

[65] Gwisai T, Hollingsworth NR, Cowles S, et al. Repurposing niclosamide as a versatile antimicrobial surface coating against device-associated, hospital-acquired bacterial infections. Biomed Mater. 2017;12:45010.

[66] Rajamuthiah R, Fuchs BB, Conery AL, et al. Repurposing salicylanilide anthelmintic drugs to combat drug resistant staphylococcus aureus. PLoS One. 2015;10:e0124595.

[67] Di Santo N, Ehrisman J. Research perspective: Potential role of nitazoxanide in ovarian cancer treatment. Old drug, new purpose? Cancers. 2013;5:1163–1176.

[68] Rossignol J-F. Nitazoxanide: A first-in-class broad-spectrum antiviral agent. Antiviral Res. 2014;110:94–103.

[69] Gau J-S, Lin W-P, Kuo L-C, et al. Nitazoxanide analogues as antimicrobial agents against nosocomial pathogens. Med Chem. 2016;12:544–552.

[70] Bailey MA, Na H, Duthie MS, et al. Nitazoxanide is active against Mycobacterium leprae. PLoS One. 2017;12:e0184107.
[71] Burnett BP, Mitchell CM. Antimicrobial activity of iodoquinol 1%-hydrocortisone acetate 2% gel against ciclopirox and clotrimazole. Cutis. 2008;82:273–280.
[72] Sawyer PR, Brogden RN, Pinder KM, et al. Clotrimazole: A review of its antifungal activity and therapeutic efficacy. Drugs. 1975;9:424–447.
[73] Frosini SM, Bond R. Activity in vitro of clotrimazole against canine methicillin-resistant and susceptible staphylococcus pseudintermedius. Antibiot. 2017;6:29.
[74] Sawyer PR, Brogden RN, Pinder RM, et al. Miconazole: A review of its antifungal activity and therapeutic efficacy. Drugs. 1975;9:406–423.
[75] Nenoff P, Koch D, Krüger C, et al. New insights on the antibacterial efficacy of miconazole in vitro. Mycoses. 2017;60:552–557.
[76] Sahoo AK, Mahajan R. Management of tinea corporis, tinea cruris, and tinea pedis: A comprehensive review. Indian Dermatol Online J. 2016;7:77–86.
[77] Chen F, Di H, Wang Y, et al. Small-molecule targeting of a diapophytoene desaturase inhibits S. aureus virulence. Nat Chem Biol. 2016;12:174–179.
[78] White NJ. Assessment of the pharmacodynamic properties of antimalarial drugs in vivo. Antimicrob Agents Chemother. 1997;41:1413–1422.
[79] White NJ, Hien TT, Nosten FH, Brief A. History of Qinghaosu. Trends Parasitol. 2015;31:607–610.
[80] Kim W-S, Choi WJ, Lee S, et al. Anti-inflammatory, antioxidant and antimicrobial effects of artemisinin extracts from artemisia annua L. KJPP. 2014;19:21–27.
[81] CDC: Artesunate,https://www.cdc.gov/malaria/diagnosis_treatment/artesunate.html.
[82] Choi WH. Novel pharmacological activity of artesunate and artemisinin: Their potential as anti-tubercular agents. J Clin Med. 2017;6:30.
[83] Gøtzsche PC. Non-steroidal anti-inflammatory drugs. Bmj. 2000;320:1058 LP –1061.
[84] Annadurai S, Basu S, Ray S, et al. Antibacterial activity of the antiinflammatory agent diclofenac sodium. Indian J Exp Biol. 1998;36:86–90.
[85] HA S, ST J, WA D, et al. Repurposing the nonsteroidal anti-inflammatory drug diflunisal as an osteoprotective, antivirulence therapy for staphylococcus aureus osteomyelitis. Antimicrob Agents Chemother. 2016;60:5322–5330.
[86] Pandey P, Verma V, Gautam G, et al. Targeting the β-clamp in Helicobacter pylori with FDA-approved drugs reveals micromolar inhibition by diflunisal. FEBS Lett. 2017;591:2311–2322.
[87] Kil J, Pierce C, Tran H, et al. Ebselen treatment reduces noise induced hearing loss via the mimicry and induction of glutathione peroxidase. Hear Res. 2007;226:44–51.
[88] MA C, KS E, L-fs J, et al. Structure, mechanism, and inhibition of aspergillus fumigatus thioredoxin reductase. Antimicrob Agents Chemother. 2019;63:e02281–18.
[89] Ngo HX, Shrestha SK, Green KD, et al. Development of ebsulfur analogues as potent antibacterials against methicillin-resistant Staphylococcus aureus. Bioorg Med Chem. 2016;24:6298–6306.
[90] Gustafsson TN, Osman H, Werngren J, et al. Ebselen and analogs as inhibitors of Bacillus anthracis thioredoxin reductase and bactericidal antibacterials targeting Bacillus species, Staphylococcus aureus and Mycobacterium tuberculosis. Biochim Biophys Acta – Gen Subj. 2016;1860:1265–1271.
[91] Zou L, Lu J, Wang J, et al. Synergistic antibacterial effect of silver and ebselen against multidrug-resistant Gram-negative bacterial infections. EMBO Mol Med. 2017;9:1165–1178.
[92] Ahmed EF, El-Baky RMA, Ahmed ABF, et al. Antibacterial activity of some non-steroidal anti-inflammatory drugs against bacteria causing urinary tract infection. Am J Infect Dis Microbiol. 2017;5:66–73.

[93] Chan EWL, Yee ZY, Raja I, et al. Synergistic effect of non-steroidal anti-inflammatory drugs (NSAIDs) on antibacterial activity of cefuroxime and chloramphenicol against methicillin-resistant Staphylococcus aureus. J Glob Antimicrob Resist. 2017;10:70–74.
[94] Liu G, Catacutan DB, Rathod K, et al. Deep learning-guided discovery of an antibiotic targeting Acinetobacter baumannii. *Nat Chem Biol 2023* https://doi.org/10.1038/s41589-023-01349-8.
[95] Corsello SM, Bittker JA, Liu Z, et al. The Drug Repurposing Hub: a next-generation drug library and information resource. *Nat Med.* 2017;23:405–408.
[96] Park K. A review of computational drug repurposing. Transl Clin Pharmacol. 2019;27:59–63.
[97] Jarada TN, Rokne JG, Alhajj R. A review of computational drug repositioning: Strategies, approaches, opportunities, challenges, and directions. J Cheminform. 2020;12:46.

Vinita Pandey, Priyanka Verma, Urooj Fatima, Iliyas Khan,
Arshad J. Ansari

7 The implication of drug repurposing in the identification of drugs for gastrointestinal disorders

Abstract: Over the past decade, advancement in the effective treatment of gastrointestinal tract (GIT) disorders are still challenging to treat and may not have complete cure. As an alternative to the traditional drug development approach, drug repurposing or "repositioning" is a time- and money-effective strategy to find new applications for already-approved medications. This study focused on critical biological targets affected by many sensitive genes and mediators essential in Inflammatory Bowel Disease (IBD). We have also focused on various FDA-approved drugs that have been repositioned to target these IBD candidate genes and mediators involved in the disease and treatment of IBD. The use of medication repurposing techniques in IBD treatment may eventually enhance IBD care. This chapter summarizes the progress of drug repurposing in identifying drugs for gastrointestinal disorders and their biological properties over the last decade.

Keywords: Drug repurposing, GIT disorder, inflammatory bowel disease (IBD)

7.1 Introduction

Discovering a new drug is a high-cost and time-consuming [1] process, so the drug discovery pipeline is dehydrating [2]. Therefore, research institutions and drug manufacturers focus on innovative drug development strategies. Drug repurposing goals are to accelerate the drug development process [3]. Drug repurposing (often referred to as drug repositioning) means finding novel indications for drugs already present in the market [2]. This strategy may lower the cost of creating new medications and increase patient access to those medications for life-threatening illnesses. Drug repositioning is based on the idea that many diseases have related molecular pathways. It is preferable to the conventional de novo drug development procedure in many ways:

i. Since existing medications have undergone human testing, extensive information on their pharmacology, dosage, formulation, and potential toxicity is accessible.
ii. The medication repurposing strategy is centered on the idea that multiple diseases share common molecular pathways that contribute to their pathogenesis.
iii. It considerably reduces the cost and time required for new therapeutic development because existing approved compounds have proven safe in humans during phase 1 clinical trials.

https://doi.org/10.1515/9783110791150-007

Many data-driven and experimental methods have been proposed to discover repurposable medication candidates, but significant technological and legal issues also need to be resolved [4].

Repurposing may involve combining an earlier drug with a recent drug to boost the effectiveness. A drug may be combined with a nondrug therapy option, such as radiation, to improve the efficacy of the nondrug treatment or the medication itself. The method also permits the combination of existing medications that are now prescribed separately but are already used to treat a particular ailment. Since medications can influence multiple targets and are connected to metabolic pathways, the total activity of other targets may alter as a result of changes to the activity of one target, which might have other impacts, which is where the idea of drug repurposing comes in. There are two types of drug repurposing [5]: on-target and off-target drug repurposing shown in Figure 7.1.

7.1.1 On-target drug repurposing

On-target drug repurposing refers to Investigating new therapeutic uses for medications that already have established targets and mechanisms. The well-known pharmacological effect of a drug molecule in on-target drug repurposing is applied to a novel therapeutic indication. In this approach, the drug's biological target is the same, but the disease is different [6].

7.1.2 Off-target drug repurposing

Off-target drug repurposing refers to investigating potential applications for a medication that acts on an unusual or unexpected target. This strategy is widely used since, on average, a medicine hits 6 to 13 targets [7]. In the off-target profile, there is no recognized pharmacological mechanism. In addition to their primary uses, drugs and drug candidates also work on new targets. Consequently, both the aims and the signals are new [6].

The practice of finding new indications for treatments that already exist is known as drug repurposing. These treatments have been approved by USFDA or shown to be safe in phase 1–2 clinical studies but were never commercialized due to factors unrelated to safety [8]. This approach to drug development might have a better chance of being successful because the repurposed medication has a demonstrated safety profile. Additionally, compared to the usual strategy, the drug repurposing approach should need much less time and money. Figure 7.2 displays the road plan for discovering new applications for the current medication.

The road to discovering new drugs has always been a twisty one. In particular, de novo drug design and development is costly and time-taking and the procedure is risky.

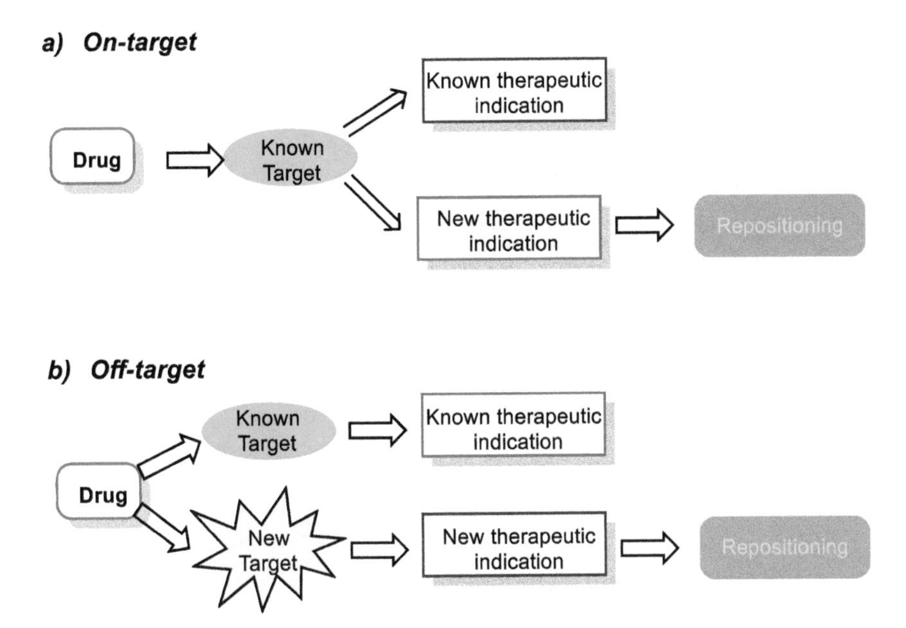

Figure 7.1: Drug repurposing: (a) on-target; (b) off-target.

For instance, to bring a medicine to market from its initial discovery to the approval stage the average cost would be around 2 to 3 billion dollars and it would take at least 13–15 years [9]. Additionally, there is a significant percentage of attrition throughout the procedure. Approximately 10% of medications tested in clinical trials are authorized by regulatory bodies [10]. As preclinical studies have a limited ability to predict future outcomes, the remaining 90% of medications fail owing to inefficiency or severe toxicity [11]. Phase 2 compound failure rates are 62%, while Phase 3 compound attrition rates are around 45% [12]. Since there are few preclinical disease models available, this attrition is caused by inadequate research and development (R&D) productivity in finding the medication response on the target, which has generated worry in the pharmaceutical business [13]. The number of newly authorized pharmaceuticals has remained constant despite tremendous advancements in technology and exponential increases in pharmaceutical R&D spending. According to the underlying theory, repositioned pharmaceuticals can enter the clinical stages more quickly in comparison to new compounds, since most authorized compounds have established bioavailability and safety profiles, validated formulation and production pathways, and fairly described pharmacology. Additionally, the 90% failure rate for therapeutic development indicates that numerous, incompletely developed therapeutic candidates may be revisited, further investigated, and possibly repurposed for a new condition, whether common or unusual. Therefore, of the pharmaceuticals that have reached the markets in recent years, 30% are repositioned medications. For instance, just seven of the 113 new pharmaceuticals and biologics that were introduced in 2017 were first-in-class agents (a drug that was approved and introduced first with a unique mechanism of ac-

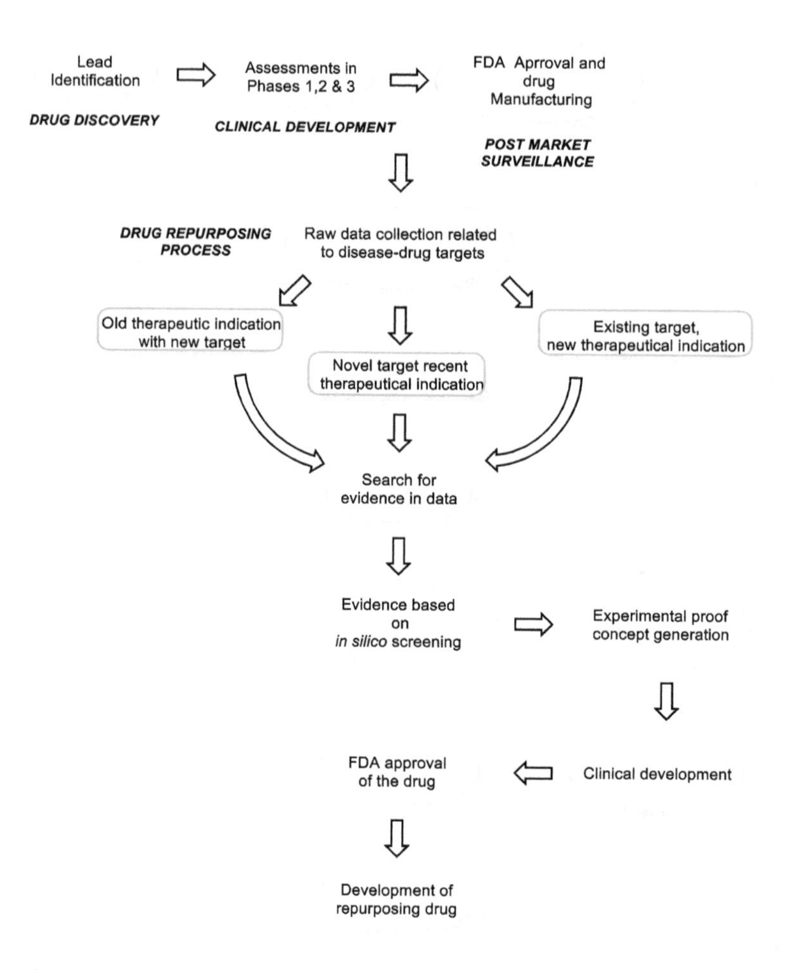

Figure 7.2: Summary of drug repurposing.

tion), whereas 36 were repositioned medications [14]. According to estimates, this bypassing may speed up the release of medicine for patient usage by 3–12 years at a total projected cost of 40–80 million dollars [15–17].

7.2 History of drug repurposing

In the midst of 2000, sildenafil, which had been used to treat angina pectoris, was repositioned to treat erectile dysfunction, and thalidomide was switched from treating morning sickness to multiple myeloma. The achievement triggered significant interest in repurposing, which lead to the establishment of other startup businesses with repurposing as their primary focus [18]. Repurposing is important in the life cycle management of products with 10–50% R&D expenditure, according to numerous evaluations and market research reports. Nancy et al. carried out a bibliometric study by carefully examining a few drugs in order to comprehend the background of the various drug repurposing practices. Among a few instances, chlorpromazine was first produced in 1950 and was intended for use in treating mental illnesses and as a preoperative drug; it was later employed in treating a variety of diseases, such as whooping cough, as symptoms first appeared in 1972 during radiation treatment for cancer patients [19–21]. A well-known antimalarial drug, chloroquine, was synthesized in 1934. Afterward, it has been used to treat many other illnesses such as fever, lupus, skin rashes, parasite disorders (before 1960), and fever [22]. A few repurposed medications are shown in Figure 7.3.

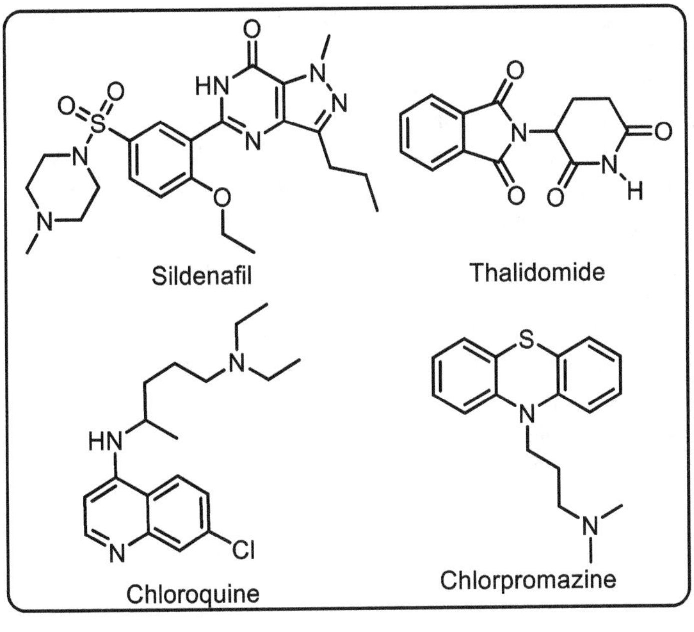

Figure 7.3: Few historical drugs repurposing.

7.3 Advantages of drug repurposing

Drug repurposing is sometimes referred to as rediscovering, rescuing, retasking, reprofiling, and repositioning. All of these terms are synonyms when describing the same procedure that aims to find new uses for drugs or compounds that are already on the market. De novo drug development is the expansive, lengthy, and risky process of discovering and creating completely new treatments. The overall average expanse is between $2 to $3 billion, and it takes a minimum time of 13 to 15 years to complete [23]. Repurposing older drugs has been recommended as an alternative approach to producing novel therapeutics that have less risk, and cost, and takes shorter time as compared to new drugs. The approval rates for repurposed medications are close to 30%, and they are often approved more quickly (3–12 years) and a cheaper cost (50–60%), and with less risk. [22]. In fact, the chemical used as the starting point has well-understood critical features in humans, including a high safety profile and predictable pharmacokinetics. It describes both medicines that have been approved and substances that have been proven safe in phase 1–2 clinical trials but were never commercialized for reasons unrelated to safety. This implies that it may be feasible to bypass the expensive and time-consuming discovery and early-stage development research necessary for completely new medications and proceed directly to clinical trials for a new indication [24]. A repurposed medicine may have roughly the same regulatory and phase 3 costs as a new drug for the same application, but there may be significant preclinical and phase 1 and 2 cost savings. According to estimates, the expenses of commercializing a medicine repurposed to treat a different condition equal those of an active pharmacological ingredient [25]. Another way to think of drug repositioning is as the process through which new biological effects for well-known medications are discovered, resulting in suggestions for new therapeutic applications [26]. Last but not least, drug rescue has been employed to draw attention to the fact that the drug has failed for its main indication [27]. The phrase "drug repurposing" will be used throughout this paper to refer to all terms because a PubMed search that only looked at titles and keywords of articles published after January 2019 shows that it predominates over the term "drug" [28]. All terms used in this article shall be referred to as "drug repurposing" for the sake of consistency. Repositioning entails the creation of an existing, previously assessed, but unapproved medicine, whereas repurposing refers to using authorized pharmaceuticals for new indications, according to certain researchers.

7.4 GIT disorders

7.4.1 Inflammatory bowel disease

The GI tract (GIT) is affected by a chronic inflammatory condition known as inflammatory bowel disease (IBD), which is characterized by alternating episodes of remission and relapse (Figure 7.4). Crohn's disease (CD) and ulcerative colitis (UC) are the two main subtypes according to clinical classification [29]. Table 7.1 lists the parts of the GIT that CD and UC affect. IBD is increasing in prevalence across the globe, particularly in recently industrialized countries where it was previously nonexistent [28]. Similar increases in childhood IBD cases are indicative of shifting environmental risk factors [30]. In 2017, there were 6.8 million IBD cases worldwide. From 79.5 per 100,000 people in 1990 to 84.3 per 100,000 people in 2017, the age-standardized prevalence rate increased.

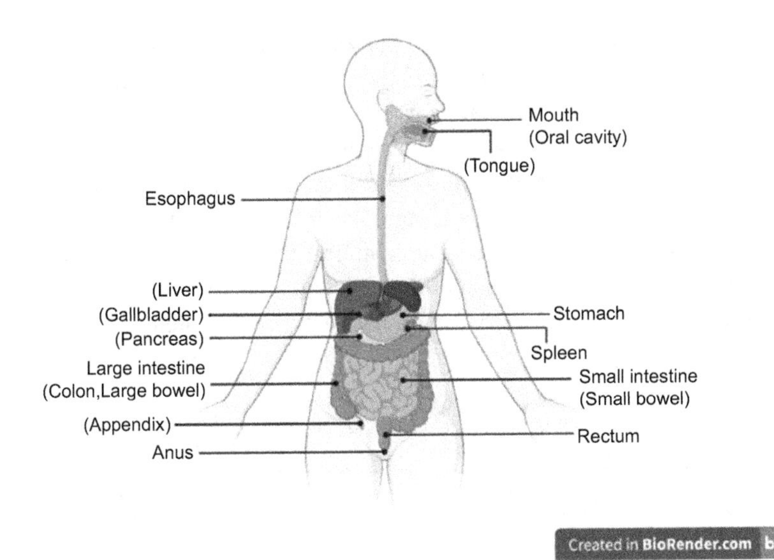

Figure 7.4: Graphical representation of gastrointestinal tract.

Table 7.1: Types of inflammatory bowel disease.

S. no.	Affected parts	Crohn's disease	Ulcerative colitis	References
1.	**Affected location**	Can impact any GIT portion, including the mouth and antrum. The section of the small intestine before the large intestine/colon is most frequently affected	Occurs in the rectum and the large intestine (colon)	[29]
2.	**Damaged areas**	Patches of damaged tissue can be seen next to areas of healthy tissue	Continuous (not patchy) damage typically begins at the rectum and progresses into the colon	[29]
3.	**Inflammation**	The multiple layers of the GIT's walls may be penetrated by inflammation	Only the innermost layer of the colon's lining exhibits inflammation	[29]

As regards global burden of disease (GBD), North America had the highest age-standardized prevalence rate in 2017 (422 per 100,000 population), and the Caribbean had the lowest age-standardized prevalence rate (6.7 per 100,000 population). Age-standardized prevalence rates were highest in areas with high sociodemographic indexes (SDIs) and lowest in areas with low SDIs [31]. The complicated pathogenesis of IBD makes prevention and therapy difficult. Both innate and adaptive immunity are dysregulated in part due to genetic predisposition [32]. Diet, infection, antibiotic use, and exposure to toxins are environmental triggers that modify immune regulation by altering the intestinal microbiome and influencing epigenetic alterations. Immunosuppressive therapy and surgery are frequently used during treatment, which is typically lifelong, especially in patients with more severe diseases. To lessen the burden of IBD, it is imperative to [1] identify and eliminate risk factors for the disease and [2] find new medical treatments with higher efficacy and fewer potential adverse effects. Current IBD medications work to bring about a profound remission of symptoms. Older IBD treatments only attempted to keep the symptoms under control. Several medication classes are used, some of which are grouped according to their biological activity in Table 7.2.

Table 7.2: Recently discovered potential treatment for inflammatory bowel disease.

S. no.	Drug	Drug type	Functions of drugs	References
1.	Balsalazide,Mesalazine, Olsalazine, and Sulfasalazine	Aminosalicylates	Anti-inflammatory (for mild to moderate IBD).	[33]
2.	Prednisone, Methylprednisolone, Hydrocortisone, Betamethasone, Triamcinolone, Dexamethasone, Budesonide, Betamethasone, Cortisone acetate, Desonide, and Hydrocortisone	Corticosteroids	Anti-inflammatory (for moderate to severe IBD)	[33]
3.	Methotrexate, 6-mercaptopurine, and azathioprine	Thiopurines	Immunosuppression (for aminosalicylate resistance patients with chronic corticosteroid use)	[33]
4.	Metronidazole, Ciprofloxacin	Antibiotics	Altering the gut microbiota, treatment of bacterial infections	[33]
5.	Adalimumab, Certolizumab, Golumumab, Infliximab	TNF blockers	TNF signaling is downregulated to reduce inflammation (for moderate to severe IBD in patients resistant to other treatments)	[33]
6.	Vedolizumab, Natalizumab	Leucocytes adhesion inhibitors	By preventing white blood cells from sticking to blood arteries and invading target tissue, anti-inflammatory effects are achieved (for CD patients whose condition is unresponsive to any other treatments)	[33]

Table 7.2 (continued)

S. no.	Drug	Drug type	Functions of drugs	References
7.	Ustekinumab	Anti-interleukin -12/23	Stop IL-12 and IL-23 from acting as pro-inflammatory cytokines in CD. Used with people who have moderate-to-severe CD	[34]
8.	Tofacitinib	JAK inhibitors	Inhibits the JAK-STAT signaling pathway, which limits transcription of several inflammatory cytokines involved in IBD development. Approved for US in-patients with UC	[34]

7.4.2 Ulcerative colitis (UC)

Inflammatory bowel illnesses are made up of two primary categories of chronic idiopathic intestinal disorders, including ulcerative colitis (UC). About 1 million people in the US alone are affected by UC, and both its incidence and prevalence are rising globally [34, 35]. The distinguishing characteristic of UC is mucosal inflammation that begins in the rectum and spreads to the entire colon. Clinical and endoscopic remissions are the primary goals of UC therapeutic management. But only roughly two-thirds of patients benefit from corticosteroids [35], and one-third of patients who get antitumor necrosis factor (anti-TNF) therapy do not respond [36]. A colectomy is frequently necessary to control the condition in patients who are resistant to pharmaceutical therapy. Between 2.4 to 10.4% is estimated to be the 10-year cumulative colectomy rate in UC patients [33, 37]. The only known curative treatment is total colectomy; however, it is only used as a last resort because of the adverse effects it might cause, including surgery-related complications such as clots, pouchitis, and intestinal blockages or strictures. The unfavorableness of colectomies highlights the demand for additional UC patient treatment alternatives that can lower colectomy rates.

7.4.3 Gastrointestinal stromal tumor (GIST)

Despite being rare, gastrointestinal stromal tumor (GIST) is the most prevalent mesenchymal malignancy of the digestive tract, with an estimated 6,000 cases per year in the United States [3]. The discovery of activating mutations in roughly 85% of GISTs in the tyrosine kinase receptors, KIT and PDGFRA, has changed how the disease is managed

[38, 39]. The various repurposed medications on GI disorders are listed in Table 7.3 and are grouped according to their novel indication.

Table 7.3: Drug repurposing on GIT disorders.

S. no.	Drug	Original indication	Novel indication	References
1.	Topiramate	Convulsion	IBD	[38]
2.	Quinacrine	Malaria	IBD	[39]
3.	Thioguanine	Leukemia	IBD	[40]
4.	Aldesleukin	Metastatic renal cell carcinoma	IBD	[41]
5.	Adalimumab	Rheumatoid arthritis, psoriatic arthritis	IBD	[42]
6.	Interferon	Autoimmune disorder; multiple sclerosis	IBD	[43]
7.	Thalidomide	Morning sickness	IBD	[44]
8.	Angiotensin-converting Enzyme 2	Hypertension	IBD	[45]
9.	Sunitinib	Renal cell carcinoma	Gastrointestinal tumor, pancreatic tumors	[46]
10.	Imatinib	BCR-ABL	GIST	[47–49]
11.	Salidroside	Aging, antioxidant, cancer, inflammation, and neuroprotective	Ulcerative colitis	[50]
12.	D-Pinitol	Inflammation and diabetes	Ulcerative colitis	[51]
13.	Troxerutin	Coagulation and thrombosis	Ulcerative colitis	[52]
14.	Poria Cocos	Antioxidant, antitumor and immunomodulation activities	IBD	[53]

7.5 Effects of the drug repurposing approach in IBD

7.5.1 Topiramate

A common indication for the anticonvulsant drug topiramate is seizure disorders [40]. Gamma-aminobutyric acid (GABA)-A receptor activity is increased by topiramate, which also acts as an antagonist for AMPA/kinate glutamate receptor subtypes and a weak inhibitor of carbonic anhydrase isozymes II. Topiramate significantly reduced gross pathophysiological and histological outcomes in the experimental animal model of IBD. The positive effects of topiramate in animal models of IBD have also been supported by computational calculations. The analysis shows that topiramate has an impact on genes related to NF-kB signaling, inflammatory response, antigen presentation, and other pathophysiological IBD-related processes. In a different study, gene expression profiles of 164 drug compounds were compared to those of IBD patients. By comparing the genomic fingerprint of topiramate with the disease signature, the researchers concluded that topiramate might be an effective management and treatment option for IBD [41].

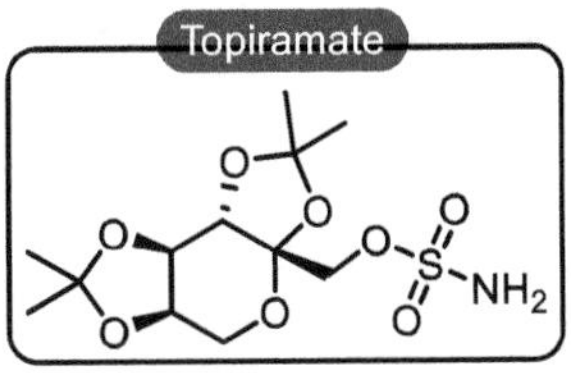

Topiramate

7.5.2 Quinacrine

Quinacrine has a variety of antiprotozoal, anticancer, antirheumatic, antiprion, and sclerosing effects inside the pleura [42]. Quinacrine successfully suppressed the clinical disease index (CDI), histological changes of the colon, and levels of inflammatory markers (iNOS, Cox-2, and p53) in vivo in experiments using the dextran sulfate sodium and oxazolone mouse models of UC. No toxic or negative effects were noticed. Prostaglandins and the arachidonic acid cascade are two more inflammatory mediators that quinacrine has been shown to block the synthesis of. Additionally, it inhibits NF-B, TNF, and IL-1, suggesting the potential use of quinacrine in the management of mild-to-severe forms of IBD and providing information on potential mechanisms of anti-TNF therapy [43].

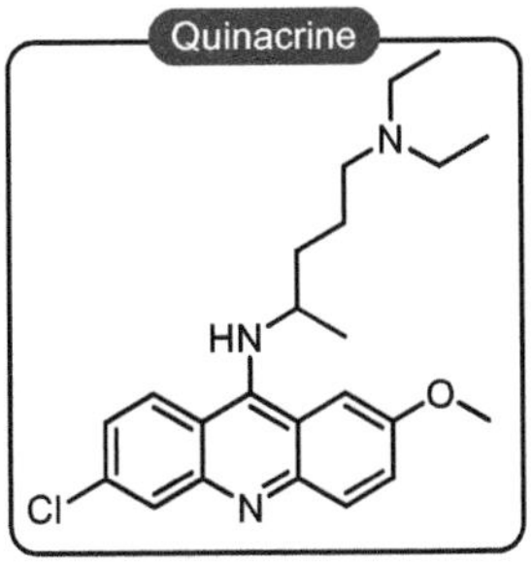

Quinacrine

7.5.3 Thioguanine

The FDA has authorized the use of thioguanine in leukemia. In the chronic stage of IBD, mucosal immune system activation has been proven to be crucial. Thiopurines are immunosuppressive medications that reduce inflammation by deactivating T cells. The effectiveness of thioguanine in treating IBD was evaluated through extensive clinical trials, and it was found that about 80% of patients benefited clinically from its use. However, this repurposed clinical benefit was overshadowed by the nodular regenerative hyperplasia advancement in the liver associated with the high dose of thioguanine. Thioguanine was subsequently therapeutically repositioned as an immunosuppressive agent for the treatment of patients with IBD in 2001, after multiple trials were conducted to show that it does not induce any liver problems at a sufficient dose of 0.2–0.3 mg/kg (not exceeding 25 mg per day) [44].

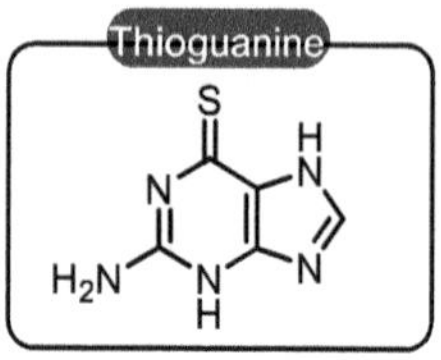

Thioguanine

7.5.4 Aldesleukin

Metastatic renal cell cancer is treated with aldesleukin, an IL2 mimic. IL2 modulates the immune system and is one of the key mediators involved in the development of IBD and other autoimmune disorders, according to earlier studies. To ascertain the likely effects of modest doses of aldesleukin, clinical trials are now being conducted in individuals with autoimmune illnesses such as IBD, rheumatoid arthritis, and primary sclerosing cholangitis [45].

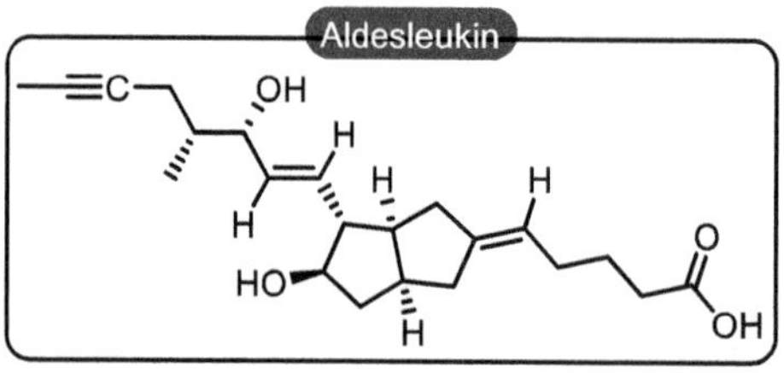

Aldesleukin

7.5.5 Adalimumab

Rheumatoid arthritis, psoriatic arthritis, and ankylosing spondylitis were the first conditions for which adalimumab received clinical approval. Adalimumab binds TNF and has been shown in randomized, double-blind studies to be safe and effective in generating and sustaining remission in up to 36% of patients with moderate-to-severe UC, who do not react to standard IBD treatments such as steroids or immunosuppressants [46].

7.5.6 Interferon

The autoimmune disease known as multiple sclerosis was the initial target of interferon beta 1b therapy. Numerous studies have revealed that interferon beta 1b has anti-inflammatory properties and may have potential results in various inflammatory illnesses, including IBD, even if the therapeutic profile of this medication is not yet apparent [47].

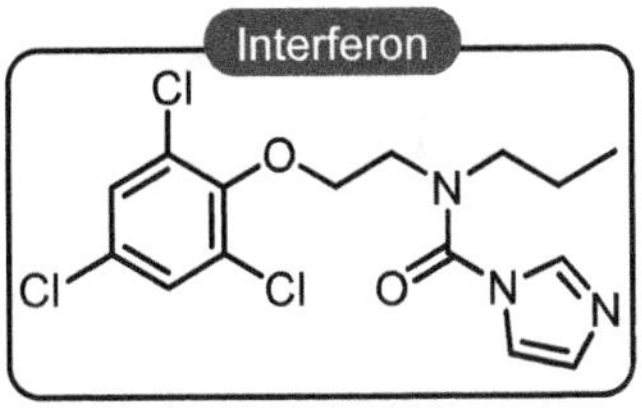

7.5.7 Thalidomide

Thalidomide is a controversial drug because it has a serious side effect called polyneuropathy. The IBD treatment plan does not include the medication. Thalidomide is nevertheless prescribed by medical professionals when a patient does not respond to steroids or other widely used treatments for IBD. Thalidomide also showed efficacy in CD patients in a small open-label clinical trial study. Consequently, a thorough examination of the genetic evidence in favor of thalidomide's use in the treatment of IBD is necessary [48].

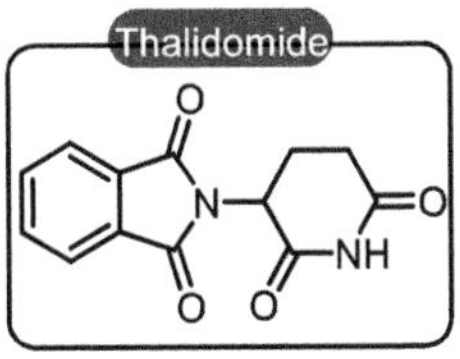

7.5.8 Angiotensin-converting enzyme 2

Angiotensin-converting enzyme 2 (ACE2), which has been shown to be significant in gastrointestinal tissue, is expressed more frequently in IBD patients. The potent and selective ACE2 inhibitor GL1001 significantly decreased colon pathology and myeloperoxidase activity while also demonstrating anti-inflammatory activity that may have therapeutic benefits for IBD [49].

7.5.9 Sunitinib

New oral multitargeted tyrosine kinase inhibitor sunitinib malate has anticancer and antiangiogenic properties. Recent studies have shown that sunitinib effectively treats individuals with advanced renal cell carcinoma (RCC) and GIST, if imatinib mesylate therapy has failed or the patient's condition has worsened [50].

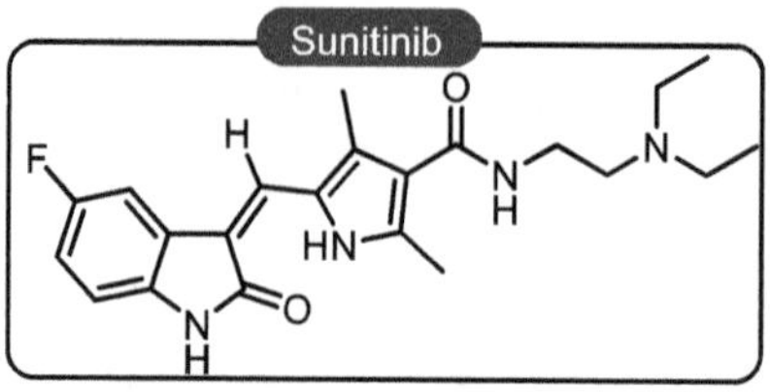

Sunitinib

7.5.10 Imatinib mesylate

Imatinib mesylate's prognosis for treating GIST has significantly improved since the FDA approved it as a drug in 2002. GIST, which typically resists conventional chemotherapy and radiation, may now be controlled by imatinib mesylate. However, the median time for GISTs to progress is about two years, so these effects are frequently fleeting [51–53]. As a result, other medications or medications used in conjunction with imatinib mesylate should be considered for GIST therapy.

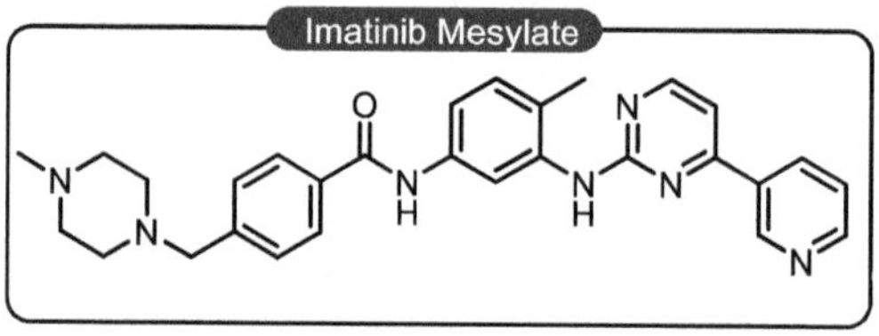

Imatinib Mesylate

7.5.11 Salidroside

As a primary glycoside isolated from Rhodiola rosea L., salidroside (Sal) has demonstrated its potent anticancer, neuroprotective, anti-inflammatory, antioxidant, and anti-aging benefits in a variety of disorders. The SIRT1/FoxOs pathway has recently been activated in colitis mouse models to demonstrate its protective effects. However, it is unknown if Sal has any further defenses against dextran sulfate sodium (DSS)-induced colitis in mice. In this study, we examined the mechanisms by which Sal shields mice from colitis brought on by DSS. Sal met the requirements to receive treatment for ulcerative colitis, per the findings [51].

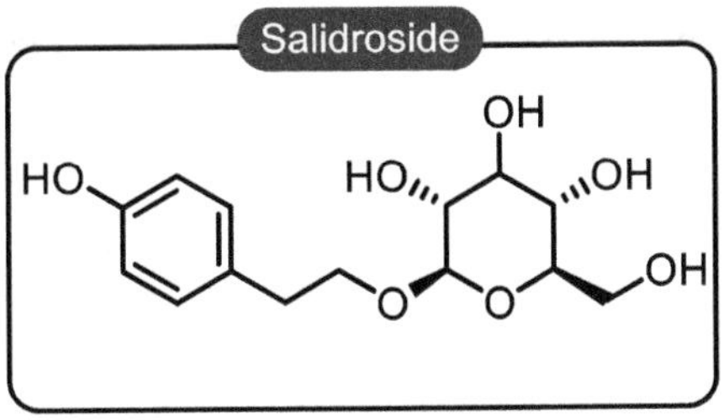

Salidroside

7.5.12 D-pinitol (3-O-methyl-chiro-inositol)

R monosperma aerial portions were used to isolate the chemical D-pinitol (3-O-methyl-chiro-inositol), which has anti-inflammatory and antidiabetic properties. Current research on D-probable pinitol's mechanism is still lacking. In the current study, the possible therapeutic effect and mechanism of D-pinitol against colitis were investigated. D-pinitol demonstrated a striking anticolitis activity by activating the Nrf2/ARE pathway and PPAR-. D-pinitol may therefore, one day, prove to be a successful UC treatment [54].

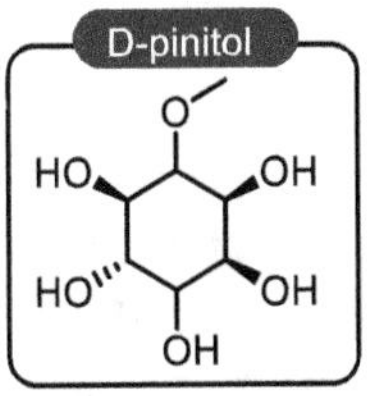

D-pinitol

7.5.13 Troxerutin

Troxerutin is a flavonoid that is used in medicine as a thrombolytic and anticoagulant. In this investigation, troxerutin was revealed to have a novel pharmacological effect; specifically, it significantly improved mice with UC. Troxerutin was demonstrated to effectively reduce the level of oxidative stress that harmed intestinal epithelial cells and colonic tissue, protect the distribution and expression of proteins linked to tight junctions, and safeguard the barrier function of colon tissue. The severe inflammatory response is another significant pathogenic component that aggravates UC in addition to oxidative stress. The production of inflammatory-related proteins and proinflammatory cytokines, as well as the infiltration of inflammatory cells in the colon tissue, may both be reduced by roxerutin, though. Due to its antioxidant and anti-inflammatory properties, roxerutin decreases the degree of colonic fibrosis by preventing the process of cell death in the colon tissue. According to a bioinformatics study, Troxerutin's ability to treat UC was likely attributed to its network control of signaling pathways. In conclusion, we identified a novel pharmacological action of the flavonoid troxerutin against UC, paving the way for the development and use of flavonoids in treating human illnesses [55].

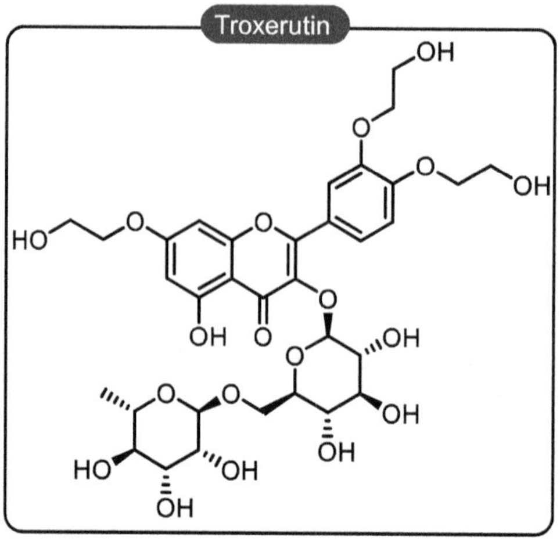

7.5.14 Poria cocos

In the formulation of cosmetics, functional meals, and tea supplements, Poria cocos, a kind of edible and medicinal fungus is frequently employed. From the cultivated mycelia of Poria cocos, Lu et al. extracted a water-soluble 1,6-branched 1,3-D-galactan (PC-II). It has been demonstrated that PC-II inhibits IP-10 synthesis, an IFN-g-induced inflammatory marker, in a dose-dependent way. Generally, the antioxidant, anticancer, and immunomodulation properties of Poria cocos polysaccharides predominate. The study uses mouse models of TNBS-induced colitis to demonstrate the anti-inflammatory properties of the carboxymethyl polysaccharide CMP33 from Poria cocos. The findings showed that CMP33 significantly lessened the severity of colitis when compared to TNBS administration, including a twofold decrease in death rate, a 50% decrease in disease activity index, and a 36–44% decrease in macro- or microscopic histopathological score. Additionally, CMP33 increased the levels of anti-inflammatory cytokines and decreased the levels of proinflammatory cytokines in the colon tissue and serum of colitic mice [56].

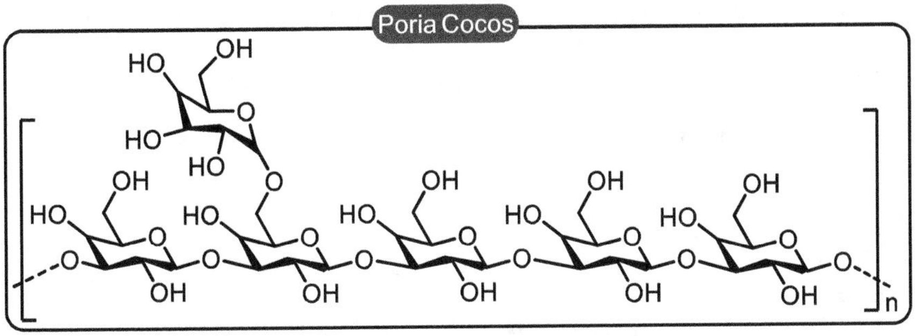

7.6 Conclusion

IBD is a growing global concern in health industry. Nowadays, chemotherapies and clinical management of IBD are unsuccessful due to divergent etiology. For the long-term treatment and maintenance of IBD, a chronic inflammatory GIT condition, few safe and efficient medications are available. In order to find medications to cure a range of ailments, drug repurposing – the process of finding new applications for already-approved medications –is a time- and money-effective strategy. It has been observed that a number of sensitive genes and mediators are important in IBD. Other FDA approved medications may be repositioned to target these IBD candidate genes and mediators involved in the disease etiology and treat IBD. The use of medication repurposing techniques in IBD treatment may eventually enhance IBD care.

Reference

[1] Dimasi, J.A., Feldman, L., Seckler, A., and Wilson, A. (2010) Trends in risks associated with new drug development: Success rates for investigational drugs. *Clin Pharmacol Ther*, **87** (3).

[2] Gil, C., and Martinez, A. (2021) Is drug repurposing really the future of drug discovery or is new innovation truly the way forward? *Expert Opin Drug Discov*, **16** (8).

[3] Parvathaneni, V., Kulkarni, N.S., Muth, A., and Gupta, V. (2019) Drug repurposing: a promising tool to accelerate the drug discovery process. *Drug Discov Today*, **24** (10).

[4] Pushpakom, S., Iorio, F., Eyers, P.A., Escott, K.J., Hopper, S., Wells, A., Doig, A., Guilliams, T., Latimer, J., McNamee, C., Norris, A., Sanseau, P., Cavalla, D., and Pirmohamed, M. (2018) Drug repurposing: Progress, challenges and recommendations. *Nat Rev Drug Discov*, **18** (1).

[5] Rudrapal, M., J. Khairnar, S., and G. Jadhav, A. (2020) Drug Repurposing (DR): An Emerging Approach in Drug Discovery, in *Drug Repurposing – Hypothesis, Molecular Aspects and Therapeutic Applications*.

[6] Ferreira, L.G., and Andricopulo, A.D. (2016) Drug repositioning approaches to parasitic diseases: a medicinal chemistry perspective. *Drug Discov Today*, **21** (10).

[7] Dhaneshwar, S., and Roy, S. (2018) Drug Repurposing Opportunities for Inflammatory Bowel Disease Inflammatory Bowel Disease.

[8] Ashburn, T.T., and Thor, K.B. (2004) Drug repositioning: Identifying and developing new uses for existing drugs. *Nat Rev Drug Discov*, **3** (8).

[9] Scannell, J.W., Blanckley, A., Boldon, H., and Warrington, B. (2012) Diagnosing the decline in pharmaceutical R&D efficiency. *Nat Rev Drug Discov*, **11** (3).

[10] Akhondzadeh, S. (2016) The importance of clinical trials in drug development. *Avicenna J Med Biotechnol*, **8** (4).

[11] Plenge, R.M., Scolnick, E.M., and Altshuler, D. (2013) Validating therapeutic targets through human genetics. *Nat Rev Drug Discov*, **12** (8).

[12] Kola, I., and Landis, J. (2004) Can the pharmaceutical industry reduce attrition rates? *Nat Rev Drug Discov*, **3** (8).

[13] Paul, S.M., Mytelka, D.S., Dunwiddie, C.T., Persinger, C.C., et al., (2010) How to improve RD productivity: The pharmaceutical industry's grand challenge. *Nat Rev Drug Discov*, **9** (3).

[14] Graul, A.I., Cruces, E., and Stringer, M. (2014) The year's new drugs & biologics, 2013: Part I. *Drugs of Today*, **50** (1).

[15] Papapetropoulos, A., and Szabo, C. (2018) Inventing new therapies without reinventing the wheel: the power of drug repurposing. *Br J Pharmacol*, **175** (2).

[16] Yella, J.K., Yaddanapudi, S., Wang, Y., and Jegga, A.G. (2018) Changing trends in computational drug repositioning. *Pharmaceuticals*, **11** (2).

[17] Hurle, M.R., Yang, L., Xie, Q., Rajpal, D.K., Sanseau, P., and Agarwal, P. (2013) Computational drug repositioning: From data to therapeutics. *Clin Pharmacol Ther*, **93** (4).

[18] Novac, N. (2013) Challenges and opportunities of drug repositioning. *Trends Pharmacol Sci*, **34** (5), 267–272.

[19] López-Muñoz, F., Alamo, C., Cuenca, E., Shen, W.W., Clervoy, P., and Rubio, G. (2005) History of the discovery and clinical introduction of chlorpromazine. *Annals of Clinical Psychiatry*, **17** (3).

[20] Frankenburg, F.R., and Baldessarini, R.J. (2008) Neurosyphilis, malaria, and the discovery of antipsychotic agents. *Harv Rev Psychiatry*, **16** (5).

[21] CAMPBELL, M., and GORDON, R.A. (1956) The use of chlorpromazine in intractable pain associated with terminal carcinoma. *Can Med Assoc J*, **75** (5).

[22] Schlitzer, M. (2007) Malaria chemotherapeutics part I: History of antimalarial drug development, currently used therapeutics, and drugs in clinical development. *ChemMedChem*, **2** (7).

[23] Fetro, C., and Scherman, D. (2020) Drug repurposing in rare diseases: Myths and reality. *Therapie*, **75** (2).

[24] Pulley, J.M., Rhoads, J.P., Jerome, R.N., Challa, A.P., Erreger, K.B., Joly, M.M., et al., (2020) Using what we already have: Uncovering new drug repurposing strategies in existing omics data. *Annu Rev Pharmacol Toxicol*, **60**.

[25] Walker, N. (2017) Accelerating drug development through repurposing, repositioning and rescue. *Pharm Outsourcing*, **18** (7).

[26] Barbosa, E.J., Löbenberg, R., de Araujo, G.L.B., and Bou-Chacra, N.A. (2019) Niclosamide repositioning for treating cancer: Challenges and nano-based drug delivery opportunities. *European Journal of Pharmaceutics and Biopharmaceutics*, **141**.

[27] Jourdan, J.P., Bureau, R., Rochais, C., and Dallemagne, P. (2020) Drug repositioning: a brief overview. *Journal of Pharmacy and Pharmacology*, **72** (9).

[28] Murteira, S., Ghezaiel, Z., Karray, S., and Lamure, M. (2013) Drug reformulations and repositioning in pharmaceutical industry and its impact on market access: reassessment of nomenclature. *J Mark Access Health Policy*, **1** (1).

[29] Ahluwalia, B., Moraes, L., Magnusson, M.K., and Öhman, L. (2018) Immunopathogenesis of inflammatory bowel disease and mechanisms of biological therapies. *Scand J Gastroenterol*, **53** (4).

[30] Sýkora, J., Pomahačová, R., Kreslová, M., Cvalínová, D., Štych, P., and Schwarz, J. (2018) Current global trends in the incidence of pediatric-onset inflammatory bowel disease. *World J Gastroenterol*, **24** (25).
[31] Alatab, S., Sepanlou, S.G., Ikuta, K., Vahedi, H., et al., (2020) The global, regional, and national burden of inflammatory bowel disease in 195 countries and territories, 1990–2017: a systematic analysis for the Global Burden of Disease Study 2017. *Lancet Gastroenterol Hepatol*, **5** (1).
[32] Kim, D.H., and Cheon, J.H. (2017) Pathogenesis of inflammatory bowel disease and recent advances in biologic therapies. *Immune Netw*, **17** (1).
[33] Grenier, L., and Hu, P. (2019) Computational drug repurposing for inflammatory bowel disease using genetic information. *Comput. Struct. Biotechnol. J.***17** .
[34] Weisshof, R., El Jurdi, K., Zmeter, N., and Rubin, D.T. (2018) Emerging Therapies for Inflammatory Bowel Disease. *Adv Ther*, **35** (11).
[35] Rubin, D.T., Ananthakrishnan, A.N., Siegel, C.A., Sauer, B.G., and Long, M.D. (2019) ACG Clinical Guideline: Ulcerative Colitis in Adults. *American Journal of Gastroenterology*, **114** (3).
[36] Paramsothy, S., Rosenstein, A.K., Mehandru, S., and Colombel, J.F. (2018) The current state of the art for biological therapies and new small molecules in inflammatory bowel disease. *Mucosal Immunol*, **11** (6).
[37] Parragi, L., Fournier, N., Zeitz, J., Scharl, M., et al. (2018) Colectomy rates in ulcerative colitis are low and decreasing: 10-year follow-up data from the Swiss IBD cohort study. *J Crohns Colitis*, **12** (7).
[38] Kedia, S. (2014) Management of acute severe ulcerative colitis. *World J Gastrointest Pathophysiol*, **5** (4), 579.
[39] Heinrich, M.C., Corless, C.L., Duensing, A., McGreevey, L., Chen, C.J., Joseph, N., Singer, S., Griffith, D. J., Haley, A., Town, A., Demetri, G.D., Fletcher, C.D.M., and Fletcher, J.A. (2003) PDGFRA activating mutations in gastrointestinal stromal tumors. *Science (1979)*, **299** (5607).
[40] Wauquier, A., and Zhou, S. (1996). Topiramate: a potent anticonvulsant in the amygdala-kindled rat. *Epilepsy Research*, **24** (2).
[41] Dudley, J.T., Sirota, M., Shenoy, M., Pai, R.K., Roedder, S., Chiang, A.P., Morgan, A.A., Sarwal, M.M., Pasricha, P.J., and Butte, A.J. (2011) Computational repositioning of the anticonvulsant topiramate for inflammatory bowel disease. *Sci Transl Med*, **3** (96).
[42] Sabzichi, M., and Mohammadian., et al. (2022) Surface functionalization of lipidic core nanoparticles with albumin: A great opportunity for quinacrine in lung cancer therapy. *J Drug Deliv Sci Technol*,**75** (103632).
[43] Chumanevich, A.A., Witalison, E.E., Chaparala, A., Chumanevich, A., Nagarkatti, P., Nagarkatti, M., and Hofseth, L.J. (2016) Repurposing the anti-malarial drug, quinacrine: New anti-colitis properties. *Oncotarget*, **7** (33).
[44] Ward, M.G., Patel, K. V., Kariyawasam, V.C., Goel, R., Warner, B., Elliott, T.R., Blaker, P.A., Irving, P.M., Marinaki, A.M., and Sanderson, J.D. (2017) Thioguanine in inflammatory bowel disease: Long-term efficacy and safety. *United European Gastroenterol J*, **5** (4).
[45] Grigorian, A., Mkhikian, H., and Demetriou, M. (2012) Interleukin-2, Interleukin-7, T cell-mediated autoimmunity, and N-glycosylation. *Ann N Y Acad Sci*, **1253** (1).
[46] Wasan, S.K., and Kane, S. V. (2011) Adalimumab for the treatment of inflammatory bowel disease. *Expert Rev Gastroenterol Hepatol*, **5** (6).
[47] Rossi, C.P., Hanauer, S.B., Tomasevic, R., Hunter, J.O., Shafran, I., and Graffner, H. (2009) Interferon beta-1a for the maintenance of remission in patients with Crohn's disease: Results of a phase II dose-finding study. *BMC Gastroenterol*, **9**.
[48] Bramuzzo, M., Ventura, A., Martelossi, S., and Lazzerini, M. (2016) Thalidomide for inflammatory bowel disease: Systematic review. *Medicine (United States)*, **95** (30).

[49] Byrnes, J.J., Gross, S., Ellard, C., Connolly, K., Donahue, S., and Picarella, D. (2009) Effects of the ACE2 inhibitor GL1001 on acute dextran sodium sulfate-induced colitis in mice. *Inflammation Research*, **58** (11).

[50] Chatziathanasiadou, M. V., Stylos, E.K., Giannopoulou, E., Spyridaki, M.H., et al., (2019) Development of a validated LC-MS/MS method for the in vitro and in vivo quantitation of sunitinib in glioblastoma cells and cancer patients. *J Pharm Biomed Anal*, **164**.

[51] Liu, J., Cai, J., Fan, P., Zhang, N., and Cao, Y. (2019) The Abilities of Salidroside on Ameliorating Inflammation, Skewing the Imbalanced Nucleotide Oligomerization Domain-Like Receptor Family Pyrin Domain Containing 3/Autophagy and Maintaining Intestinal Barrier Are Profitable in Colitis. *Front. Pharmacol*, **10** (1385).

[52] Pessetto, Z. Y., Ma, Y., Hirst, J. J., von Mehren, M., Weir, S. J., and Godwin, A. K. (2014). Drug Repurposing Identifies a Synergistic Combination Therapy with Imatinib Mesylate for Gastrointestinal Stromal Tumor. *Molecular Cancer Therapeutics*, **13** (10).

[53] Rink, L., Skorobogatko, Y., Kossenkov, A. V., Belinsky, M.G., Pajak, T., Heinrich, M.C., Blanke, C.D., Von Mehren, M., Ochs, M.F., Eisenberg, B., and Godwin, A.K. (2009) Gene expression signatures and response to imatinib mesylate in gastrointestinal stromal tumor. *Mol Cancer Ther*, **8** (8).

[54] Lin, Y., Wu, Y., Su, J., Wang, M., Wu, X., Su, Z., Yi, X., Wei, L., Cai, J., and Sun, Z. (2021) Therapeutic role of d-pinitol on experimental colitis: Via activating Nrf2/ARE and PPAR-γ/NF-κB signaling pathways. *Food Funct*, **12** (6).

[55] Wang, X., Gao, Y., Wang, L., Yang, D., Bu, W., Gou, L., Huang, J., Duan, X., Pan, Y., Cao, S., Gao, Z., Cheng, C., Feng, Z., Xie, J., and Yao, R. (2021) Troxerutin Improves Dextran Sulfate Sodium-Induced Ulcerative Colitis in Mice. *J Agric Food Chem*, **69** (9).

[56] Liu, X., Yu, X., Xu, X., Zheng, X., and Zhang, X. (2018) The protective effects of Poria cocos-derived polysaccharide CMP33 against IBD in mice and its molecular mechanism. *Food Funct*, **9** (11).

Mayank Joshi, Mayank, Sandeep Rathor, Sumeet Gupta

8 Implication of drug repurposing in the identification of drugs for renal disorders

Abstract: Drug repurposing seeks to give already-available, authorized medications new applications. Now, we will give a quick rundown of the recent advances in drug repurposing and highlight a few uses for both acute and long-term renal problems. Renal disorders can be divided into two categories: chronic and acute diseases, both of which are linked to cardiovascular disease, a major cause of high mortality. As a result, by relocating the drug, cardiovascular receptors also become a source of interest in the drug development process for renal illnesses. The medication candidate, vitamin D, and the vasodilator, levosimendan, can be added here in the event of acute renal problems. The drug candidate for chronic renal disease is provided by autosomal dominant polycystic kidney disease, focal segmental glomerulosclerosis, and diabetic renal disease. Both diabetic and non-diabetic patients may benefit from taking the glucose-lowering drugs metformin, glucagon-like peptide 1 agonists, and sodium glucose co-transporter 2 inhibitors to prevent kidney and heart damage. Atrasentan (a selective endothelin receptor antagonist), febuxostat, allopurinol, abatacept (a selective co-stimulation modulator), baricitinib (a Janus kinase inhibitor), pentoxifylline (a phosphodiesterase inhibitor), and hydralazine (a DNA demethylating agent/vasodilator) are some additional examples of potential drug candidates. These medication molecules are reintroduced as a novel therapeutic purpose using the drug repurposing approach.

Keywords: Acute renal injuries, chronic kidney diseases, drug repositioning, novel drug targets

8.1 Introduction

Introducing a new drug molecule into the market is very costly and tedious process. A lot of money, around $700 million to $2.7 billion as per recent estimation, and a long time period are involved in the drug discovery process [1, 2]. The process of finding new drugs can be accelerated, thanks to the recent technological and scientific advances, which also drastically cut costs. Here, we can introduce the concept of drug repurposing, with significant evidence of its promise for drug discovery in the past, and thousands of drug molecules have been attempted for use in more than one disease condition [3]. The primary goal of drug repurposing is to identify a new application for a medication candidate, other than its initial targeted therapy. The Food and Drug Administration (FDA) has approved thousands of medicinal compounds, and aca-

https://doi.org/10.1515/9783110791150-008

demic institutions and businesses can access these libraries for high throughput screening [4]. Many cases have been identified for the new uses of the existing drug molecule by drug repurposing, and they have reached the stage of clinical trials. As a result, it motivates drug companies to concentrate on drug repurposing of existing or failed drug compounds, for other indications. Drug repurposing provides advantages over conventional drug development in terms of known pharmacokinetic and safety profiles. We can identify drug candidates via unexpected observations, knowledge discovery in data, or better understanding of a mechanism for a disease condition. We are trying to encompass the attempts made to discover drugs for renal disorders through drug repurposing.

Drug discovery is a multistep process; however, it may be classified into three steps: lead identification, preclinical testing on animals, and human clinical trials. Reverse drug discovery process has become more a popular approach in order to bypass the conventional drug development and discovery phase. Another option for a drug development strategy is to investigate medications that have already been approved for the treatment of other diseases and/or whose targets have been discovered. Finding new pharmacological applications for FDA-approved drugs and for biological products that have already been commercialized is a process known as drug repurposing. It also refers to the use of recently developed pharmaceuticals to treat conditions other than those for which they were originally developed [5]. The drug repurposing approaches are classified in Figure 8.1.

Renal disorders may be classified into two types: chronic and acute disease conditions, which are further associated with cardiovascular disease, a cause of high mortality rate. Thus, cardiovascular receptors also become a cause of interest in the drug discovery for renal disorders, by repositioning of drug. Here, we can include the drug candidate, vitamin D, and the levosimendan, vasodilator, in the case of acute renal disorders. Diabetic renal sickness, focal segmental glomerulosclerosis, and autosomal dominant polycystic kidney disorder provide a drug candidate for chronic renal disease. The glucose-lowering medications, metformin, glucagon-like peptide 1 agonists, and sodium glucose co-transporter 2 inhibitors, also have the potential to protect the kidneys and the heart in both diabetic and non-diabetic patients [6]. The other examples of a possible drug candidate that can be enlisted are atrasentan (selective endothelin receptor antagonist), febuxostat and allopurinol (xanthine oxidase inhibitors), abatacept (selective co-stimulation modulator), baricitinib (Janus kinase inhibitor), pentoxifylline (phosphodiesterase inhibitor), and hydralazine (the DNA demethylating agent/vasodilator). Drug repurposing concept is used to reintroduce these drug molecules as a fresh therapeutic use. The drug repurposing approach provides some advantages as discussed below.

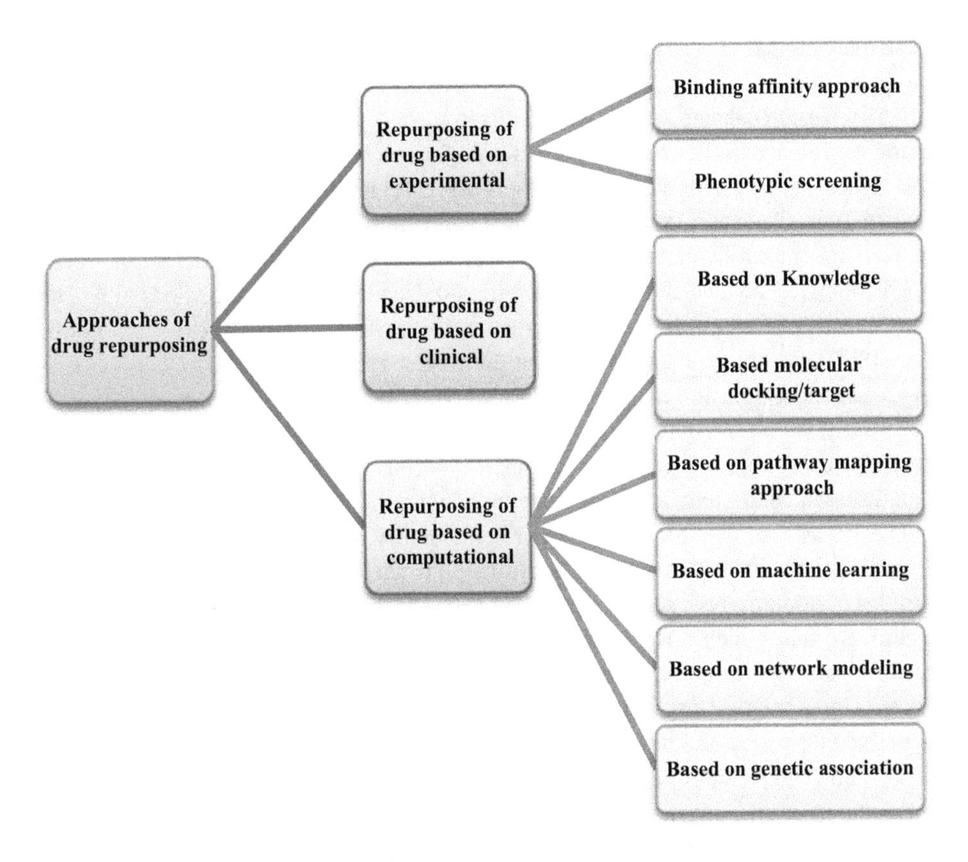

Figure 8.1: Approaches of drug repurposing.

8.2 Advantages of drug repurposing

- Significantly reduces research and development expenses
- Decreases the timeline for drug development
- Easily uses the safety data of existing drug molecules; it cuts off Phase 1 clinical trials
- Allows reuse of the adverse effects and failed efficacy of drugs for new indications
- Lowers chances of failure
- Offers new target or pathway for discovery

A rational use of drug repurposing approach can be utilized for the treatment of patients with renal disease. It offers an upper hand in terms of safety profile, pharmacokinetics, and pharmacodynamics of the compounds, which are previously established by others. Here, we are trying to focus on identifying compounds, by drug repositioning, which have the potential to become drug candidates for acute and chronic conditions of renal diseases [7].

The World Diabetes Federation predicted in 2007 that the number of people with diabetes will rise in the coming years and might reach 380 million by 2025. However, this figure was reached in 2015 when there were 415 million people worldwide. Renal failure and cardiovascular diseases, which contribute to 12% of the global health expenses, are caused by diabetes in most cases, and it is more likely observed in developed countries. At the same time, conventional drug development methods and drug repurposing processes are used to overcome the barriers in drug development in this area [8]. These approaches may reduce the cost of research, offer a convenient herapy, reduce the time and increase focus on targets.

Firstly, we have to do a huge amount of data mining of the genes or protein molecules for screening of fresh drug targets for differentiating between diseased and non-diseased conditions. Then, we need to validate the identified targets to recognize the variances that may be associative or contributory. After the establishment of preclinical concept by evidences, we determine pharmacokinetic and pharmacodynamic profiles in humans, with toxicity testing in phase 1 clinical trials. Next, it enters phase 2 clinical trials, which offers provisional endpoints to backing an expensive phase 3 clinical trial. The drug enters the commercial market after positive results in clinical trials. A new medication molecule's total development and approval costs are estimated to be roughly $2.6 billion [9].

An analogous kind of approach is adopted in case of drug repurposing. Here, we mine the data and use bioinformatics and preclinical studies. These data are then used to determine whether the new indication can be applied to an existing drug or not. An unexpected observation during preclinical or clinical studies can be a motivation for the repurposing of drug. These new indications need to be tested and strictly validated during clinical trials. Drug repurposing provides the advantages that the toxicities and side effects are already known; thus, it reduces the time, cost, and risk of failure, by avoiding the repetition of experiments [10].

Thalidomide is a one of the best cases of drug repurposing. In 1950s, it was an over-the-counter drug used for nausea and vomiting but it was withdrawn from the market due to phocomelia in newborn babies. Thalidomide was repurposed and reinstated for the treatment of leprosy in 1998 after adhering to all new medication indication regulations and received US Food and Drug Administration's approval. Once again, thalidomide was repurposed for multiple myeloma treatment in 2006 [11, 12]. In this chapter, we make an effort to highlight the medications that can be used to treat both acute and chronic renal problems. The primary cause of death in chronic renal disease is cardiovascular disease as already described. So, it becomes rational to consider renal endpoint along with cardiovascular prospect. In clinical trials too, most experiments are designed for safety and foremost results, with consideration for adverse cardiac actions. So, we have included the drugs that encompass action on kidney disease and offer safety from cardiovascular events. Drug dosage is the main factor to be taken into consideration because the kidneys are crucial in the excretion of medications. A drug should follow some features while being repurposed for kidney disease. These features are as follows: (i) should be safe from renal impairment; (ii) provide renal protection; (iii) provide cardiac protection (iv) be cheap and; (v) globally accessible. We discuss the drug candidates for drug repurposing based on the classification of renal diseases. These examples are classified in two categories: (I) Acute renal disease and (II) Chronic renal disease.

8.3 Acute renal disease

We have enough evidence to say that acute kidney injury is individually related to risk of mortality, cardiovascular events, progressive or episodes of chronic kidney disease, and it is considered as a variety of chronic renal diseases [13, 14]. As a priority, it has become a global problem and we need to decrease or avoid acute renal diseases [15]. Some agents may be used for drug repurposing, like vitamin D and levosimendan.

8.3.1 Vitamin D

The body uses vitamin D to maintain calcium homeostasis and bone metabolism. Immunity, and cardiovascular and endothelial systems are related to vitamin D, which further plays a significant role in the kidney's functions [16, 17]. The inflammatory and profibrotic pathways are related to renal disorders and can be understood by various mechanisms; these risks may be eliminated through vitamin D. According to preclinical investigations in rats, vitamin D deficiency is connected to worsening ischemic acute renal injury and vascular damage [18]. Vitamin D inhibits the transforming growth factor β-mediated tubular epithelial to mesenchymal conversion and renal fibrosis, which rely on the vitamin D receptors [19]. Nuclear factor kβ is present in kidney cells, and is activated and translocated by the activated vitamin D receptor [20]. Lipopolysaccharide causes acute kidney injury and inflammation in kidney, which is regulated by the pretreatment of vitamin D, and it acts by suppressing the nuclear factor kβ in the renal tubules. Nuclear factor kβ in the kidney cells is activated by lipopolysaccharide; simultaneously, it inhibits the receptors of vitamin D and the targeted gene [21]. When patients are diagnosed with acute renal injuries, their vitamin D level is found to be significantly low and this condition worsens as the severity of acute kidney injury increases. As we can see at the time of diagnosis, there is no direct link between the serum level of vitamin D and death, but in clinical studies, it is found that this event causes mortality in 3 months [22].

The level of vitamin D is studied in critically affected persons with acute renal injuries to see the difference in vitamin D levels in healthy and severely ill adult volunteers [23]. Other evidences are collected from phase 2 clinical study in critically suffering patients with acute kidney disease, and as per that evidence, we use vitamin D for preventing or treating acute renal injury. They have determined the composition of serum creatinine amount on a daily basis, which is measured in case of kidney replacement therapy; otherwise, we see the primary outcome of kidney failure and the patient dies within a week [24]. Though, all evidences and clinical trials conducted indicate that minimum alteration in the acute renal injury can be caused by several factors, vitamin D shows the possibility of drug repurposing.

8.3.2 Levosimendan

Levosimendan is drug for the treatment of acute heart failure and shows vasodilation and inotropic effects through acting as calcium sensitizer and adenosine triphosphate-dependent potassium channel agonist. We have seen that levosimendan does show pleiotropic effects along with betterment in heart's functioning. One paper has been published in 2016 by Farmakis et al. and they have indicated the potency of levosimendan beyond heart diseases, and they have included the application of drug in renal disease [25].

Some preclinical experiments have been carried out in mice and it suggested that the levosimendan provides protection to kidney against ischemia reperfusion disease. Experimentally, the protection is provided in the acute renal failure generated in mice by endotoxins and it activates nitric oxide synthase and potassium channel, which is opened by ATP of mitochondria [26]. This activation process dilates the vessels along with renal protection [27].

Levosimendan has been studied in a larger population and meta-analyses are performed [28]. These results indicate that levosimendan has properties of vasodilation, and this helps increase the blood supply in the kidney, which improves oxygenation and the glomerular filtration rate. This decreases the chance of kidney injury in patient with cardiac disease, thus avoiding surgery [29]. Clinical trials have been carried out to understand the vasodilatory effects of levosimendan and scientists are continuously working to improve the results for acute renal injury along with cardiac surgery and learn new things from the negative results.

8.4 Chronic renal disease

Chronic renal diseases can be encountered during the treatment of cardiovascular diseases; so we need to improve the treatment to reduce the risk of kidney diseases that can eventually develop over time. Many attempts have been made individually and by government funds to develop new drugs for chronic renal injuries, and many patents have expired in doing so. There is an emerging cause for chronic kidney disease known as diabetic nephropathy and we need to concentrate upon it. Herein, we discuss many strategies to make new drugs by repurposing.

8.4.1 Glucose-lowering drugs

Many antidiabetic drugs mainly decrease the glucose concentration in blood, but these drugs offer benefit to cardiovascular and renal diseases, apart from glycemic control. New drug indications have been developed for kidney disease and heart disorders in the nondiabetic population. These data are evaluated on the basis of clinical trials being carried out. Here, we discuss the potential compounds for drug repurposing.

8.4.1.1 Metformin

A biguanide drug, metformin, is generally used for the treatment of diabetes for many years. Metformin has been found effective on age-related diseases and reduce the mortality rate, apart from its antidiabetic action [30]. There are some advantages of metformin such as low cost, and an effective and safer drug. It is a hypoglycemic drug and causes weight gain; these symptoms are undesirable. Gluconeogenesis in the liver is inhibited by metformin and it opposes glucagon efficacy. The complete mechanism of action is very complicated and under research, but it understood to work at the mitochondrion level. Metabolism through energy governs the signaling pathways, for example, inhibition of the mammalian target of rapamycin and the activation of protein kinase by adenosine monophosphate.

Age generally worsens disease conditions such as renal fibrosis and causes a decline in the estimated glomerular filtration rate. Metformin shows significant results in the treatment of renal proximal tubules in humans via the mammalian target of rapamycin and adenosine monophosphate activating protein kinase, signaling activation [31]. But, polycystic renal disease in adults is caused by the mammalian target of rapamycin, adenosine monophosphate activating protein kinase, and sirtuins. It is reported that in the animal model of adult polycystic kidney disease, metformin reduces kidney cystogenesis by activating adenosine monophosphate-activating protein kinase [32]. Clinical trials are in progress for studying the effect of metformin in the therapeutic use for polycystic kidney disease [33].

There is a long lasting issue of lactic acidosis in patients with renal impairment,. The recommendations of dose-adjusted use of metformin for chronic renal disease and the estimated glomerular filtration rate within the range of 30 to 45 ml/min is relaxed by US Food and Drug Administration [34]. The effect of metformin in overweight patients is dose-independent and has seen a decline in cases of myocardial infarction [35]. Metformin shows its potential in plasma lipids and a decrease in HbA1c. Metformin can show several benefits in diseases such as cardiovascular, chronic renal diseases, and cancers by various mechanisms of signaling pathways, insulin reductions, and increase energetic stress. Here, metformin can be repurposed for many new indications like adult polycystic kidney disease and many other chronic renal disorders [36].

8.5 Glucagon-like peptide-1 receptor agonists

Glucagon-like peptide-1 (GLP-1) regulates post-meal blood sugar elevations in the pancreas by binding to GLP-1 receptor sites in a glucose-dependent manner to stimulate the release of insulin and to suppress the secretion of glucagon. Due to dipeptidyl-peptidase-4 metabolizing activity, the glucagon-like peptide-1 has a limited half-life. It cleaves the peptide bond at the amino terminal of amino acids, generally proline, and sometimes alanine at the penultimate position. During this metabolic process, GLP-1 (7–36) amide is converted to GLP-1 (9–36) amide. The body largely circulates GLP-1 (9–36) amide; however, it has no insulinotropic effects. Neprilysin continues to break down GLP-1 (9–36) amide into GLP-1 (28–36) amide [37]. Both these metabolites show pleiotropic actions on the cardiovascular system and the renal system and are a choice of study for drug repurposing for new indications [38].

In animal study, it is noticed that agonist drug molecules act on GLP-1 receptor, exerting protective effects on the kidney for type-1 and type-1 diabetes. These agents provide treatment by action on the endothelial cells of glomerular and inflammatory infiltrate cells [39]. Anti-inflammatory action of GLP-1 on mesangial cells is produced by GLP-1 receptor [40]. It can also be accomplished by enhancing the cyclic adenosine monophosphate pathway while reducing the expression of the receptor for advanced glycation end products. Recombinant GLP-1 and exendin-4 show antihypertensive action by weakening the effect of angiotensin II, which helps in the disease conditions of the heart vessels, the kidney, and proteinuria [41]. Liu et al. also reported the analogous properties of exanetide analog [42]. Sodium homeostasis modulation is observed in the proximal tubules of the kidney and act by the sodium–hydrogen antiporter isoform 3 [43]. For the same effect in an overweight patient, GLP-1 intravenous infusions enhance sodium ion excretion, reduce proton secretion, and reduce glomerular hyperfiltration. These facts back up its impact on kidney patients [44].

In preclinical and clinical studies, we have observed that cardiovascular protection is provided by GLP-1 receptor agonists [45]. GLP-1 treats pre-ischemia and shrinks the infarct size in both isolated highly vascular rat heart as well as the whole model organisms of ischemia reperfusion and left ventricular dysfunction [46]. Improvement has been noticed in patients with ischemia or left ventricular dysfunction alone as myocardial salvage or left ventricular wall motion, when treated with GLP-1 infusion. GLP-1 and its analogs show the effects on cardiovascular outcome [47]. It has an effect on cardiomyocytes and this is interceded by two actions. Firstly, GLP-1 improves myocardial insulin sensitivity, which is followed by glucose uptake. Secondly, GLP-1 binds to the GLP-1 receptor and modifies the apoptotic control for the persistence of cardiomyocyte cells [48].

GLP-1 receptor exerts its action on the endothelial cells and also on the vascular smooth muscle cells [49]. According to the animal study conducted, GLP-1 has potential to bring NO-facilitated endothelial-dependent relaxation of the respiratory artery vessel rings [50]. While some other studies have detected independency of vasodilation from NO, it can be explained by a straight action on vascular smooth muscle cells via the GLP-1 receptor [51]. To know the reason behind this, we extended the study to the molecular level and found that increase in flow through the coronary arteries is encountered through the action of GLP-1 (9–36), a derivative of GLP-1, and it is by the autonomous activity, apart from GLP-1 receptor [52]. Vascular injury models have been developed to find out the growth in vascular calls and it is found that a GLP-1 analog, exendin-4 infusion, decreases neointimal growth, which is sovereign from other metabolic processes. It acts by the suppression of platelet-derived growth factor, which promotes the creation of vascular smooth muscle cells [53, 54]. Along with it, endothelial cells of the human coronary artery proliferated by exendin-4 action via protein kinase A, nitric oxide synthase in endothelium, and phosphatidylinositol-3-kinase/Akt- driven routes [53]. Vascular endothelial growth factor enhances the growth of progenitor endothelial cells for the protection of vessels. So, the application of the GLP-1 analogue can replace the formation of proinflammatory markers, derived by endothelial cells [55]. Atherosclerosis is promoted by the reduction of monocyte chemoattractant protein-1 and tumor necrosis factor-α mRNA. These are related to exenatide. Thus, vascular cell growth and its functions can be altered by GLP-1 and its analogs, and these are connected to modifications in the inflammatory situation for various types of cells.

In the completed clinical trials, the agonist activity of GLP-1 receptor was studied in consideration of the cardiovascular endpoint. Persons having type-2 diabetes and severe coronary syndrome were administered with the drug lixisenatide and effects are seen for myocardial infarction, cardiovascular mortality, stroke, and unstable angina. The results with lixisenatide are not significantly improved in these cases. Another study has been performed with exenatide on cardiovascular events, but it also failed to show significant improvement in disease conditions, while liraglutide and semaglutide offered noticeable improvement in the heart and kidney safety. In this study, patients with high cardiovascular risk and associated with disease conditions like type-2 diabetes were selected and split into two study groups. These groups were administered with either liraglutide (GLP-1 analog) or a placebo [56]. After this, all events like first the incidence of death due to cardiovascular disease, nonfatal stroke, and myocardial infarction related to type-2 diabetes were analyzed. The effectiveness of liraglutide was found better than placebo. Moreover, it was evident from the study that liraglutide reduced the onset and progression of persistent proteinuria and diabetic nephropathy [57]. Semaglutide showed similar results in clinical trials.

8.6 Sodium and glucose co-transporter 2 inhibitors

Sodium and glucose co-transporter 2 inhibitors act at the proximal tubule located in the kidney by blocking the reuptake of sodium and glucose into the proximal tubule. Subsequently, it causes natriuresis and glycosuria. In the clinical trial conducted on type-2 diabetes mellitus patients for cardiovascular activity with the drug empagliflozin, which acts through inhibition of sodium and glucose co-transporter 2 receptor [58, 59], it reduced the risk of cardiovascular disease-related death by 35%. Improvement by 38% was also observed in the renal microvascular terminals, which act as a protective agent for the kidney. These results could be understood by the advantageous metabolic events like lessening in HbA1c, blood pressure, and body weight [60]. In the proximal tubule, empagliflozin reduces the glucose level and additionally, it directly reduces the sodium reuptake in the kidney. As a result of this event, it activates the tubuloglomerular feedback mechanism and produces a decreased volume in the vessels, resulting in a decline of the glomerular pressure. Furthermore, this causes variations in the activation of intrarenal hormonal, transportation of uric acid, and in the metabolism of the lipid. These findings can be used as evidence for the further assessment of empagliflozin treatment in heart failure associated with and without diabetes [61]. By a thorough study, empagliflozin was repurposed for a new indication in patients with type-2 diabetes for reduced mortality by cardiovascular diseases [62]. This new indication was approved by the US Food and Drug Administration in 2016.

Canagliflozin was also studied for the analogues situations and it promises to slowdown the decrease of the estimated glomerular filtration rate. The results of canagliflozin treatment showed decreased cases of myocardial infarction along with albuminuria and the estimated glomerular filtration rate in kidney [63]. These results have similarities with the empagliflozin study. Empagliflozin shows better results than canagliflozin in terms of cardiovascular deaths reduction due to the higher cardiovascular baseline. But at the point of renal protection, both drugs have shown similar benefits with some minor differences in the clinical study [64]. Current trials have been conducted for canagliflozin with patients having diabetic nephropathy to detect kidney and cardiovascular mortality, end-stage renal injuries, and enhancement of the creatinine level of serum.

Only the drugs that result in the protection of heart and kidney despite lowering the glucose levels are significant in nondiabetic patients with chronic renal disease. In a clinical study with empagliflozin for chronic kidney treatment, disordered patients show inhibition of angiotensin-converting enzyme, with remarkably upsurges of Ang-1-7 levels, by taking the petameters for kidney as a subsidiary endpoint [65]. A current clinical study is being carried out on dapagliflozin in patients with chronic renal disease for renal consequences and mortality by cardiovascular complications [66]. Dapagliflozin shows significant results in proteinuria in nondiabetic patients [67]. Another study suggested the dapagliflozin function to treat albuminuria and for the smooth functioning of hemodynamic activity. One clinical study was carried out to determine the addition of Na^+ and glucose co-transporter 2 inhibitors with normal care results in the reduction of intraglomerular pressure and destruction of proteinuria in focal segmental glomerular sclerosis patients [68].

Sodium and glucose co-transporter 2 inhibitors exert their effects not only for glycemia control but it shows benefit for heart and kidney disorders too. Additionally, this drug family may exhibit molecule-specific effects that could account for negative side effects like amputations. Currently, clinical research is looking at how cardiorenal protection is seen in those with chronic kidney disease but who are not diabetic.

8.7 Selective endothelin receptor A antagonist: atrasentan

A powerful vasoactive peptide, called endothelin-1, is responsible for the development of high blood pressure, albuminuria, insulin intolerance, fibrosis, swelling, and endothelial disfunction [69]. Endothelin-1 mediates its effects through endothelin receptor A and endothelin receptor B, two receptor classes. Atrasentan, a specific endothelin receptor A agonist, was developed in the middle of the 1990s. An antagonist was developed after it was found that patients with metastatic hormone-resistant prostate cancer had much higher plasma endothelin-1 levels than healthy people or those with localized prostate cancer [70]. This specific endothelin receptor A antagonist was simultaneously discovered to lower blood pressure in animal models of hypertension. Atrasentan has been reevaluated as a possible treatment for diabetic nephropathy even though phase 3 clinical trials found no evidence to support its use in prostate cancer [71]. Renal tubular development is accelerated by endothelin receptors, which also have pro-inflammatory and pro-fibrotic effects [72]. Glomerular filtration rate is decreased by endothelin receptor A because it causes more afferent than efferent vasoconstriction [73]. It was believed that the addition of endothelin receptor A and endothelin receptor B as inhibitors to the treatment of kidney disease would be promising. Clinical findings of reduced albuminuria in studies discussed below may have a molecular explanation related to the repair of the glomerular endothelial glycocalyx barrier [74].

According to the research, an endothelin receptor A antagonist, avosentan, decreased albuminuria in patients with type-2 diabetes and overt diabetic nephropathy when it was added to the standard therapy. Congestive heart failure and substantial fluid overload caused it to end prematurely; this was most likely caused by a lack of specificity for the endothelin receptor A and endothelin receptor B [75]. At a dose of 0.75 mg, the more selective endothelin receptor A antagonist atrasentan had a 40% decrease in proteinuria in 50% of the patients, when added to the standard therapy, even though the estimated glomerular filtration rate did not change substantially. There was still a noticeable and dose-dependent peripheral edema [76]. So, at lesser doses, atrasentan reduced albuminuria without evident indications of retaining fluid.

The effectiveness and safety of two modest dosages of atrasentan, 0.75 and 1.25 mg/d, on albuminuria and other renal hazard indications in patients with diabetic nephropathy receiving conventional treatment were evaluated as a result of this. Research demonstrated that 0.75 mg of atrasentan per day had tolerable adverse effects related to fluid overload, while dramatically reducing albuminuria in type-2 diabetes with diabetic nephropathy. Patients having type-2 diabetes and chronic renal disorders of 2nd to 4th stage participated in the study of diabetic nephropathy with atrasentan, a phase 3 clinical study [77]. According to the enhancement protocol, patients with eligibility got atrasentan, 0.75 mg/day, for 6 weeks to measure their urinary albumin-to-creatinine ratio response and tolerance. The time until serum creatinine doubled or critical-stage kidney disease, as indicated through the requirement for ongoing dialysis, kidney transplant, or demise from renal failure, which were the main endpoints. Urinary albumin elimination, estimated glomerular filtration rate, and cardiovascular events like myocardial infarction, stroke, and cardiovascular mortality were considered secondary endpoints.

8.8 Xanthine oxidase inhibitors

There is now mounting data that suggests a causal connection between hyperuricemia and the advancement of long-lasting renal disorders, in addition to gout, high blood pressure, kidney stones, and cardiovascular disorder [78, 79]. A cheap, efficient treatment for reducing serum uric acid, allopurinol, is a xanthine oxidase inhibitor with a known toxicity profile. Febuxostat, a more recent nonpurine xanthine oxidase inhibitor, is also therapeutically accessible to reduce uric acid. The renin-angiotensin system can be activated by uric acid, a byproduct of purine metabolism, and this can lead to a rise in reactive oxygen species [80]. The remarkable decrease in serum uric acid levels, induced by sodium and glucose co-transporter 2 inhibitors, is thought to be mediated by SLC2A9b, which is responsible for the secretion of urate back into the urinary output, in exchange for luminal glucose [81]. One of the many ways sodium and glucose co-transporter 2 inhibitors cause uricosuric effects to give renoprotection may be through these effects.

8.8.1 Febuxostat

Newer nonpurine xanthine oxidase inhibitors like febuxostat have demonstrated comparable renoprotective effects in preclinical studies in animals with type-1 and type-2 diabetes [82]. In comparison to placebo, febuxostat slowed a decline in the estimated glomerular filtration rate in chronic renal disease stages 3 and 4 in a single center study of 93 individuals over a 6-month period [83]. Despite effectively lowering plasma uric acid, febuxostat had no discernible effects on the predetermined primary (urinary transforming growth factor, adiponectin levels, and thiobarbituric acid-reactive compound in adipose tissue) or secondary (C-reactive protein, IL-6, and TNF-α in the plasma) endpoints, according to a minor randomized controlled test of the drug's effects on adipokines and markers of renal fibrosis in asymptomatic hyperuricemic patients with diabetes [84]. Difficult renal endpoints, however, were not taken into account in this minor study. Research is now being done on how febuxostat affects the kidney and the systemic hemodynamic abnormalities in young patients with type-1 diabetes [85].

8.8.2 Allopurinol

The key conclusion was that properly powered randomized controlled trials are required to give definitive data despite the fact that quantitative methods and meta-analyses typically suggest a reno-protective benefit in lowering uric acid [86]. In a randomly selected concurrent intervention, allopurinol treatment improved the kidney function in type-2 diabetic patients [83]. Also, a prospective, randomized investigation of individuals with the estimated glomerular filtration rate of less than 60 ml/min discovered that allopurinol lowers C-reactive protein and slows the advancement of renal illness in patients with chronic renal disease, at two years. Five years later, a post hoc analysis of this unit revealed that long-term allopurinol use may slow the progression of renal disorder and lower the chance of cardiovascular disease [87]. This research was not double blind and had limitations due to its small sample size and single center. Allopurinol is being tested in a multicenter clinical experiment to maintain kidney function in people with type 1 diabetes. Additionally, research using allopurinol and a randomized controlled trial to slow the progression of kidney disease caused by xanthine oxidase inhibition are ongoing [88].

8.9 Selective co-stimulation modulator: abatacept

By directly interacting with B7-1 (CD80) and B7-2 (CD86), which simultaneously stimulate signaling molecules on antigen-presenting cells, abatacept inhibits T-cell-mediated responses. For the treatment of rheumatoid arthritis, a genetically modified protein was authorized in 2005. Reiser et al. reported their preclinical findings in 2004 that various disease conditions in mice cause the induction of podocyte B7-1 [89]. The potential of repurposing this medication for proteinuria kidney diseases was raised by directing B7-1 toward the medical use of abatacept in patients with focal segmental glomerular sclerosis.

Proteinuric kidney disease, called focal segmental glomerular sclerosis, frequently leads to end-stage renal illness. Recurrence after transplantation is difficult to manage, and results in transplant failure. Staining of B7-1 was detected in renal tissue samples removed from patients having proteinuric kidney disease and with positive focal segmental glomerular sclerosis. Podocytes from normal kidneys do not stain B7-1. Five patients with treatment-resistant focal segmental glomerular sclerosis were given abatacept treatment; four received abatacept post transplant; and one received abatacept primary, based on these results and supporting in vitro mechanistic studies. The fact that all five individuals had partial or full remission is significant. It has been questioned, among other things, whether or not podocytes generate B7, though it is uncertain whether this is simply a technical issue [90]. To determine the effectiveness and feasibility of abatacept in decreasing proteinuria in patients with focal segmental glomerular sclerosis or minimal change disease, phase 2 pilot trials are presently being conducted [91].

Betalacept is a second-generation drug formulation that is only used in transplantation and has received alterations to lower the rate of dissociation from B7-1 and B7-2.

8.10 JAK1/JAK2 inhibitor: baricitinib

Signal transducer and activator of transcription are activated by Janus kinases (JAKs) in experimental renal disease models, including diabetes [92]. In diabetic rats, proteinuria was substantially reduced by the administration of a nonselective JAK inhibitor (AG490). According to Brosius et al., dysregulation of this pathway is associated with higher diabetic nephropathy in people [93]. The functional significance of this pathway in diabetic nephropathy was verified by mechanistic studies conducted in vivo [94]. By chance, Eli Lilly, Co. had applied to the FDA for permission to practice baricitinib for the treatment of rheumatoid arthritis. By working together, the academia and the industry, they recognized the possibility for using this medication for diabetic nephropathy. There was a decrease in albuminuria through the energy-dependent inhibition with respect to placebo in a second phase clinical trial. Here, we evaluate the effectiveness of baricitinib in type-2 diabetes participants having proteinuric diabetic nephropathy already, on the renin-angiotensin system blockade [95].

8.11 Phosphodiesterase inhibitor: pentoxyfylline

An anti-inflammatory nonselective phosphodiesterase inhibitor, pentoxyfylline, is a xanthine derivative. It was initially authorized for the treatment of intermittent claudication due to its hemorheological outcome of declining blood viscosity and enhancing microcirculatory stream. Renoprotection has been demonstrated in preclinical studies as a result of its anti-inflammatory and antifibrotic effects [96]. The clinical study demonstrated that the addition of pentoxyfylline to the renin-angiotensin system blockade slowed the progression of nephropathy, as indicated by a slower decline in the estimated glomerular filtration rate and albuminuria over a two-year period [97]. The results of the recent meta-analyses confirm this point.

8.12 DNA demethylating compound and vasodilator: hydralazine

Vasodilator hydralazine received approval for hypertension earlier and was repurposed as a treatment for heart failure. Hydralazine has recently gained recognition as a DNA demethylating agent, or a substance that can remove an attached methyl group from DNA nucleotides, when used at a reduced dose. This happens as a result of Tet3, a natural demethylating enzyme, being induced [98].

The word "epigenetics" refers to heritable chromatin structure changes that do not involve a change in the DNA sequence. Chronic stimuli like high blood sugar cause histone deacetylation or DNA methylation-based epigenetic changes that can alter gene regulation and eventually result in renal disease. In population studies of people with chronic kidney disease, DNA methylation has been thoroughly documented [99]. The practice of low-dose hydralazine in renal disease as a means of modifying the expression of pathological gene is supported by preclinical research. Independent of its blood pressure-lowering effects, hydralazine encouraged demethylation in a murine model of acute renal injury-to-chronic renal disease development and reduced renal fibrosis, and maintained the excretory function of the kidney [100].

There are not many clinical studies that use hydralazine as a demethylating agent in chronic renal disease with strict cardiorenal outcomes. Given the connection between histone modification and chronic kidney disease, epigenetic modification is a topic worth researching.

8.13 Conclusions

Due to high failure rates, the process of discovering and developing new drugs is very costly. The majority of drug failures happen late in the clinical stages as a result of subpar efficacy results from tests conducted on bigger populations. Inappropriately, the costliest phase of the entire drug-discovery procedure is the clinical phase. The costs and development times are increasing as a consequence of the need to include more a diverse human population in clinical trials, raising safety concerns. Additionally, the safety measures of the drug compound must be re-validated if enhanced doses are required for the new drug usage than for the initial one. An adequate quantity of the repositioned drug must also be able to approach its target when delivered using a specific delivery method or formulation, according to the discovery team's evidence. Before conducting drug repurposing research, it is crucial to consider how well the intellectual property coverage for the target molecule could be specified. Over the past years, government support for chemical probe development, clinical and translational research, and university and minor research institutes' drug discovery and development programs have helped them grow. Furthermore, scientists can now use new bioinformatics tools that are being developed concurrently with new technologies to analyze, interpret, and connect the accelerated expansion of openly accessible data that can be deposited in large cloud computing systems. A molecule with a recognized bioavailability and safety profile would likely be the starting point for the development of a repurposed drug candidate, lowering the chance of failure in preliminary testing. Researchers primarily employ knowledge-based and experiment-oriented methods as their two main drug repurposing strategies.

Drug repurposing gives the chance to get around problems with planning for novel drug development planning, like the unaffordable prices and length of time. For effective results to be achieved quickly, a completely operational translational pipeline, built on joint energies between academia and industry is essential. With patent expiration, this is frequently constrained by weak economic incentives, leaving charity organizations or governments to connect stakeholders. Various classes of drugs, including those with concurrent cardiovascular and renal benefits are being repositioned or are appropriate for drug repurposing in renal disorders.

References

[1] Medical Association A. Erratum. Estimated research and development investment needed to bring a new medicine to market, 2009–2018. Jama. 2020;323(9):844–853.

[2] Paul SM, Mytelka DS, Dunwiddie CT, Persinger CC, Munos BH, Lindborg SR, et al. How to improve RD productivity: The pharmaceutical industry's grand challenge. Nat Rev Drug Discov. 2010;9:203–214.

[3] Baker NC, Ekins S, Williams AJ, Tropsha A. A bibliometric review of drug repurposing. Drug Discov Today. 2018;23:661–672.

[4] Anderson E, Havener TM, Zorn KM, Foil DH, Lane TR, Capuzzi SJ, et al. Synergistic drug combinations and machine learning for drug repurposing in chordoma. Sci Rep. 2020:10.

[5] Emig D, Ivliev A, Pustovalova O, Lancashire L, Bureeva S, Nikolsky Y, et al. Drug target prediction and repositioning using an integrated network-based approach. PLoS One. 2013;8:e60618.

[6] Zhang M, Luo H, Xi Z, Rogaeva E. Drug repositioning for diabetes based on "omics" data mining. PLoS One. 2015;10:e0126082.

[7] Joshua Swamidass S. Mining small-molecule screens to repurpose drugs. Brief Bioinform. 2011;12:327–335.

[8] Ogurtsova K, Da Rocha Fernandes JD, Huang Y, Linnenkamp U, Guariguata L, Cho NH, et al. IDF diabetes atlas: Global estimates for the prevalence of diabetes for 2015 and 2040. Diabetes Res Clin Pract. 2017;128:40–50.

[9] DiMasi JA, Grabowski HG, Hansen RW. Innovation in the pharmaceutical industry: New estimates of R&D costs. J Health Econ. 2016;47:20–33.

[10] Jadamba E, Shin M, Systematic A. Framework for drug repositioning from integrated omics and drug phenotype profiles using pathway-drug network. Biomed Res Int. 2016:2016.

[11] Antoszczak M, Markowska A, Markowska J, Huczyński A. Old wine in new bottles: Drug repurposing in oncology. Eur J Pharmacol. 2020;866:172784.

[12] Schein CH. Repurposing approved drugs for cancer therapy. Br Med Bull. 2021;137:13–27.

[13] Coca SG, Singanamala S, Parikh CR. Chronic kidney disease after acute kidney injury: A systematic review and meta-analysis. Kidney Int. 2012;81:442–448.

[14] Chawla LS, Eggers PW, Star RA, Kimmel PL. Acute kidney injury and chronic kidney disease as interconnected syndromes. N Engl J Med. 2014;371:58–66.

[15] Levin A, Tonelli M, Bonventre J, Coresh J, Donner JA, Fogo AB, et al. Global kidney health 2017 and beyond: A roadmap for closing gaps in care, research, and policy. Lancet. 2017;390:1888–1917.

[16] Dusso AS. Vitamin D receptor: Mechanisms for vitamin D resistance in renal failure. Kidney Int Suppl. 2003;63. Blackwell Publishing Inc.

[17] Zoccali C, Curatola G, Panuccio V, Tripepi R, Pizzini P, Versace M, et al. Paricalcitol and endothelial function in chronic kidney disease trial. Hypertens. 2014;64:1005–1011.

[18] De Bragança AC, Volpini RA, Canale D, Gonçalves JG, Heloisa M, Shimizu M, et al. Vitamin D deficiency aggravates ischemic acute kidney injury in rats. Physiol Rep. 2015;3:e12331.

[19] Ito I, Waku T, Aoki M, Abe R, Nagai Y, Watanabe T, et al. A nonclassical vitamin D receptor pathway suppresses renal fibrosis. J Clin Invest. 2013;123:4579–4594.

[20] Chen Y, Zhang J, Ge X, Du J, Deb DK, Li YC. Vitamin D receptor inhibits nuclear factor κb activation by interacting with IκB kinase β protein. J Biol Chem. 2013;288:19450–19458.

[21] Xu S, Chen YH, Tan ZX, Xie DD, Zhang C, Zhang ZH, et al. Vitamin D3 pretreatment regulates renal inflammatory responses during lipopolysaccharide-induced acute kidney injury. Sci Rep. 2015:5.

[22] Lai L, Qian J, Yang Y, Xie Q, You H, Zhou Y, et al. Is the Serum Vitamin D Level at the Time of Hospital-Acquired Acute Kidney Injury Diagnosis Associated with Prognosis?. PLoS One. 2013;8: e64964.

[23] Cameron LK, Lei K, Smith S, Doyle NL, Doyle JF, Flynn K,et al. Vitamin D levels in critically ill patients with acute kidney injury: A protocol for a prospective cohort study (VID-AKI). BMJ Open. 2017;7.

[24] NCT02962102. Activated Vitamin D for the prevention and treatment of acute kidney injury. https://clinicaltrials.gov/ct2/show/NCT02962102 (accessed March 22, 2023).

[25] Farmakis D, Alvarez J, Ben GT, Brito D, Fedele F, Fonseca C, et al. Levosimendan beyond inotropy and acute heart failure: Evidence of pleiotropic effects on the heart and other organs: An expert panel position paper. Int J Cardiol. 2016;222:303–312.

[26] Grossini E, Molinari C, Pollesello P, Bellomo G, Valente G, Mary D, et al. Levosimendan protection against kidney ischemia/reperfusion injuries in anesthetized pigs. J Pharmacol Exp Ther. 2012;342:376–388.

[27] Zager RA, Johnson AC, Lund S, Hanson SY, Abrass CK. Levosimendan protects against experimental endotoxemic acute renal failure. Am J Physiol – Ren Physiol. 2006;290:1453–1462.

[28] Zhou C, Gong J, Chen D, Wang W, Liu M, Liu B. Levosimendan for prevention of acute kidney injury after cardiac surgery: A meta-analysis of randomized controlled trials. Am J Kidney Dis. 2016;67:408–416.

[29] NCT02531724. Effects of Levosimendan in Acute Kidney Injury After Cardiac Surgery. https://clinicaltrials.gov/ct2/show/NCT02531724 (accessed March 28, 2023).

[30] Campbell JM, Bellman SM, Stephenson MD, Lisy K. Metformin reduces all-cause mortality and diseases of ageing independent of its effect on diabetes control: A systematic review and meta-analysis. Ageing Res Rev. 2017;40:31–44.

[31] Dong D, Cai GY, Ning YC, Wang JC, Lv Y, Hong Q, et al. Alleviation of senescence and epithelial-mesenchymal transition in aging kidney by short-term caloric restriction and caloric restriction mimetics via modulation of AMPK/mTOR signaling. Oncotarget. 2017;8:16109–16121.

[32] Takiar V, Nishio S, Seo-Mayer P, King JD, Li H, Zhang L, et al. Activating AMP-activated protein kinase (AMPK) slows renal cystogenesis. Proc Natl Acad Sci U S A. 2011;108:2462–2467.

[33] Seliger SL, Watnick T, Althouse AD, Perrone RD, Abebe KZ, Hallows KR, et al. Baseline characteristics and patient-reported outcomes of ADPKD patients in the multicenter TAME-PKD clinical trial. Kidney360. 2020;1:1363–1372.

[34] Bloomgarden Z,Diabetes. Metformin and renal insufficiency – Is 45, or even 30, the new 60?. Nat Rev Endocrinol. 2015;11:693–694.

[35] Crowley MJ, Diamantidis CJ, McDuffie JR, Cameron CB, Stanifer JW, Mock CK, et al. Clinical outcomes of metformin use in populations with chronic kidney disease, congestive heart failure, or chronic liver disease: A systematic review. Ann Intern Med. 2017;166:191–200.

[36] Holman RR, Paul SK, Bethel MA, Matthews DR, Neil HAW. 10-year follow-up of intensive glucose control in type 2 diabetes. N Engl J Med. 2008;359:1577–1589.

[37] Li J, Besada JA, Bernardos AM, Tarrío P, Casar JR. A novel system for object pose estimation using fused vision and inertial data. Inf Fusion. 2017;33:15–28.

[38] Guglielmi V, Sbraccia P. GLP-1 receptor independent pathways: Emerging beneficial effects of GLP-1 breakdown products. Eat Weight Disord. 2017;22:231–240.

[39] Kodera R, Shikata K, Kataoka HU, Takatsuka T, Miyamoto S, Sasaki M, et al. Glucagon-like peptide-1 receptor agonist ameliorates renal injury through its anti-inflammatory action without lowering blood glucose level in a rat model of type 1 diabetes. Diabetologia. 2011;54:965–978.

[40] Park CW, Kim HW, Ko SH, Lim JH, Ryu GR, Chung HW, et al. Long-term treatment of glucagon-like peptide-1 analog exendin-4 ameliorates diabetic nephropathy through improving metabolic anomalies in db/db mice. J Am Soc Nephrol. 2007;18:1227–1238.

[41] Hirata K, Kume S, Araki SI, Sakaguchi M, Chin-Kanasaki M, Isshiki K, et al. Exendin-4 has an anti-hypertensive effect in salt-sensitive mice model. Biochem Biophys Res Commun. 2009;380:44–49.

[42] Liu Q, Adams L, Broyde A, Fernandez R, Baron AD, Parkes DG. The exenatide analogue AC3174 attenuates hypertension, insulin resistance, and renal dysfunction in Dahl salt-sensitive rats. Cardiovasc Diabetol. 2010;9:1–10.

[43] Carraro-Lacroix LR, Malnic G, Girardi ACC. Regulation of Na+/H+ exchanger NHE3 by glucagon-like peptide 1 receptor agonist exendin-4 in renal proximal tubule cells. Am J Physiol – Ren Physiol. 2009;297:1647–1655.

[44] Crajoinas RO, Oricchio FT, Pessoa TD, Pacheco BPM, Lessa LMA, Malnic G, et al. Mechanisms mediating the diuretic and natriuretic actions of the incretin hormone glucagon-like peptide-1. Am J Physiol – Ren Physiol. 2011;301:355–363.

[45] Gutzwiller JP, Tschopp S, Bock A, Zehnder CE, Huber AR, Kreyenbuehl M, et al. Glucagon-like peptide 1 induces natriuresis in healthy subjects and in insulin-resistant obese men. J Clin Endocrinol Metab. 2004;89:3055–3061.

[46] Bose AK, Mocanu MM, Carr RD, Brand CL, Yellon DM. Glucagon-like peptide 1 can directly protect the heart against ischemia/reperfusion injury. Diabetes. 2005;54:146–151.

[47] Nikolaidis LA, Elahi D, Hentosz T, Doverspike A, Huerbin R, Zourelias L, et al. Recombinant glucagon-like peptide-1 increases myocardial glucose uptake and improves left ventricular performance in conscious dogs with pacing-induced dilated cardiomyopathy. Circulation. 2004;110:955–961.

[48] Timmers L, Henriques JPS, De Kleijn DPV, DeVries JH, Kemperman H, Steendijk P, et al. Exenatide reduces infarct size and improves cardiac function in a porcine model of ischemia and reperfusion injury. J Am Coll Cardiol. 2009;53:501–510.

[49] Arakawa M, Mita T, Azuma K, Ebato C, Goto H, Nomiyama T, et al. Inhibition of monocyte adhesion to endothelial cells and attenuation of atherosclerotic lesion by a glucagon-like peptide-1 receptor agonist, exendin-4. Diabetes. 2010;59:1030–1037.
[50] Golpon HA, Puechner A, Welte T, Wichert PV, Feddersen CO. Vasorelaxant effect of glucagon-like peptide-(7–36)amide and amylin on the pulmonary circulation of the rat. Regul Pept. 2001;102:81–86.
[51] Nyström T, Gonon AT, Sjöholm Å, Pernow J. Glucagon-like peptide-1 relaxes rat conduit arteries via an endothelium-independent mechanism. Regul Pept. 2005;125:173–177.
[52] Ban K, Hui S, Drucker DJ, Husain M. Cardiovascular consequences of drugs used for the treatment of diabetes: Potential promise of incretin-based therapies. J Am Soc Hypertens. 2009;3:245–259.
[53] Erdogdu Ö, Nathanson D, Sjöholm Å, Nyström T, Zhang Q. Exendin-4 stimulates proliferation of human coronary artery endothelial cells through eNOS-, PKA- and PI3K/Akt-dependent pathways and requires GLP-1 receptor. Mol Cell Endocrinol. 2010;325:26–35.
[54] Goto H, Nomiyama T, Mita T, Yasunari E, Azuma K, Komiya K, et al. Exendin-4, a glucagon-like peptide-1 receptor agonist, reduces intimal thickening after vascular injury. Biochem Biophys Res Commun. 2011;405:79–84.
[55] Xie XY, Mo ZH, Chen K, He HH, Xie YH. Glucagon-like peptide-1 improves proliferation and differentiation of endothelial progenitor cells via upregulating VEGF generation. Med Sci Monit. 2011;17:BR35–41.
[56] Pfeffer MA, Claggett B, Diaz R, Dickstein K, Gerstein HC, Køber LV, et al. Lixisenatide in patients with type 2 diabetes and acute coronary syndrome. N Engl J Med. 2015;373:2247–2257.
[57] Holman RR, Bethel MA, Mentz RJ, Thompson VP, Lokhnygina Y, Buse JB, et al. Effects of once-weekly exenatide on cardiovascular outcomes in type 2 diabetes. N Engl J Med. 2017;377:1228–1239.
[58] Ingelheim B. BI 10773 (Empagliflozin) cardiovascular outcome event trial in type 2 diabetes mellitus patients (EMPA-REG OUTCOME). Boehringer Ingelheim 2015;10773. https://clinicaltrials.gov/ct2/show/NCT01131676 (accessed March 29, 2023).
[59] Steiner S. Empagliflozin, cardiovascular outcomes, and mortality in type 2 diabetes. Z Fur Gefassmedizin. 2016;13:17–18.
[60] Wanner C, Inzucchi SE, Lachin JM, Fitchett D, Von Eynatten M, Mattheus M, et al. Empagliflozin and progression of kidney disease in type 2 diabetes. N Engl J Med. 2016;375:323–334.
[61] Griffin M, Rao VS, Ivey-Miranda J, Fleming J, Mahoney D, Maulion C, et al. Empagliflozin in heart failure: Diuretic and cardiorenal effects. Circulation. 2020;142:1028–1039.
[62] Pham D, Albuquerque Rocha ND, Dk M, IJ N. Impact of empagliflozin in patients with diabetes and heart failure. Trends Cardiovasc Med. 2017;27:144–151.
[63] Heerspink HJL, Desai M, Jardine M, Balis D, Meininger G, Perkovic V. Canagliflozin slows progression of renal function decline independently of glycemic effects. J Am Soc Nephrol. 2017;28:368–375.
[64] Neal B, Perkovic V, Mahaffey KW, De Zeeuw D, Fulcher G, Erondu N, et al. Canagliflozin and cardiovascular and renal events in type 2 diabetes. N Engl J Med. 2017;377:644–657.
[65] NCT03078101. EMPRA (EMPagliflozin and RAs in Kidney Disease). https://clinicaltrials.gov/ct2/show/NCT03078101 (accessed March 29, 2023).
[66] A Study to Evaluate the Effect of Dapagliflozin on Renal Outcomes and Cardiovascular Mortality in Patients With Chronic Kidney Disease – Full Text View –ClinicalTrials.gov https://clinicaltrials.gov/ct2/show/NCT03036150 (accessed March 29, 2023).
[67] NCT03190694. Effects of dapagliflozin in non-diabetic patients with proteinuria. https://clinicaltrials.gov/ct2/show/NCT03190694 (accessed March 29, 2023).
[68] Treating to reduce albuminuria and normalize hemodynamic function in focal sclerosis with dapagliflozin trial effects (translate). https://clinicaltrials.gov/ct2/show/NCT02585804 (accessed March 29, 2023).,

[69] Kohan DE, Pollock DM. Endothelin antagonists for diabetic and non-diabetic chronic kidney disease. Br J Clin Pharmacol. 2013;76:573–579.
[70] Nelson JB, Hedican SP, George DJ, Reddi AH, Piantadosi S, Eisenberger MA, et al. Identification of endothelin-1 in the pathophysiology of metastatic adenocarcinoma of the prostate. Nat Med. 1995;1:944–949.
[71] Carducci MA, Saad F, Abrahamsson PA, Dearnaley DP, Schulman CC, North SA, et al. A phase 3 randomized controlled trial of the efficacy and safety of atrasentan in men with metastatic hormone-refractory prostate cancer. Cancer. 2007;110:1959–1966.
[72] Ong ACM, Jowett TP, Firth JD, Burton S, Karet FE, Fine LG. An endothelin-1 mediated autocrine growth loop involved in human renal tubular regeneration. Kidney Int. 1995;48:390–401.
[73] Inscho EW, Imig JD, Cook AK, Pollock DM. ET A and ET B receptors differentially modulate afferent and efferent arteriolar responses to endothelin. Br J Pharmacol. 2005;146:1019–1026.
[74] Boels MGS, Avramut MC, Koudijs A, Dane MJC, Lee DH, Van Der Vlag J, et al. Atrasentan reduces albuminuria by restoring the glomerular endothelial glycocalyx barrier in diabetic nephropathy. Diabetes. 2016;65:2429–2439.
[75] Mann JFE, Green D, Jamerson K, Ruilope LM, Kuranoff SJ, Littke T, et al. Avosentan for overt diabetic nephropathy. J Am Soc Nephrol. 2010;21:527–535.
[76] Kohan DE, Pritchett Y, Molitch M, Wen S, Garimella T, Audhya P, et al. Addition of atrasentan to renin-angiotensin system blockade reduces albuminuria in diabetic nephropathy. J Am Soc Nephrol. 2011;22:763–772.
[77] Study of diabetic nephropathy with atrasentan – full text view – clinicaltrials.gov 2017. https://clinicaltrials.gov/ct2/show/NCT01858532 (accessed March 29, 2023).
[78] Hovind P, Rossing P, Tarnow L, Johnson RJ, Parving HH. Serum uric acid as a predictor for development of diabetic nephropathy in type 1 diabetes: An inception cohort study. Diabetes. 2009;58:1668–1671.
[79] Zoppini G, Targher G, Chonchol M, Ortalda V, Abaterusso C, Pichiri I, et al. Serum uric acid levels and incident chronic kidney disease in patients with type 2 diabetes and preserved kidney function. Diabetes Care. 2012;35:99–104.
[80] Corry DB, Eslami P, Yamamoto K, Nyby MD, Makino H, Tuck ML. Uric acid stimulates vascular smooth muscle cell proliferation and oxidative stress via the vascular renin-angiotensin system. J Hypertens. 2008;26:269–275.
[81] Cheeseman C. Solute carrier family 2, member 9 and uric acid homeostasis. Curr Opin Nephrol Hypertens. 2009;18:428–432.
[82] Nakamura T, Murase T, Nampei M, Morimoto N, Ashizawa N, Iwanaga T, et al. Effects of topiroxostat and febuxostat on urinary albumin excretion and plasma xanthine oxidoreductase activity in db/db mice. Eur J Pharmacol. 2016;780:224–231.
[83] Sircar D, Chatterjee S, Waikhom R, Golay V, Raychaudhury A, Chatterjee S, et al. Efficacy of febuxostat for slowing the GFR decline in patients with CKD and asymptomatic hyperuricemia: A 6-month, double-blind, randomized, placebo-controlled trial. Am J Kidney Dis. 2015;66:945–950.
[84] NCT01350388. Effects of febuxostat on adipokines and kidney disease in diabetic chronic kidney disease. https://clinicaltrials.gov/ct2/show/NCT01350388 (accessed March 29, 2023).
[85] The effect of uric acid lowering in type 1 diabetes. Clin Trials. 2015. https://clinicaltrials.gov/ct2/show/NCT02344602 (accessed March 29, 2023).
[86] Kanji T, Gandhi M, Clase CM, Yang R. Urate lowering therapy to improve renal outcomes in patients with chronic kidney disease: Systematic review and meta-analysis. BMC Nephrol. 2015:16.
[87] Goicoechea M, Garcia De Vinuesa S, Verdalles U, Verde E, Macias N, Santos A, et al. Allopurinol and progression of CKD and cardiovascular events: Long-term follow-up of a randomized clinical trial. Am J Kidney Dis. 2015;65:543–549.

[88] NCT04929379. A pilot study of fenofibrate to prevent kidney function loss in type 1 diabetes. https://ClinicaltrialsGov/Show/NCT04929379 2021. (accessed March 29, 2023).
[89] Reiser J, Von Gersdorff G, Loos M, Oh J, Asanuma K, Giardino L, et al. Induction of B7-1 in podocytes is associated with nephrotic syndrome. J Clin Invest. 2004;113:1390–1397.
[90] Benigni A, Gagliardini ERG. Abatacept in B7-1–positive proteinuric kidney disease. N Engl J Med. 2014;370:1261–1266.
[91] Pilot study to evaluate the safety and efficacy of abatacept in adults and children 6 years and older with excessive loss of protein in the urine due to either focal segmental Glomerulosclerosis (FSGS) or Minimal Change Disease (MCD) – Full Text View – Cl. https://clinicaltrials.gov/ct2/show/NCT02592798 (accessed March 29, 2023).
[92] Berthier CC, Zhang H, Schin M, Henger A, Nelson RG, Yee B, et al. Enhanced expression of janus kinase-signal transducer and activator of transcription pathway members in human diabetic nephropathy. Diabetes. 2009;58:469–477.
[93] Marrero MB, Banes-Berceli AK, Stern DM, Eaton DC. Role of the JAK/STAT signaling pathway in diabetic nephropathy. Am J Physiol – Ren Physiol. 2006;290:762–768.
[94] Brosius FC, Tuttle KR, Kretzler M. JAK inhibition in the treatment of diabetic kidney disease. Diabetologia. 2016;59:1624–1627.
[95] Lin SL, Chen RH, Chen YM, Chiang WC, Lai CF, Wu KD, et al. Pentoxifylline attenuates tubulointerstitial fibrosis by blocking Smad3/4-activated transcription and profibrogenic effects of connective tissue growth factor. J Am Soc Nephrol. 2005;16:2702–2713.
[96] Gunduz Z, Canoz O, Per H, Dusunsel R, Poyrazoglu MH, Tez C, et al. The effects of pentoxifylline on diabetic renal changes in streptozotocin-induced diabetes mellitus. Ren Fail. 2004;26:597–605.
[97] Navarro-González JF, Mora-Fernández C, Muros De Fuentes M, Chahin J, Méndez ML, Gallego E, et al. Effect of pentoxifylline on renal function and urinary albumin excretion in patients with diabetic kidney disease: The PREDIAN trial. J Am Soc Nephrol. 2015;26:220–229.
[98] Tampe B, Tampe D, Zeisberg EM, Müller GA, Bechtel-Walz W, Koziolek M, et al. Induction of Tet3-dependent Epigenetic Remodeling by Low-dose Hydralazine Attenuates Progression of Chronic Kidney Disease. EBioMedicine. 2015;2:19–36.
[99] Smyth LJ, McKay GJ, Maxwell AP, McKnight AJ. DNA hypermethylation and DNA hypomethylation is present at different loci in chronic kidney disease. Epigenetics. 2013;9:366–376.
[100] Tampe B, Steinle U, Tampe D, Carstens JL, Korsten P, Zeisberg EM, et al. Low-dose hydralazine prevents fibrosis in a murine model of acute kidney injury–to–chronic kidney disease progression. Kidney Int. 2017;91:157–176.

Firdoos Ahmad Sofi, Mubashir H. Masoodi

9 Drug repurposing strategy in the identification of drugs for cardiovascular disorders

Abstract: Cardiovascular disorders are considered a major health concern and the leading cause of death worldwide. New drug approvals for cardiovascular disorders have witnessed a sharp decline over the last few decades. The rationale development of new cardiovascular drugs, with improved therapeutic efficacy and lesser side effects, has been in progress over the years. Of note, the Drug repurposing strategy has proven to be the most effective and safe means of identification of new drugs. The drug repurposing strategy for cardiovascular disease has been of great value due to the reduction in time investment, lesser chances of failure, and the greater probability of getting an approved drug. A few reviews on the general topic of drug repurposing have been published. However, no comprehensive book chapter based on the drug repurposing strategy for cardiovascular disorders is reviewed. In the present book chapter, we have attempted to highlight the use of a drug repurposing strategy for the identification of new cardiovascular drugs.

Keywords: Cardiovascular disorders, drug repurposing, Metformin, IL antagonists, TNF-α antagonists, randomized cohort studies

9.1 Introduction

Drug repurposing is considered an emerging field in medicinal chemistry for the identification of new therapeutic indications for available/existing drugs. It is an effective means for developing drug molecules with new therapeutic or pharmacological implications. Drugs that may be either clinically approved marketed drugs or those withdrawn from the market, due to the presence of potential toxicities, could be repositioned [1, 2]. Besides, the investigational agents that failed to exert the desired pharmacological effect may also be repositioned for some alternate indications. Drug repositioning is the means for expanding the therapeutic domain of the currently established drugs and offers a great opportunity for the reversal of failures [3]. It has a great success rate compared to the conventional drug discovery programs in which the chances of development of new

Acknowledgments: F.A.S. thanks UGC, New Delhi for the award of the Dr. D. S. Kothari Post-Doctoral Fellowship (file no. F.4-2/2006 (BSR)/CH/20-21/0256).
Funding: University Grants Commission (UGC) New Delhi.
Conflicts of Interest: The authors declare no conflict of interest.

https://doi.org/10.1515/9783110791150-009

chemical entities (NCES) is approximately 2.01% [4] and a decline in the approval of new drugs has been witnessed over the last few decades.

The conventional approaches to drug discovery are laborious, time consuming, and involve a huge expenditure of money. The use of modern drug repurposing strategy has been of great significance in terms of low monetary cost, short duration of drug development, and low risk of failure [5]. The development of drugs by virtue of the drug repurposing approach has the potential to cut down the additional steps required in the case of the traditional drug development strategy. This involves only four stages of drug development, which include the selection of the target disease, clinical development stage, registration, and post-marketing safety surveillance [6] (Figure 9.1). The availability of various pharmacoinformatic tools has accelerated the drug development process through the drug repurposing strategy. Further, the use of structure-based drug design and the availability of 3d structures of target proteins has simplified the drug purposing process [7, 8].

In the process of drug development by the traditional approach, it takes approximately 10–16 years for the development of a new drug. In the case of the drug repurposing approach, it is estimated that 3–12 years are being spent on the development of a new drug. Drug discovery researchers usually identify new targets in approximately 2 years and discover a new repurposed drug in an average of 8 years [9, 10]. A repositioned drug usually skips the initial 6–9 years and directly enters into clinical settings, thereby reducing the overall time, risk factor, and the cost of drug development.

The drug that can be repurposed for new indications has all the initial parameters tested, including preclinical testing, clinical efficacy, and safety profiles. The drug can itself be an approved drug for some conditions and hence considered pharmacological effective, without any toxicity problem. Hence, it is a preferred approach adopted for the rational development of new therapeutic indications, which confers a high success rate and a significant reduction in the time investment, of up to 5–7 years [9–12]. The drug repurposing approach has gained considerable importance over the last few decades as one-third of the new drugs approved belong to the repurposed category, and contributes to generating approximately 25% of the revenue in the pharmaceutical industry [13]. The identification of new indications that are clinically approved has been in progress since the 1920s and has gained considerable momentum due to the availability of various pharmacoinformatic tools. A few of the representative examples of drugs developed by the drug repurposing approach include sildenafil, minoxidil, aspirin, valproic acid, methotrexate, etc. (Figure 9.2) [14] For example, sildenafil citrate was developed for the treatment of hypertension and is now approved by the FDA for the treatment of erectile dysfunction.

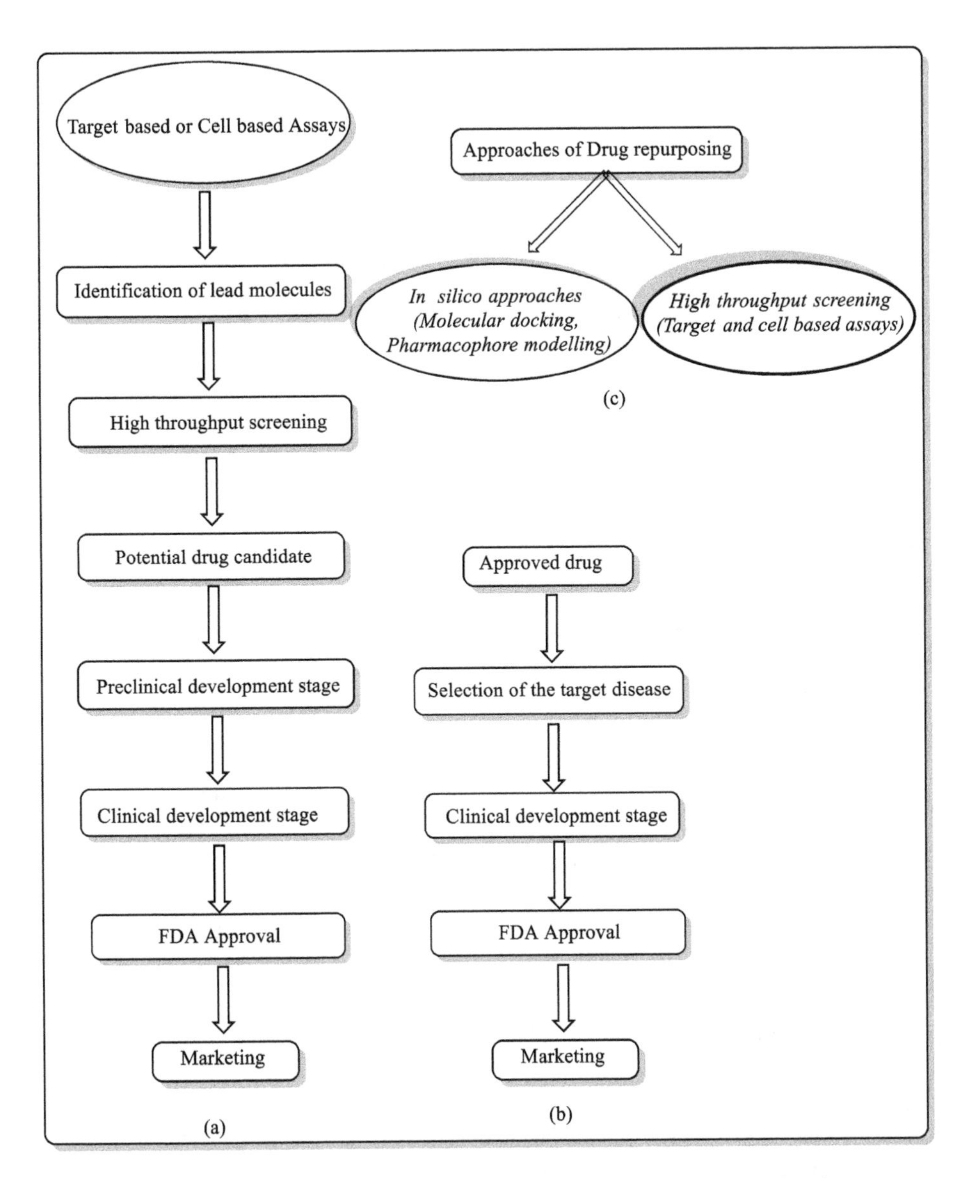

Figure 9.1: (a) Conventional Drug discovery approach. (b) Drug repurposing strategy. (c) various approaches of drug repurposing.

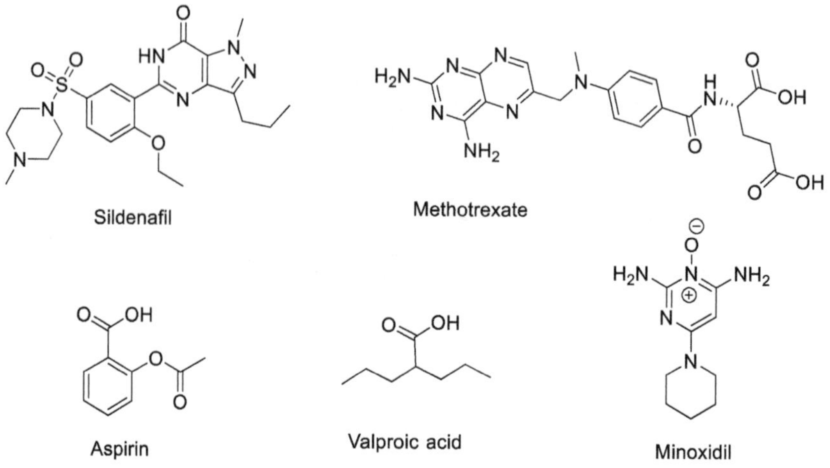

Figure 9.2: Representative structures of repurposed drugs.

9.2 Various strategies adopted for general drug repurposing

The two main strategies employed for the repurposing of drugs include on-target and off-target strategies (Figure 9.3). In an on-target drug repurposing approach, the pharmacological mechanism of a drug can be used for the prediction of a new therapeutic indication. In an on-target drug repurposing strategy, the drug target remains the same but is used for different therapeutic disorders [13]. For example, minoxidil causes vasodilation of smooth blood vessels and exhibits an antihypertensive effect. On the other hand, minoxidil sulphate also stimulates the growth of hair follicles through the opening of potassium channels (KATP). This process helps to supply a constant proportion of nutrients to hair follicles and therefore can be used for the treatment of male pattern baldness (androgenic alopecia).

In the off-target drug repurposing strategy, the pharmacological mechanism of the drug is unknown and acts on a new target to establish a new therapeutic indication. Therefore, in the off targeting approach, both the target on which the drug acts and the indications are new [1]. For example, the antiplatelet effect of aspirin drug (prevents blood coagulation by blocking the normal functioning of platelets) was established by the off-targeting drug approach. It was originally used as an NSAID in the treatment of various pain and inflammatory disorders. This off-targeting strategy has led to the development of aspirin as a repurposed drug for the treatment of heart attacks and strokes.

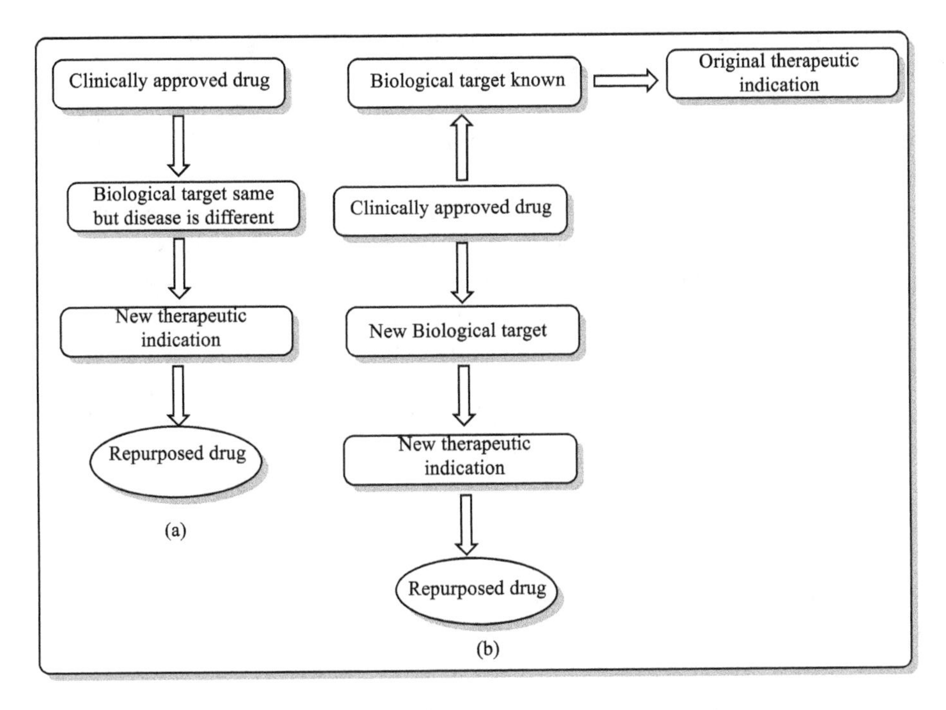

Figure 9.3: (a) On-target drug repurposing strategy (b) off-target drug repurposing strategy.

9.3 Modern methods of drug repurposing

The drug repurposing strategy has been employed for the identification of new drugs with approved therapeutic indications and for establishing the correlation between them. This method of drug development has accelerated due to the employment of various computational methods and the use of biological screening methods [12]. Computational tools have been especially used for the rational development of new orphan drugs for diseases with unknown pathophysiology [13]. The modern computational drug repurposing strategies can be classified into three major domains based on the drug, disease, and the genetic profile of the diseases. Drug repurposing strategy involves the use of pharmacophore modeling of the target enzyme and the presence of pharmacodynamic as well pharmacokinetic features of the preexisting drugs for establishing the drug structural similarity. Drug-based approach was adopted for repurposing the antitubercular drug ethambutol and the sympathomimetic agent metaraminol for treating the African sleeping sickness [14].

The disease-based strategy of repurposing has been employed for the identification of new drug indications by virtue of the disease similarity predicted, based on the genetic as well as the phenotypic expressions data. In a nutshell, the repurposing based on drug and disease use the similarities between the two drugs and the similarities between the two diseases for the establishment of new drug-disease correlations [13]. Drug repositioning in the case of profile strategy uses the genetic expressions of the diseases and the changes in expressions upon drug treatment. This strategy has been used for the repurposing of drugs in the case of inflammatory bowel disease (IBS) [14] and small-cell lung cancer [15].

Two main approaches to drug repurposing include high throughput screening and in silico methods. The high throughput screening method of drug repurposing involves the screening of original drugs on the basis of target- and cell-based assays [13, 14]. In the case of the in silico method of drug repurposing, the use of various drug/chemical databases is virtually screened with the help of various filters, which allows for the identification of the lead drug candidate. The presence of key residual interactions between the drug and the target enzyme helps to identify the bioactive conformation of the drug [15]. The in silico-based methods of drug repurposing has gained considerable importance over the experimental methods due to the involvement of less time, low expenditure, and low risks of failure. However, an in silico method possesses certain limitations in the case of the non-availability of the crystal structure of the drug target [16]. The use of a combination strategy i.e., in silico and experimental approaches for the identification of new indications of clinically approved drugs has become the need of the hour for drug discovery scientists in the modern drug discovery era [17].

9.4 Representative class of repurposed drugs for cardiovascular diseases (CVDs)

Various classes of clinically approved drugs have been repurposed for cardiovascular disorders. The identification of new repurposed drugs for cardiovascular disease has gained considerable importance over the last decade. Both in silico and in vivo experimental approaches have been adopted to discover a new repurposed drug for cardiovascular disorder. Some of the few clinically significant drugs that are being considered for the development of potential repurposed drugs against cardiovascular disorders are briefly summarized here.

9.4.1 LTR antagonists as potential candidates for drug repurposing

Cysteinyl-LTs have been implicated in the generation of various pathophysiological changes, including contraction of bronchial smooth muscles, sequestration of leukocytes at the site of inflammation, and formation of edema due to increased capillary permeability for fluids and cells [13]. Of note, the cysteinyl-LTs LTC4, LTD4, and LTE4 are key mediators in inflammation and asthma, [14, 15] and are held responsible for various cardiovascular (CV) risks [16–19]. With reference to the cardiovascular system, cysteinyl-LTs are considered potent coronary artery vasoconstrictors [20] that cause the proliferation of arterial smooth muscle cells [21], enhance the platelet-activating factor (PAF) synthesis [22], and cause an overall increase in von Willebrand factor secretion [23], as evidenced from various experimental outcomes. Leukotriene receptor antagonists, such as montelukast, zafirlukast, or pranlukast, are primarily used for the treatment of allergic rhinitis [24] urticaria, allergic fungal conjunctivitis, paranasal sinus disease, and atopic dermatitis [25].

Leukotriene receptor antagonists such as montelukast and zafirlukast displayed a protective effect in the case of cerebral ischemia [13] and caused a reduction in the blood-brain barrier permeability, beneficial in brain injuries, evidenced in vivo through several experimental models [14–20]. Saad et al. reported the neuroprotective effect of montelukast by virtue of its antioxidant, antiapoptotic, and anti-inflammatory action. It was further demonstrated to exhibit a neuroprotective effect in cerebral ischemia/reperfusion (I/R) injury in rats [21]. Montelukast, as a repurposed drug, reduced the generation of vascular reactive oxygen species (ROS), inhibited atherosclerosis, and hyperplasia [22, 23] as well as decreased the atherosclerotic plaque generation in experimental in vivo animal models [24]. Becher et al. reported that montelukast reduces oxidative stress and apoptosis in heart cells in mice by the inhibition of leukotriene C4. This can prove beneficial in the case of heart remodeling after left ventricular injury [25].

These on-target effects of cysteinyl-LTs could be exploited for the development of a potential repurposed drug against atherosclerosis (Figure 9.4). All these in vivo results suggest that leukotriene receptor antagonists could be explored as potential drugs for cardiovascular disorders. However, due to insufficient clinical data, additional efforts need to be exerted in the validation of the use of these agents as clinically approved cardiovascular agents.

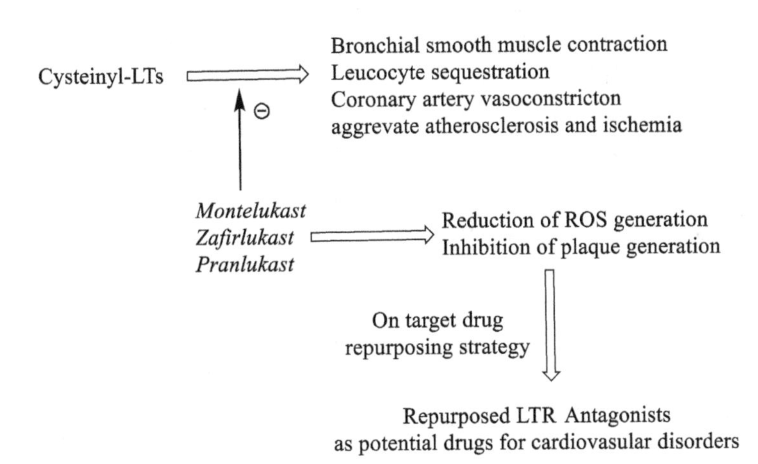

Figure 9.4: On-target strategy adopted for the repurposing of LTs against cardiovascular disorders.

9.4.2 TNF-α antagonists as potential candidates for drug repurposing

Tumor necrosis factor (TNF-α) inhibitors such as infliximab, adalimumab, certolizumab, golimumab, and etanercept are used for the treatment of rheumatic diseases. The use of TNF-α antagonists in rheumatoid arthritis has been associated with a reduced risk of cardiovascular disorders, including myocardial infarction (MI) and cerebrovascular accident (CVA) [26]. Further, anti-TNF-α treatment in rheumatoid arthritis patients has resulted in a reduced risk of acute coronary syndrome (ACS) [27]. In recent studies, the anti-TNF-α monoclonal antibodies have been associated with a reduced occurrence of carotid atherosclerotic plaques in psoriatic arthritis patients [28]. Similarly, the use of golimumab controlled the development of mean intima-media thickness (IMT), compared to patients treated with a placebo [29].

The preclinical data suggested the significance of TNF-α antagonists for the inhibition of the progression of heart failure. However, a proper justification for the involvement of TNF-α antagonists in suppressing the progression of heart failure could not be made due to a lack of support from clinical shreds of evidence [30]. Etanercept, a TNF-α blocking agent treatment by single intravenous infusion, has improved left ventricular function in patients with advanced heart failure [31]. Similarly, etanercept in suitable dosage was found to cause dose-dependent improvement of LV ejection fraction (EF) and LV remodeling in patients with advanced heart failure [32]. However, the presence of these beneficial effects could not be supported by clinical data from a large group of patients. In a study, the use of infliximab has failed to improve the symptoms of heart failure, in spite of it exaggerating the risk of death and re-hospitalization [33].

There remains ambiguity about the use of TNF-α inhibitors in chronic heart failure because of the presence of either beneficial or deleterious effects of TNF-α [34]. Randomized clinical trials of a large group of patients do not support the use of TNF-α antagonists, whereas in small group of patients, these may exert beneficial effects in subclinical atherosclerosis, artery stiffness, and KD. Therefore, continuous efforts need to be made for the rationalization of the use of TNF-α inhibitors in the treatment of various cardiovascular disorders (Figure 9.5). Further, more clinical data need to

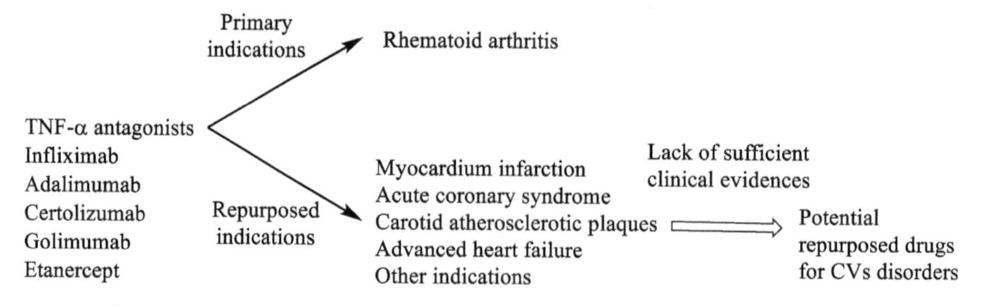

Figure 9.5: Use of TNF-α inhibitors in the development of repurposed drugs.

be generated in favor of the use of TNF-α antagonists as potential repurposed drugs for the various cardiovascular risks.

9.4.3 Interleukin receptor antagonists as potential candidates for drug repurposing

Interleukin receptor antagonists are considered immunosuppressive agents that are known to inhibit the action of interleukins. They modify the action of interleukins, generated in response to any kind of infection and inflammatory conditions. Drugs targeting the IL-1β isoform have been used in cryopyrin-associated periodic syndromes (CAPS), whereas canakinumab and rilonacept have been approved for rare auto-inflammatory diseases. Anakinra, a recombinant form of human interleukin-1 receptor antagonist, is a US FDA-approved drug for the treatment of rheumatoid arthritis and neonatal-onset multisystem inflammatory disease. It competitively inhibits both the isoforms of interleukin receptor, IL-1α and IL-1β. The acute and chronic treatment of anakinra in rheumatoid arthritis patients provided initial evidence that inhibiting IL-1 could have beneficial effects, in terms of higher myocardial contractility and relaxation thresholds, coronary flow reserve, and endothelial functioning [35].

In recent studies, anakinra has been demonstrated to produce encouraging results in decompensated systolic heart failure patients with enhanced C-reactive Protein (CRP) levels. It has resulted in an improvement in the peak Vo2, which is a measure of cardiorespiratory fitness [36]. In another study, Anakinra further demonstrated a significant increase in Vo2 in heart failure patients with preserved ejection fraction (HFpEF) [37, 38]. Anakinra also demonstrated a favorable reduction of NTproBNP, around 50%, and of high-sensitivity CRP (hs-CRP), approximately 60–70%, in HF patients [36–38]. Similarly, anakinra significantly decreased the plasma levels of IL-6 and hs-CRP in patients with acute stroke. In another study conducted on patients with acute stroke, anakinra demonstrated a significant reduction of IL-6 and highly sensitive CRP plasma levels, although not considered a favorable outcome on the Rankin Scale [39].

In a recent study, canakinumab, in suitable doses, exhibited promising inhibitory effects on inflammation in patients with diabetes mellitus at high cardiovascular risk, with plasma levels of CRP and IL-6 decreasing by more than 50% [40]. This was attributed to the presence of dose-dependent lower rates of hospitalization in heart failure patients with a history of myocardium infarction and elevated levels of biomarker (CRP) [41]. Further, canakinumab, in proper dosage, can lead to decreased levels of CRP, which can have greater clinical benefits in patients with cardiovascular disorders [42]. This became the first-ever evidence that IL-1 targeting agents could be potential agents in the management of heart failure. However, more data need to be generated in support of the use of canakinumab for the reduction of cardiovascular risks, keeping in view the cost-effectiveness of the therapy.

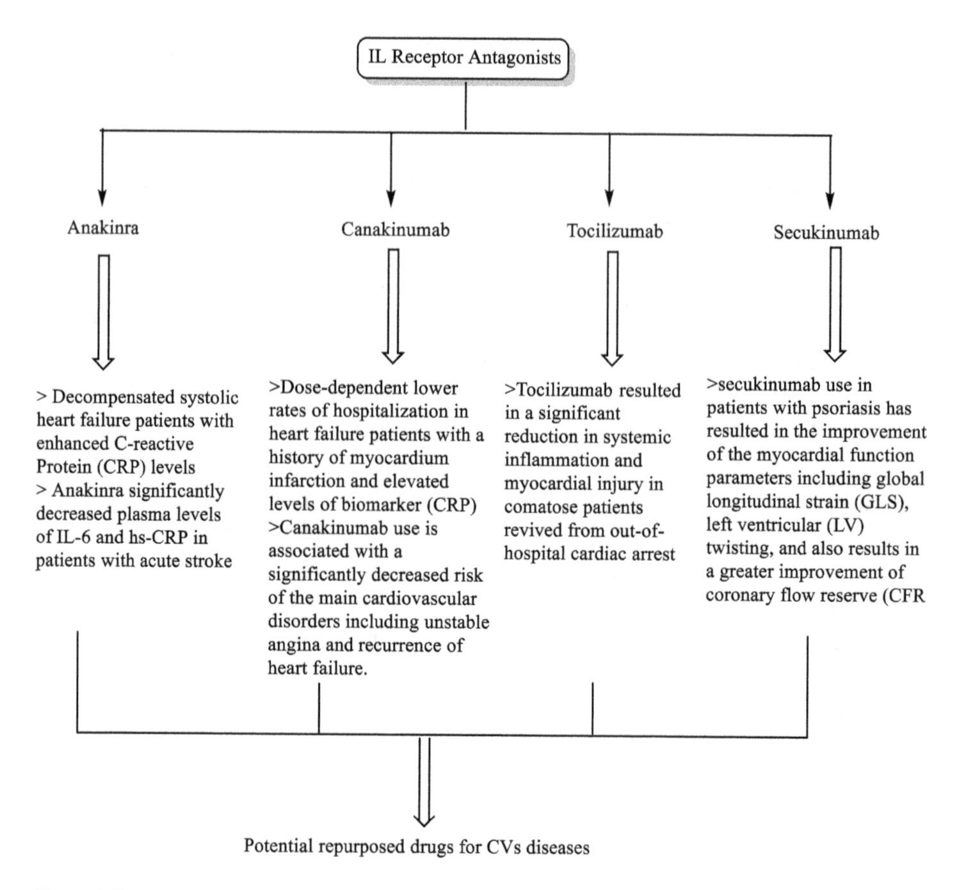

Figure 9.6: IL antagonists as potential repurposed drugs for CV disorders.

In a meta-analysis conducted on a large group of patients using both anakinra and canakinumab, IL-1 antagonists were responsible for reduced risk of the main cardiovascular disorders, including unstable angina and recurrence of heart failure. However, no correlation was established between the usage of IL-1 antagonists and acute myocardium infarction, and the deaths resulting from the use of IL-1 blockade agents [43]. Anakinra also demonstrated reduced risks in the case of recurrent pericarditis [44]. In a recent study, anakinra treatment failed to improve aerobic exercise performance (peak VO2) in a group of obese heart failure patients with preserved ejection fraction [38, 45]. Tocilizumab, an interleukin-6 (IL-6) antagonist, is an FDA approved drug for rheumatoid arthritis (RA), juvenile idiopathic arthritis, and giant cell arteritis. In a recent study, it was found effective as an abatacept or tofacitinib in the occurrence of main cardiovascular risks in rheumatoid arthritis patients [46]. In another study, tocilizumab was recommended to be used in combination therapy for myocardium infarction. Further, in Non-ST-Elevation Myocardial Infarction (NSTEMI or a non-STEMI) patients experiencing percutaneous coronary intervention, tocilizumab demonstrated reduced peri-procedural myocardial injury, assessed by the presence of less serum high-sensitive Troponin-T and CRP [47].

Tocilizumab might also be considered for the treatment of arteritis. In one study of patients with giant-cell arteritis, tocilizumab in intravenous infusion caused complete remission after a long treatment of approximately 52 weeks [48]. This observation was further supported by other studies [49, 50]. In a recent clinical study, treatment with tocilizumab resulted in a significant reduction in systemic inflammation and myocardial injury in comatose patients, revived from out-of-hospital cardiac arrest [51, 52]. Similarly, tocilizumab increased myocardial recovery in patients with acute ST-Elevation Myocardial Infarction (STEMI) [53, 54]. Tocilizumab treatment demonstrated a constant decrease in intima-media thickness (IMT) of the temporal arteries (Tas) and exhibited a little and delayed effect on the axillary and subclavian arteries (AAs/SAs) [55].

Secukinumab, an IL-17 antagonist, has been approved for the treatment of arthritis or psoriasis. The use of secukinumab in patients with arthritis or psoriasis did not demonstrate any beneficial effects on the occurrence of cardiovascular events in the case of psoriatic arthritis and ankylosing spondylitis [56, 57]. In a recent study, the use of secukinumab in patients with psoriasis has resulted in the improvement of parameters, including global longitudinal strain (GLS), left ventricular (LV) twisting, and also results in the improvement of coronary flow reserve (CFR) [58]. The presence of interferon-gamma and interleukin (IL)-17 as key pathological biomarkers in giant cell arteritis (GCA) leads to the generation of T helper (Th) 1 and Th17 cells, respectively, in the affected arteries. The monoclonal anti-IL-17A antibody, secukinumab, has been successful in inducing and maintaining remission. Therefore, IL-17A antagonists could be considered potential therapeutic agents for the treatment of GCA (Figure 9.6). A study was carried out to evaluate the effects of secukinumab in the treatment of giant cell arteritis with reference to the remission of the disease and the control of CRP levels [59].

9.4.4 Colchicine as a potential repurposed drug for CVDs

Colchicine is a naturally occurring secondary product isolated from *Colchicum autumnale* [60]. Colchicine is an approved anti-inflammatory drug for the treatment and prophylaxis of acute gout, recurrent pericarditis, and familial Mediterranean fever (FMF). Colchicine may work through a number of different methods, including a decrease in neutrophil chemotaxis, an inhibition of inflammation signaling, and a reduction in the generation of cytokines, including interleukin-1β. Colchicine has demonstrated beneficial effects on cardiovascular disorders and it reduces the risk of cardiovascular events with coronary artery disease. The effects related to cardiovascular disorders could be attributed to the presence of broad anti-inflammatory action. These beneficial cardiovascular effects can be correlated with the disruption of microtubule assembly [61] and the downregulation of the cellular functions of leucocytes [62]. Colchicine blocks the addition of neutrophils to the endothelium [63] and suppresses the leukotriene B4 (LTB4) liberation, triggering neutrophil deformability [62]. Colchicine also prevents the activation of caspase-1 by the NALP3 inflammasome, the processing and release of IL-1, the generation of superoxides, and the release of different cytokines and pyrogens [64, 65]. It is interesting to note that colchicine has also been demonstrated to promote cell death and reduce smooth muscle cell proliferation [62].

Colchicine is used for the treatment of pericarditis and post-pericardiotomy syndrome (PPS) over the past 20 years [66]. Till 2015, the European Society of Cardiology has recommended the use of colchicine as the drug of first choice for the acute treatment of both acute and recurrent pericarditis as well as acute PPS (Figure 9.7) [67]. In a recent study, colchicine has shown to reduce the occurrence of pericardial effusion in pericarditis and recurrent pericarditis, as well as in PPS [68]. More specifically, colchicine decreased the risk of recurrence in individuals with both first acute pericarditis and recurrent pericarditis. Colchicine, on the other hand, had no effect on individuals who had undergone cardiac surgery. Additional clinical experiments need to be conducted to explore the use of colchicine for the management of PPS and postoperative PE, following cardiac surgery. Colchicine considerably reduced the occurrence of MI in patients with gout and caused a reduction of CRP levels [69]. According to a recent study, Colchicine use has resulted in approximately 49% lower risk of myocardium infarction, stroke, and transient ischemic attack. It has also resulted in a 73% lower risk in all cases of mortality, compared to control patients [70].

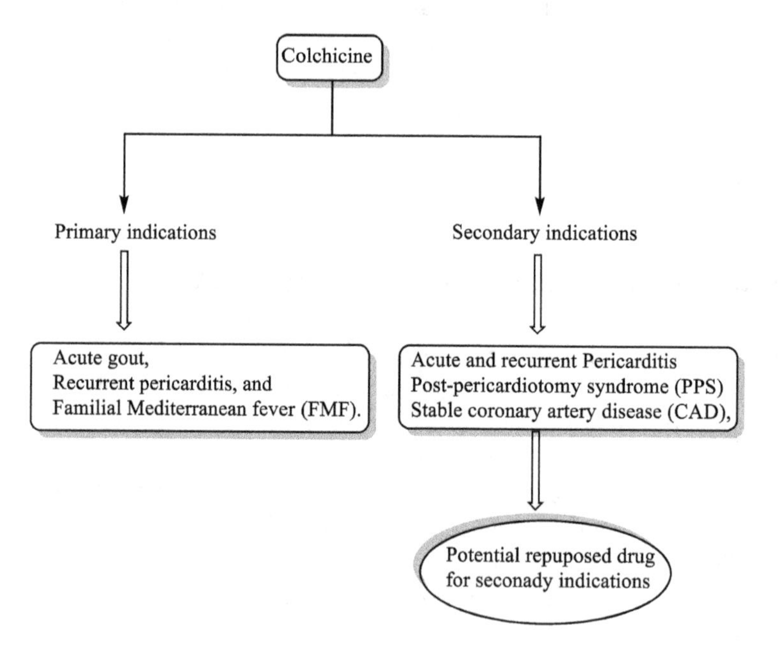

Figure 9.7: Colchicine as a potential repurposed drugs for acute and recurrent pericarditis.

Colchicine has been demonstrated to reduce the incidence of a second cardiovascular incident in the general population. In patients with stable coronary artery disease (CAD), the use of colchicine was associated with a 60% relative reduction in hs-CRP [71]. It was further associated with a decreased cardiovascular death, non-cardioembolic stroke, and acute coronary syndrome (ACS), compared to the placebo [72]. Depending on the type of coronary angioplasty performed, colchicine may also prevent re-stenosis. In fact, colchicine was found ineffective in preventing re-stenosis in patients undergoing balloon angioplasty [73]. However, colchicine was linked to less in-stent lumen area loss and a lower rate of in-stent re-stenosis in diabetic patients with bare-metal coronary stents [74]. Additionally, colchicine exhibited decreased transcoronary inflammation in patients with acute ischemic stroke (ACS). This was further supported by studies that demonstrated that colchicine can lead to a significant reduction of interleukins, including IL-1, IL-18, and IL-6, as well as chemokine ligands, including CCL2, CCL5, and CX3CL1 [75].

Colchicine treatment in appropriate dosage can significantly decrease low attenuation plaque volume (LAPV), a measure of plaque instability, and hs-CRP in patients with recent acute ischemic stroke (ACS), compared to the placebo [76]. Colchicine significantly decreased the myocardium creatine kinase-MB concentration, which is used for the diagnosis of acute myocardial infarction. It has also caused reduction of troponin-T concentration in a recent study. Notably, individuals receiving colchicine had significantly reduced absolute infarct volumes, as determined by LGE-CMR after the primary STEMI [77]. However, no effect on C-Reactive protein concentration was noticed in the case of STEMI patients who had received colchicine treatment for one month [78]. Low-dose colchicine decreased the composite risk of death from cardiovascular causes, resuscitated cardiac arrest, MI, stroke, or urgent hospitalization for angina, leading to coronary revascularization in the COLCOT study. In more detail, it decreased the risk of myocardial infarction, urgent hospitalization for angina and leading to death from cardiovascular causes, resuscitated cardiac arrest, and death from cardiovascular causes [79].

In a recent study, colchicine in appropriate dosage was associated with decreased CRP as well as IL-6 levels in patients with chronic heart failure, compared to the placebo. However, it was found to be ineffective in achieving minimal improvement as per the New York Heart Association (NYHA), and in reduction of hospitalization in the case of heart failure patients [80]. Peri-procedural inflammation is associated with greater chances of post-procedural myocardial infarction (MI), which occurs in approximately 35% of PCI patients. In another recent study, acute preprocedural administration of colchicine administration has resulted in decreased interleukin-6 and high-sensitivity C-reactive protein levels after PCI. However, it did not reduce the risk of PCI-related myocardial damage compared to the placebo [81]. Colchicine administration in high doses did not reduce the myocardial infarct size and left ventricular remodeling at the acute phase of ST-segment-elevation myocardial infarction [82].

In a nutshell, colchicine has demonstrated to be beneficial in patients with CAD, ACS, or STEMI, thereby suggesting a new possibility for the use of colchicine as an alternate repurposed drug for these indications.

9.4.5 Methotrexate as a potential repurposed drug for CVDs

Methotrexate has been used extensively for more than 50 years to treat a number of malignancies and autoimmune conditions such as psoriasis and rheumatoid arthritis. Methotrexate has been demonstrated to provide cardioprotection by virtue of its anti-inflammatory and immunomodulatory effects. The cardioprotective effects of methotrexate could be attributed to the inhibition of two key enzymes, dihydrofolate reductase (DHFR) and aminoimidazole carboxamide ribonucleotide (AICAR) transformylase, which resulted in a decreased synthesis of DNA/RNA and activation of the adenosine A2A receptor, respectively [83]. Methotrexate has been further demonstrated to decrease the expression of various cytokines, including IL-1, IL-6, and TNF, as well as in the adhesion of ICAM-1 and VCAM-1 molecules on the endothelial cells. It also resulted in the decreased production of free radicals, malondialdehyde(MDA), and acetaldehyde protein adducts [83].

Methotrexate may also have preventive effects against atherosclerotic cardiovascular disease, according to epidemiological research. The use of methotrexate in patients with rheumatoid arthritis and psoriasis has resulted in a reduced risk of cardiovascular disorders by 21% and myocardium infarction by 18% [84]. In another study, methotrexate was linked to a 28% reduced risk for overall cardiovascular disorders and a 19% reduced risk in the case of MI [85]. Prospective studies, however, have not been impressed by the possibility of using methotrexate in a different way to treat people with atherosclerotic cardiovascular disease. In a recent study with STEMI patients, methotrexate in a suitable dosage was found not effective in reducing the risk of reinfarction, death rates, or plasma levels of hs-CRP, or creatine phosphokinase-MB (CPK-MB).

Further, methotrexate-treated patients had lower LVEF compared to the placebo group [86]. In another study, methotrexate in low dosage was also found to be ineffective in patients with a past history of myocardium infarction, multivessel coronary disease, type 2 diabetes, or metabolic syndrome. The risk of hospitalization for unstable angina or a composite of nonfatal MI, nonfatal stroke, or CV mortality was not decreased by methotrexate [87]. It further did not lower IL-1, IL-6, or CRP levels. Currently, methotrexate use for cardiovascular disorders is limited because of the lack of sufficient clinical evidence. The use of methotrexate as a repurposed drug for cardioprotection needs to be established by virtue of various randomized control trials.

9.4.6 Metformin as a potential repurposed drug for CVDs

Metformin is a blockbuster drug approved for the treatment of type 2 diabetes mellitus (T2DM) (Figure 9.8). Metformin has been demonstrated to possess anticancer activities and many other additional pharmacological activities. In vitro and in vivo studies demonstrated the presence of anti-inflammatory activity, reduced fibrosis, and reduced endothelial apoptosis through downregulation of FOXO3 and caspase-3, and vascular remodeling [88]. By preventing the phosphorylation of a serine/threonine kinase (PKB/Akt) and mTOR enzyme, metformin also reduces cell hypertrophy [89]. The cardiovascular protection seen in numerous experimental trials could be explained by these molecular and cellular effects. In this regard, it has been demonstrated that metformin lowers pulmonary tension and vascular remodeling, reduces injury due to myocardial ischemia and reperfusion, reduces hypertrophic cardiomyopathy, and reduces the onset of heart failure (Figure 9.9) [88, 90].

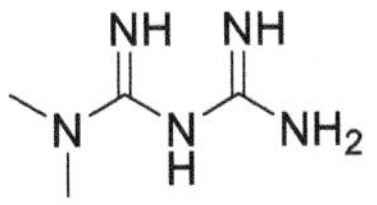

NH NH
N N NH$_2$
H

Figure 9.8: Chemical structure of metformin.

Metformin therapy for a short duration in nondiabetic patients with ischemic does not result in cardioprotection. However, it has demonstrated cardioprotection in patients suffering from diabetes mellitus. In a study, the use of metformin in patients with coronary heart disease has demonstrated negligible effects on various cardiovascular parameters, such as total cholesterol, HDL levels, LDL levels, TGA, and hs- CRP) [91]. In a piece of evidence, metformin pre-treatment in patients who had undergone coronary artery bypass surgery (CABG) did not reduce peri-procedural myocardial injury [92]. Intriguingly, metformin treatment was found to be an effective medication for lowering the risks of deaths associated with myocardial infarction after peri-procedural myocardial injury and ischemia-driven target lesion revascularization [93].

Metformin's impact on HF patients without T2DM was also assessed. The secondary outcome of VE/VCO$_2$ slope (ventilation/carbon dioxide production slope) was improved in the TAYSIDE study by treating insulin-resistant HF patients with metformin for a period of four months [91]. In the recent cohort studies of non-diabetic patients with several cardiovascular complications, the use of metformin has resulted in a significant reduction of left ventricular hypertrophy in comparison to the placebo group [mean difference (MD) [92]. In another cohort study involving T2DM and hypertensive patients, one year treatment with metformin did not reduce B-type natriuretic peptide (BNP) levels, in comparison to other antidiabetic agents [93]. In conclusion, metformin demonstrated potential efficacy for usage in treating CVD, regardless of diabetic or nondiabetic patients. However, various trials need to be performed in support metformin use in nondiabetic patients with coronary artery disease (CAD).

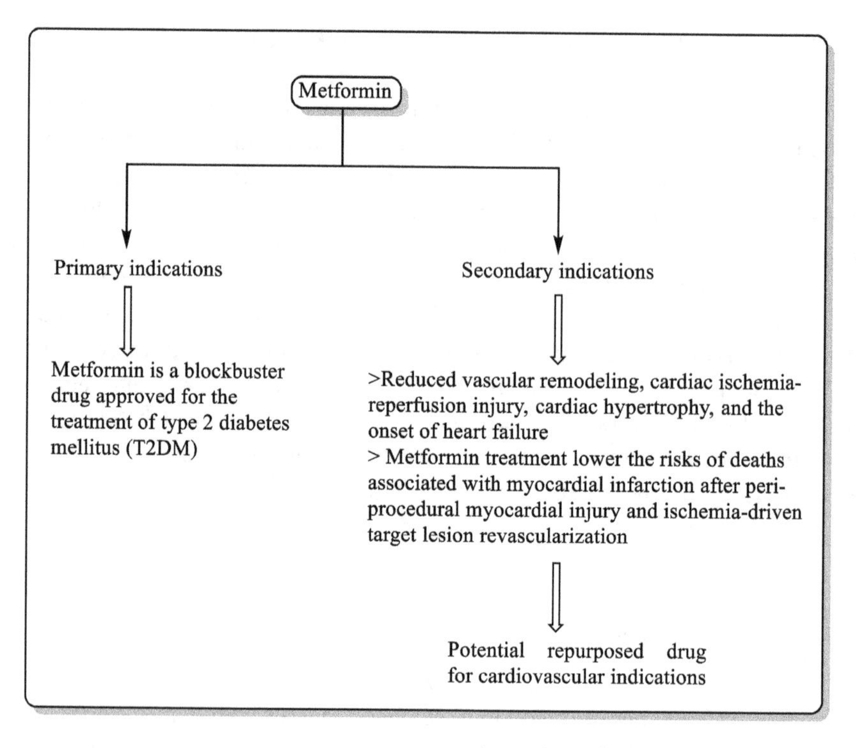

Figure 9.9: Metformin repurposing, based on secondary indications.

Many studies are in progress to evaluate the effects of metformin in nondiabetic (type-2) patients suffering from heart failure and coronary artery disease (CAD). In particular, heart failure patients with intact ejection fraction are being assessed to check for the possible effects of metformin on the changes in Left Ventricular function, BNP levels, and pulmonary artery pressure.

Apart from these drug categories, a few additional drug classes have been tested for the presence of repurposed use against cardiovascular disorders. However, additional experimental and clinical support need to be established for their use as drugs of first choice in various cardiovascular disorders. A list of drugs with their repurposed indications for cardiovascular disorders is depicted in Table 9.1.

Table 9.1: A list of drugs repurposed for cardiovascular disorders.

Drug	Original use	Repurposed use	Evidences in support for repurposed use
Colchicine	Gout	Acute coronary syndrome, Early stages of atherosclerosis [94, 95]	RCT showed a reduction in MACE after myocardial infarction [69, 79]. It leads to the inhibition of interleukin (IL)-1β production and reduction of platelet leukocytes. It further suppresses the growth of fibroblasts and osteophytes [94].
Methotrexate	Rheumatoid arthritis and psoriatic arthritis	Myocardium infarction (MI) and Endothelial dysfunction	Methotrexate causes inhibition of AICAR transformylase enzyme, which leads to increased adenosine levels, thereby exerting an anti-inflammatory effect. Methotrexate also inhibits the dihydrofolate reductase enzyme to increase the synthesis of nitic oxide (NO) [96].
Tocilizumab IL-6 antagonist	Rheumatoid Arthritis	Acute coronary syndrome and Endothelial dysfunction	Tocilizumab was recommended to be used in a combination therapy for myocardium infarction. Further, in Non-ST-Elevation Myocardial Infarction (NSTEMI or a non-STEMI) patients experiencing percutaneous coronary intervention, tocilizumab demonstrated reduced peri-procedural myocardial injury, assessed by the presence of less serum high-sensitive Troponin-T and CRP [53, 74].
Sildenafil*	Pulmonary arterial hypertension	Congestive Heart failure (heart failure with reduced ejection fraction) HFrEF	Randomized cohort studies revealed that a treatment with sildenafil improves the survival of patients after acute myocardial infarction. It was approved by the FDA in 2017 for HFrEF [97, 98].
Anakinra	US FDA-approved drug for rheumatoid arthritis and neonatal- inflammatory disease	Congestive Heart failure (heart failure with preserved ejection fraction) HFpEF	Anakinra further demonstrated a significant increase in Vo2 in heart failure patients with preserved ejection fraction (HFpEF) [64, 65].

Table 9.1 (continued)

Drug	Original use	Repurposed use	Evidences in support for repurposed use
Canakinumab	Canakinumab is an IL-1β antagonist and FDA-approved drug for the treatment of (CAPS) and (FMF). It is also used to treat active (SJIA).	Acute coronary syndrome	Randomized cohort studies revealed the reduction of hospitalizations in patients with HF. It has also resulted in the reduction of mortality in heart failure patients with a history of myocardial infarction [40, 41].
Losartan	Hypertension	Marfan syndrome	Randomized cohort studies revealed that losartan in a combination therapy with β- blockers has proved to be of significant value in patients with Marfan Syndrome [99, 100].
Dapagliflozin, Empagliflozin	Type2 diabetes mellitus	Congestive heart failure	They belong to the class of Sodium-glucose Cotransporter-2 (SGLT2) inhibitors. Both drugs are involved in the reduction of heart failure events in patients with reserved injection fraction or preserved injection fraction [101, 102].
Exenatide	Type 2 diabetes mellitus	Acute coronary syndrome	Randomized cohort studies demonstrated the use of exenatide to improve myocardial salvage index in patients with STEMI after PCI. It can also be used in patients without diabetes [103, 104].
Trametinib	Malignant melanoma	Arteriovenous Malformations (AVM)	Currently, a phase II trial is being conducted for establishing the use of trametinib in extracranial arteriovenous malformation [105–107]

*Sildenafil citrate was originally approved by the US FDA for the treatment of erectile dysfunction. (CAPS) Cryopyrin-Associated Periodic Syndromes, (FMF) Familial Mediterranean Fever, and (SJIA) Systemic Juvenile Idiopathic Arthritis.

9.5 Conclusions and future perspectives

The drug repurposing strategy has emerged as one of the effective and safe means of identification of new cardiovascular drugs with new approved therapeutic indications. The use of a combined approach for the rational development of new drugs has been in progress for the last few decades. The use of Leukotriene receptor antagonists in various cardiovascular events has been rationalized by the reduction of ROS generation and inhibition of plaque formation. Sufficient clinical evidence needs to be generated in support of the use of TNF-α antagonists in myocardium infarction, acute coronary syndrome, and many other CVs conditions. Similarly, IL antagonists have demonstrated potential for the treatment of various cardiovascular disorders due to the support of various clinical evidence. Other classes of drugs have been attempted for the repurposing approach for cardiovascular disorders. However, some of the drugs were successfully repurposed and a few examples are either in the initial phase or in the later stages of repurposing for different cardiovascular indications. The constant efforts of drug discovery scientists shall pave the way for the repurposing of clinically approved drugs for various cardiovascular disorders.

List of abbreviations

NCES	New chemical entities
KATP	ATP-sensitive potassium channel
NSAIDs	Non-Steroidal Anti-Inflammatory Drugs
IBS	Irritable bowel syndrome
LTR	Leukotriene receptor
CVDs	Cardiovascular Disorders
PAF	Platelet-activating factor
ACS	Acute coronary syndrome
CAPS	Cryopyrin-associated periodic syndromes
CRP	C-reactive Protein
HFpEF	Heart failure-Preserved ejection fraction
HFrEF	Reduced ejection fraction
NTproBNP	Brain natriuretic peptide 32
Hs-CRP	Highly sensitive-C-reactive protein
NSTEMI	Non-ST-elevation myocardial infarction
STEMI	ST-Elevation Myocardial Infarction
GCA	Giant cell arteritis
PPS	Post-pericardiotomy syndrome
CAD	Coronary artery disease
LAPV	Low attenuation plaque volume
COLCOT	Colchicine Cardiovascular Outcomes Trial
NYHA	New York Heart Association
DHFR	Dihydrofolate reductase
AICAR	Aminoimidazole carboxamide ribonucleotide transformylase
LVEF	Left ventricular ejection fraction
CABG	Coronary artery bypass surgery
MACE	Major adverse cardiovascular events
PCI	Percutaneous coronary intervention

References

[1] Ashburn TT, Thor K. Drug repositioning: Identifying and developing new uses for existing drugs. 2004;3(8):673–683.

[2] Dey G, Research C. An overview of drug repurposing: Review article. 2019;7(2):3–5.

[3] Frail DE, Barratt M, Drugs E. Opportunities and challenges associated with developing additional indications for clinical development candidates and marketed drugs. 2012:33–51.

[4] Arrowsmith J, Harrison R, Drugs E. Drug repositioning: The business case and current strategies to repurpose shelved candidates and marketed drugs. 2012;9.

[5] Hughes JP, Rees S, Kalindjian SB, Philpott K. Principles of early drug discovery. 2011;162 (6):1239–1249.

[6] Rudrapal M, DJDd C. Endoperoxide antimalarials: Development, structural diversity and pharmacodynamic aspects with reference to 1, 2, 4-trioxane-based structural scaffold. Drug Des Devel Ther. 2016;10:3575.

[7] Kalita J, Chetia D, Rudrapal MJMC. Design, synthesis, antimalarial activity and docking study of 7-chloro-4-(2-(substituted benzylidene) hydrazineyl) quinolines. 2020;16(7):928–937.

[8] Agrawal P. Artificial intelligence in drug discovery and development. 2018;6(2):1000e173.

[9] Mjdd A. Development, Therapy. Open-source approaches for the repurposing of existing or failed candidate drugs: Learning from and applying the lessons across diseases. 2013;7(753).

[10] Jin G, Wong S. Toward better drug repositioning: Prioritizing and integrating existing methods into efficient pipelines. 2014;19(5):637–644.

[11] Chong CR, Sullivan DJJN. New uses for old drugs. 2007;448(7154):645–646.

[12] Xue H, Li J, Xie H, Wang Y. Review of drug repositioning approaches and resources. 2018;14(10):1232.

[13] Bäck MJCD. Leukotriene signaling in atherosclerosis and ischemia. Cardiovasc Drug Ther. 2009;23 (1):41–48.

[14] Bıber N, Toklu HZ, Solakoglu S, Gultomruk M, Hakan T, Berkman Z, et al. Cysteinyl-leukotriene receptor antagonist montelukast decreases blood–brain barrier permeability but does not prevent oedema formation in traumatic brain injury. 2009;23(6):577–584.

[15] Yu G-L, Wei E-Q, Zhang S-H, Xu H-M, Chu L-S, Zhang W-P, et al. Montelukast, a cysteinyl leukotriene receptor-1 antagonist, dose-and time-dependently protects against focal cerebral ischemia in mice. 2005;73(1):31–40.

[16] Yu G-L, Wei E-Q, Wang M-L, Zhang W-P, Zhang S-H, Weng J-Q, et al. Pranlukast, a cysteinyl leukotriene receptor-1 antagonist, protects against chronic ischemic brain injury and inhibits the glial scar formation in mice. 2005;1053(1–2):116–125.

[17] Qian X-D, Wei E-Q, Zhang L, Sheng -W-W, Wang M-L, Zhang W-P, et al. Pranlukast, a cysteinyl leukotriene receptor 1 antagonist, protects mice against brain cold injury. 2006;549(1–3):35–40.

[18] Fang S, Wei E, Zhou Y, Wang M, Zhang W, Yu G, et al. Increased expression of cysteinyl leukotriene receptor-1 in the brain mediates neuronal damage and astrogliosis after focal cerebral ischemia in rats. 2006;140(3):969–979.

[19] Zhao R, Shi W-Z, Zhang Y-M, Fang S-H, E-qjjop W. Montelukast, a cysteinyl leukotriene receptor-1 antagonist, attenuates chronic brain injury after focal cerebral ischaemia in mice and rats. J Pharm Pharmacol. 2011;63(4):550–557.

[20] Huang X-Q, Zhang X-Y, Wang X-R, Yu S-Y, Fang S-H, Lu Y-B, et al. Transforming growth factor β1-induced astrocyte migration is mediated in part by activating 5-lipoxygenase and cysteinyl leukotriene receptor 1. 2012;9(1):1–13.

[21] Saad M, Abdelsalam R, Kenawy S, Attia A. Montelukast, a cysteinyl leukotriene receptor-1 antagonist protects against hippocampal injury induced by transient global cerebral ischemia and reperfusion in rats. 2015;40(1):139–150.

[22] Kaetsu Y, Yamamoto Y, Sugihara S, Matsuura T, Igawa G, Matsubara K, et al. Role of cysteinyl leukotrienes in the proliferation and the migration of murine vascular smooth muscle cells in vivo and in vitro. 2007;76(1):160–166.

[23] Olszanecki R, Korbut RJJPP. The effect of montelukast on atherogenesis in apoE/LDLR-double knockout mice. 2008;59:633–639.

[24] Mueller CF, Wassmann K, Widder JD, Wassmann S, Chen CH, Keuler B, et al. Multidrug resistance protein-1 affects oxidative stress, endothelial dysfunction, and atherogenesis via leukotriene C4 export. 2008;117(22):2912–2918.

[25] Becher UM, Ghanem A, Tiyerili V, Fürst DO, Nickenig G, Mueller C, et al. Inhibition of leukotriene C4 action reduces oxidative stress and apoptosis in cardiomyocytes and impedes remodeling after myocardial injury. 2011;50(3):570–577.

[26] Hyphantis T, Kotsis K, Voulgari PV, Tsifetaki N, Creed F, Drosos AAJAC, et al. Diagnostic accuracy, internal consistency, and convergent validity of the Greek version of the patient health questionnaire 9 in diagnosing depression in rheumatologic disorders. 2011;63(9):1313–1321.

[27] Ljung L, Rantapää-Dahlqvist S, Jacobsson LT, Askling J. Response to biological treatment and subsequent risk of coronary events in rheumatoid arthritis. 2016;75(12):2087–2094.

[28] Di Minno MND, Iervolino S, Peluso R, Scarpa R, Di Minno GJA. Carotid intima-media thickness in psoriatic arthritis: Differences between tumor necrosis factor-α blockers and traditional disease-modifying antirheumatic drugs. Asterioscler Thromb Vasc Biol. 2011;31(3):705–712.

[29] Tam L-S, Shang Q, Kun EW, Lee K-L, Yip M-L, Li M, et al. The effects of golimumab on subclinical atherosclerosis and arterial stiffness in ankylosing spondylitis – A randomized, placebo-controlled pilot trial. 2014;53(6):1065–1074.

[30] Bhatia R, Hare JMJCHF. Mesenchymal stem cells: Future source for reparative medicine. 2005;11 (2):87–93.

[31] Deswal A, Bozkurt B, Seta Y, Parilti-Eiswirth S, Hayes FA, Blosch C, et al. Safety and efficacy of a soluble P75 tumor necrosis factor receptor (Enbrel, etanercept) in patients with advanced heart failure. 1999;99(25):3224–3226.

[32] Bozkurt B, Torre-Amione G, Warren MS, Whitmore J, Soran OZ, Feldman AM, et al. Results of targeted anti–tumor necrosis factor therapy with etanercept (ENBREL) in patients with advanced heart failure. 2001;103(8):1044–1047.

[33] Chung ES, Packer M, Lo KH, Fasanmade AA, Willerson JTJC. Randomized, double-blind, placebo-controlled, pilot trial of infliximab, a chimeric monoclonal antibody to tumor necrosis factor-α, in patients with moderate-to-severe heart failure: Results of the anti-TNF Therapy Against Congestive Heart Failure (ATTACH) trial. 2003;107(25):3133–3140.

[34] Ferrari R. Tumor necrosis factor in CHF: A double facet cytokine. 1998;37(3):554–559.

[35] Ikonomidis I, Lekakis JP, Nikolaou M, Paraskevaidis I, Andreadou I, Kaplanoglou T, et al. Inhibition of interleukin-1 by anakinra improves vascular and left ventricular function in patients with rheumatoid arthritis. 2008;117(20):2662–2669.

[36] Van Tassell BW, Canada J, Carbone S, Trankle C, Buckley L, Oddi Erdle C, et al. Interleukin-1 blockade in recently decompensated systolic heart failure: Results from REDHART (Recently Decompensated Heart Failure Anakinra Response Trial). 2017;10(11):e004373.

[37] Van Tassell BW, Arena R, Biondi-Zoccai G, Canada JM, Oddi C, Abouzaki NA, et al. Effects of interleukin-1 blockade with anakinra on aerobic exercise capacity in patients with heart failure and preserved ejection fraction (from the D-HART pilot study). 2014;113(2):321–327.

[38] Van Tassell BW, Trankle CR, Canada JM, Carbone S, Buckley L, Kadariya D, et al. IL-1 Blockade in Patients With Heart Failure With Preserved Ejection Fraction: Results From DHART2. 2018;11(8): e005036.

[39] Smith CJ, Hulme S, Vail A, Heal C, Parry-Jones AR, Scarth S, et al. SCIL-STROKE (subcutaneous interleukin-1 receptor antagonist in ischemic stroke) a randomized controlled phase 2 trial. 2018;49 (5):1210–1216.

[40] Ridker PM, Howard CP, Walter V, Everett B, Libby P, Hensen J, et al. Effects of interleukin-1β inhibition with canakinumab on hemoglobin A1c, lipids, C-reactive protein, interleukin-6, and fibrinogen: A phase IIb randomized, placebo-controlled trial. 2012;126(23):2739–2748.

[41] Everett BM, Cornel JH, Lainscak M, Anker SD, Abbate A, Thuren T, et al. Anti-inflammatory therapy with canakinumab for the prevention of hospitalization for heart failure. 2019;139(10):1289–1299.

[42] Ridker PM, MacFadyen JG, Everett BM, Libby P, Thuren T, Glynn RJ, et al. Relationship of C-reactive protein reduction to cardiovascular event reduction following treatment with canakinumab: A secondary analysis from the CANTOS randomised controlled trial. 2018;391(10118):319–328.

[43] Bertoldi EG, Stella SF, Rohde LE, Polanczyk C. Long-term cost-effectiveness of diagnostic tests for assessing stable chest pain: Modeled analysis of anatomical and functional strategies. 2016;39 (5):249–256.

[44] Brucato A, Imazio M, Gattorno M, Lazaros G, Maestroni S, Carraro M, et al. Effect of anakinra on recurrent pericarditis among patients with colchicine resistance and corticosteroid dependence: The AIRTRIP randomized clinical trial. 2016;316(18):1906–1912.

[45] Van Tassell BW, Buckley LF, Carbone S, Trankle CR, Canada JM, Dixon DL, et al. Interleukin-1 blockade in heart failure with preserved ejection fraction: Rationale and design of the Diastolic Heart Failure Anakinra Response Trial 2 (D-HART2). 2017;40(9):626–632.

[46] Kim SC, Solomon DH, Rogers JR, Gale S, Klearman M, Sarsour K, et al. Cardiovascular safety of tocilizumab versus tumor necrosis factor inhibitors in patients with rheumatoid arthritis: A multi-database cohort study. 2017;69(6):1154–1164.

[47] Kleveland O, Kunszt G, Bratlie M, Ueland T, Broch K, Holte E, et al. Effect of a single dose of the interleukin-6 receptor antagonist tocilizumab on inflammation and troponin T release in patients with non-ST-elevation myocardial infarction: A double-blind, randomized, placebo-controlled phase 2 trial. 2016;37(30):2406–2413.

[48] Villiger PM, Adler S, Kuchen S, Wermelinger F, Dan D, Fiege V, et al. Tocilizumab for induction and maintenance of remission in giant cell arteritis: A phase 2, randomised, double-blind, placebo-controlled trial. 2016;387(10031):1921–1927.

[49] Stone JH, Tuckwell K, Dimonaco S, Klearman M, Aringer M, Blockmans D, et al. Trial of tocilizumab in giant-cell arteritis. 2017;377(4):317–328.

[50] Adler S, Reichenbach S, Gloor A, Yerly D, Cullmann JL, Villiger PMJR. Risk of relapse after discontinuation of tocilizumab therapy in giant cell arteritis. 2019;58(9):1639–1643.

[51] Meyer MAS, Wiberg S, Grand J, Meyer ASP, Obling LER, Frydland M, et al. Treatment effects of interleukin-6 receptor antibodies for modulating the systemic inflammatory response after out-of-hospital cardiac arrest (The IMICA Trial) a double-blinded, placebo-controlled, single-center, randomized, clinical trial. 2021;143(19):1841–1851.

[52] Meyer MA, Wiberg S, Grand J, Kjaergaard J, Hassager CJT. Interleukin-6 Receptor Antibodies for Modulating the Systemic Inflammatory Response after Out-of-Hospital Cardiac Arrest (IMICA): Study protocol for a double-blinded, placebo-controlled, single-center, randomized clinical trial. 2020;21(1):1–12.

[53] Broch K, Anstensrud AK, Woxholt S, Sharma K, Tøllefsen IM, Bendz B, et al. Randomized trial of interleukin-6 receptor inhibition in patients with acute ST-segment elevation myocardial infarction. 2021;77(15):1845–1855.

[54] Anstensrud AK, Woxholt S, Sharma K, Broch K, Bendz B, Aakhus S, et al. Rationale for the ASSAIL-MI-trial: A randomised controlled trial designed to assess the effect of tocilizumab on myocardial salvage in patients with acute ST-elevation myocardial infarction (STEMI). 2019;6(2):e001108.

[55] Seitz L, Christ L, Lötscher F, Scholz G, Sarbu A-C, Bütikofer L, et al. Quantitative ultrasound to monitor the vascular response to tocilizumab in giant cell arteritis. 2021;60(11):5052–5059.
[56] Mease PJ, McInnes IB, Kirkham B, Kavanaugh A, Rahman P, Van Der Heijde D, et al. Secukinumab inhibition of interleukin-17A in patients with psoriatic arthritis. 2015;373(14):1329–1339.
[57] Baeten D, Sieper J, Braun J, Baraliakos X, Dougados M, Emery P, et al. Secukinumab, an Interleukin-17A Inhibitor. 2015;373(26):2534–2548.
[58] Makavos G, Ikonomidis I, Andreadou I, Varoudi M, Kapniari I, Loukeri E, et al. Effects of interleukin 17A inhibition on myocardial deformation and vascular function in psoriasis. 2020;36(1):100–111.
[59] Venhoff N, Schmidt WA, Lamprecht P, Tony H-P, App C, Sieder C, et al. Efficacy and safety of secukinumab in patients with giant cell arteritis: Study protocol for a randomized, parallel group, double-blind, placebo-controlled phase II trial. 2021;22(1):1–15.
[60] Sakurada M, Yoshioka N, Kuse A, Nakagawa K, Morichika M, Takahashi M, et al. Rapid identification of Gloriosa superba and Colchicum autumnale by melting curve analysis: Application to a suicide case involving massive ingestion of G. superba. 2019;133(4):1065–1073.
[61] Lu Y, Chen J, Xiao M, Li W, Miller D. An overview of tubulin inhibitors that interact with the colchicine binding site. 2012;29(11):2943–2971.
[62] Leung YY, Hui LLY, Kraus VB, editors. Colchicine – Update on mechanisms of action and therapeutic uses. Semin Arthritis Rheum. 2015:Elsevier.
[63] Asahina A, Tada Y, Nakamura K, Tamaki K. Colchicine and griseofulvin inhibit VCAM-1 expression on human vascular endothelial cells – Evidence for the association of VCAM-1 expression with microtubules. 2001;25(1):1–9.
[64] Martinon F, Pétrilli V, Mayor A, Tardivel A, Tschopp JJN. Gout-associated uric acid crystals activate the NALP3 inflammasome. 2006;440(7081):237–241.
[65] Marques-da-silva C, Chaves M, Castro N, Coutinho-Silva R, Guimaraes M. Colchicine inhibits cationic dye uptake induced by ATP in P2X2 and P2X7 receptor-expressing cells: Implications for its therapeutic action. 2011;163(5):912–926.
[66] Imazio M. Colchicine for pericarditis. 2015;25(2):129–136.
[67] Adler Y, Charron P, Imazio M, Badano L, Barón-Esquivias G, Bogaert J, et al. ESC Guidelines for the diagnosis and management of pericardial diseases. 2015;73(11):1028–1091.
[68] Lutschinger LL, Rigopoulos AG, Schlattmann P, Matiakis M, Sedding D, Schulze PC, et al. Meta-analysis for the value of colchicine for the therapy of pericarditis and of postpericardiotomy syndrome. 2019;19(1):1–11.
[69] Crittenden DB, Lehmann RA, Schneck L, Keenan RT, Shah B, Greenberg JD, et al. Colchicine use is associated with decreased prevalence of myocardial infarction in patients with gout. 2012;39 (7):1458–1464.
[70] Solomon DH, Liu -C-C, Kuo I-H, Zak A, Kim S. Effects of colchicine on risk of cardiovascular events and mortality among patients with gout: A cohort study using electronic medical records linked with Medicare claims. 2016;75(9):1674–1679.
[71] Nidorf M, Thompson P. Effect of colchicine (0.5 mg twice daily) on high-sensitivity C-reactive protein independent of aspirin and atorvastatin in patients with stable coronary artery disease. 2007;99 (6):805–807.
[72] Nidorf SM, Eikelboom JW, Budgeon CA, Thompson P. Low-dose colchicine for secondary prevention of cardiovascular disease. 2013;61(4):404–410.
[73] O'Keefe JJH, McCallister BD, Bateman TM, Kuhnlein DL, Ligon RW, GOJJotACoC H. Ineffectiveness of colchicine for the prevention of restenosis after coronary angioplasty. 1992;19(7):1597–1600.
[74] Deftereos S, Giannopoulos G, Raisakis K, Kossyvakis C, Kaoukis A, Panagopoulou V, et al. Colchicine treatment for the prevention of bare-metal stent restenosis in diabetic patients. 2013;61 (16):1679–1685.

[75] Tucker B, Kurup R, Barraclough J, Henriquez R, Cartland S, Arnott C, et al. Colchicine as a novel therapy for suppressing chemokine production in patients with an acute coronary syndrome: A pilot study. 2019;41(10):2172–2181.
[76] Vaidya K, Arnott C, Martínez GJ, Ng B, McCormack S, Sullivan DR, et al. Colchicine therapy and plaque stabilization in patients with acute coronary syndrome: A CT coronary angiography study. 2018;11(2Part 2):305–316.
[77] Deftereos S, Giannopoulos G, Angelidis C, Alexopoulos N, Filippatos G, Papoutsidakis N, et al. Anti-inflammatory treatment with colchicine in acute myocardial infarction: A pilot study. 2015;132 (15):1395–1403.
[78] Akodad M, Lattuca B, Nagot N, Georgescu V, Buisson M, Cristol J-P, et al. COLIN trial: Value of colchicine in the treatment of patients with acute myocardial infarction and inflammatory response. 2017;110(6–7):395–402.
[79] Tardif J-C, Kouz S, Waters DD, Bertrand OF, Diaz R, Maggioni AP, et al. Efficacy and safety of low-dose colchicine after myocardial infarction. 2019;381(26):2497–2505.
[80] Deftereos S, Giannopoulos G, Panagopoulou V, Bouras G, Raisakis K, Kossyvakis C, et al. Anti-inflammatory treatment with colchicine in stable chronic heart failure: A prospective, randomized study. 2014;2(2):131–137.
[81] Shah B, Pillinger M, Zhong H, Cronstein B, Xia Y, Lorin JD, et al. Effects of acute colchicine administration prior to percutaneous coronary intervention: COLCHICINE-PCI randomized trial. 2020;13(4):e008717.
[82] Mewton N, Roubille F, Bresson D, Prieur C, Bouleti C, Bochaton T, et al. Effect of colchicine on myocardial injury in acute myocardial infarction. 2021;144(11):859–869.
[83] Bălănescu AR, Bojincă VC, Bojincă M, Donisan T, Bălănescu SMJE. Cardiovascular effects of methotrexate in immune-mediated inflammatory diseases. Exp Ther Med. 2019;17(2):1024–1029.
[84] Micha R, Imamura F, Von Ballmoos MW, Solomon DH, Hernán MA, Ridker PM, et al. Systematic review and meta-analysis of methotrexate use and risk of cardiovascular disease. 2011;108 (9):1362–1370.
[85] Roubille C, Richer V, Starnino T, McCourt C, McFarlane A, Fleming P, et al. The effects of tumour necrosis factor inhibitors, methotrexate, non-steroidal anti-inflammatory drugs and corticosteroids on cardiovascular events in rheumatoid arthritis, psoriasis and psoriatic arthritis: A systematic review and meta-analysis. 2015;74(3):480–489.
[86] Moreira DM, Lueneberg ME, Da Silva RL, Fattah T, Gottschall C. MethotrexaTE Therapy in ST-Segment Elevation Myocardial Infarctions: A randomized double-blind, placebo-controlled trial (TETHYS Trial). J Cardiovasc Pharmacol Ther. 2017;22(6):538–545.
[87] Ridker PM, Everett BM, Pradhan A, MacFadyen JG, Solomon DH, Zaharris E, et al. Low-dose methotrexate for the prevention of atherosclerotic events. 2019;380(8):752–762.
[88] Nesti L, Natali AJN. Metabolism, Diseases C. Metformin effects on the heart and the cardiovascular system: A review of experimental and clinical data. 2017;27(8):657–669.
[89] Fu Y-N, Xiao H, Ma X-W, Jiang S-Y, Xu M, Zhang Y-Y. Metformin attenuates pressure overload-induced cardiac hypertrophy via AMPK activation. 2011;32(7):879–887.
[90] Zhou J, Massey S, Story D, LJIjoms. L. Metformin: An old drug with new applications. 2018;19 (10):2863.
[91] Wong AK, Symon R, AlZadjali MA, Ang DS, Ogston S, Choy A, et al. The effect of metformin on insulin resistance and exercise parameters in patients with heart failure. 2012;14(11):1303–1310.
[92] Mohan M, Al-Talabany S, McKinnie A, Mordi IR, Singh JS, Gandy SJ, et al. A randomized controlled trial of metformin on left ventricular hypertrophy in patients with coronary artery disease without diabetes: The MET-REMODEL trial. 2019;40(41):3409–3417.

[93] Ono K, Wada H, Satoh-Asahara N, Inoue H, Uehara K, Funada J, et al. Effects of metformin on left ventricular size and function in hypertensive patients with type 2 diabetes mellitus: Results of a randomized, controlled, multicenter, phase IV trial. 2020;20(3):283–293.
[94] Nidorf SM, Thompson PLJCT. Why colchicine should be considered for secondary prevention of atherosclerosis: An overview. 2019;41(1):41–48.
[95] Fiolet AT, Nidorf SM, Mosterd A, Cornel J. Colchicine in stable coronary artery disease. 2019;41 (1):30–40.
[96] Aquilante CL, Niemi M, Gong L, Altman RB, Klein TEJP. PharmGKB summary: Very important pharmacogene information for cytochrome P450, family 2, subfamily C, polypeptide 8. Pharmacogenet Genom. 2013;23(12):721.
[97] Hwang I-C, Kim Y-J, Park J-B, Yoon YE, Lee S-P, Kim H-K, et al. Pulmonary hemodynamics and effects of phosphodiesterase type 5 inhibition in heart failure: A meta-analysis of randomized trials. 2017;17 (1):1–12.
[98] Andersson DP, Lagerros YT, Grotta A, Bellocco R, Lehtihet M, Holzmann MJJH. Association between treatment for erectile dysfunction and death or cardiovascular outcomes after myocardial infarction. 2017;103(16):1264–1270.
[99] Habashi JP, Judge DP, Holm TM, Cohn RD, Loeys BL, Cooper TK, et al. Losartan, an AT1 antagonist, prevents aortic aneurysm in a mouse model of Marfan syndrome. 2006;312(5770):117–121.
[100] Van Andel MM, Indrakusuma R, Jalalzadeh H, Balm R, Timmermans J, Scholte AJ, et al. Long-term clinical outcomes of losartan in patients with Marfan syndrome: Follow-up of the multicentre randomized controlled COMPARE trial. 2020;41(43):4181–4187.
[101] Zannad F, Ferreira JP, Pocock SJ, Anker SD, Butler J, Filippatos G, et al. SGLT2 inhibitors in patients with heart failure with reduced ejection fraction: A meta-analysis of the EMPEROR-Reduced and DAPA-HF trials. 2020;396(10254):819–829.
[102] Nassif ME, Windsor SL, Borlaug BA, Kitzman DW, Shah SJ, Tang F, et al. The SGLT2 inhibitor dapagliflozin in heart failure with preserved ejection fraction: A multicenter randomized trial. 2021;27(11):1954–1960.
[103] Lønborg J, Vejlstrup N, Kelbæk H, Bøtker HE, Kim WY, Mathiasen AB, et al. Exenatide reduces reperfusion injury in patients with ST-segment elevation myocardial infarction. 2012;33 (12):1491–1499.
[104] Ratner R, Han J, Nicewarner D, Yushmanova I, Hoogwerf BJ, Shen LJCD. Cardiovascular safety of exenatide BID: An integrated analysis from controlled clinical trials in participants with type 2 diabetes. 2011;10(1):1–10.
[105] Li D, March ME, Gutierrez-Uzquiza A, Kao C, Seiler C, Pinto E, et al. ARAF recurrent mutation causes central conducting lymphatic anomaly treatable with a MEK inhibitor. 2019;25(7):1116–1122.
[106] Al-Olabi L, Polubothu S, Dowsett K, Andrews KA, Stadnik P, Joseph AP, et al. Mosaic RAS/MAPK variants cause sporadic vascular malformations which respond to targeted therapy. 2018;128 (4):1496–1508.
[107] Medicine UNLo. Clinical Trials. gov. 2022.

Hitesh Kumar, Yirivinti Hayagreeva Dinakar, Arshad J. Ansari, Iliyas Khan

10 Successful examples of blockbusters drugs for drug repurposing

Abstract: The development of new chemical entities or drugs is a time-taking and costly process. The unfulfilled need for the development of efficacious drugs led to a shift in the paradigm of drug repurposing against cancer. Drug repurposing means the utilization of known drugs for the treatment of new diseases. Because of the several advantages such as the lower cost, short development times, and better safety profiles that drug repurposing offers, it emerged as a hot topic and an excellent strategy for drug development in recent times. Various drugs, such as antiviral, antibiotics, antimicrobial, antibacterial, cardiovascular, non-steroidal anti-inflammatory, etc. are being explored for drug repurposing in cancer treatment. This chapter provides an overview of the repurposing of the various blockbuster drugs approved by the FDA, which exist to treat various diseases. We have summarized the repurposing of these drugs for treating various cancers. Furthermore, we have also elaborated on the possible studied mechanisms involved in demonstrating their anticancer activities.

Keywords: Drug repurposing, clinical trial, cancer, therapeutics

10.1 Introduction

Drug repurposing is an innovative approach to drug development that involves using existing drugs for new indications. This strategy has gained significant interest in recent years as it offers several advantages, including reduced costs, shorter development timelines, and improved safety profiles, over traditional drug development. Several drugs initially developed for other indications have been found to have anticancer properties and have been repurposed for cancer treatment. Several FDA-approved molecules/pharmaceuticals that are used in the treatment of various kinds of illnesses have been proposed for drug repurposing to treat several diseases such as diabetes, viral infections, covid-19, cancer, etc. [1]. Drug repurposing in cancer was introduced in 1972 to treat leukemia [2]. Arsenic trioxide, used in skin erosion, showed effective results against acute promyelocytic leukemia and has now been approved by FDA as the first-line treatment, in combination with tretinoin [3]. Similarly, Thalidomide is another example of the classic drug used to treat morning sickness in a stage of pregnancy, which is repurposed for myeloma but restricted for use in pregnant women due to adverse effects [4]. Various classes of drugs such as nonsteroidal anti-inflammatory, cardiovascular, antimicrobial, antimalarial, antifungal, antibacterial,

https://doi.org/10.1515/9783110791150-010

antiviral, antibiotics, antineurogenerative, antidiabetic, and antipsychotic are being studied for repurposing in cancer therapy. Most drugs failed due to their side/adverse effects and were withdrawn from the clinical trial stages; some are still being investigated because of their extraordinary ability and effects against cancer.

10.2 Successful blockbuster drugs

10.2.1 Metformin

Metformin, a well-known antidiabetic and diuretic drug, is being studied for repurposing in the treatment of several cancers such as breast, ovarian, prostate, colon, and endocrine [5, 6]. Metformin, a widely prescribed oral medication for type 2 diabetes, has gained interest in recent years due to its potential anticancer effects. Several preclinical and clinical studies have shown promising results in using Metformin as an anticancer agent, with its ability to target multiple pathways involved in cancer development and progression. One of the primary mechanisms by which metformin may inhibit cancer growth is targeting cancer cells' energy metabolism. Cancer cells have a unique energy metabolism, relying more heavily on glycolysis for energy production than normal cells, known as the Warburg effect. Metformin blocks the respiratory chain of the mitochondria complex I, causing AMP-activated protein kinase to be activated (AMPK), subsequent inhibiting the mTOR signaling pathway. This results in decreased glucose uptake and reduced energy supply, which inhibits cancer cell growth [7]. Several studies have also shown that metformin reduces insulin levels, another potential mechanism for its anticancer effects. High insulin levels have been linked to increased cancer risk and poor outcomes in cancer patients [8].

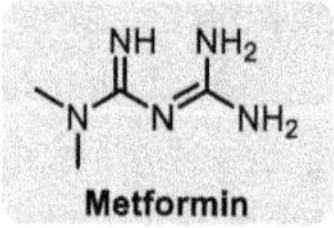

Metformin

Numerous preclinical studies have shown that metformin has anticancer effects in various cancer types, including breast, prostate, colorectal, pancreatic, and lung [9]. For example, a meta-analysis of several observational studies found that metformin use was associated with a significant reduction in breast cancer incidence and improved survival in patients with breast cancer [10]. In prostate cancer, metformin has been shown to reduce the risk of developing prostate cancer and in improving outcomes in patients with prostate cancer [11]. Several ongoing clinical trials are investigating the use of metformin as an anticancer agent. For example, the Phase III TAME (Targeting Aging with Metformin) trial evaluates the use of metformin to prevent age-related diseases, including cancer in older adults [12]. Another Phase II trial is investigating the use of metformin in combination with chemotherapy in patients with pancreatic cancer [13].

10.2.2 Thiazolidinediones

Similarly, Thiazolidinediones (TZDs) are drugs commonly used to treat type 2 diabetes. In recent years, there has been growing interest in the potential anticancer effects of TZDs. Several studies have suggested that TZDs may have antitumor activity in vitro and in vivo. One proposed mechanism of TZD anticancer activity is the induction of apoptosis (programmed cell death) in cancer cells. In one study, treatment with the TZD Rosiglitazone induced apoptosis in human pancreatic cancer cells via activation of PPARγ-dependent and -independent pathways [14]. Other studies have shown that TZDs can induce apoptosis in breast cancer cells [15] and gastric cancer cells [16]. Another proposed mechanism of TZD's anticancer activity is by inhibiting cancer cell proliferation. In one study, treatment with the TZD, Pioglitazone, inhibited the proliferation of human pancreatic cancer cells via activation of the PPARγ pathway [17]. In addition to their direct anticancer effects, TZDs may indirectly affect cancer development and progression through their effects on insulin resistance and inflammation. Insulin resistance and chronic inflammation have been implicated in the development and progression of several types of cancer. TZDs have improved insulin sensitivity and reduced inflammation in patients with type 2 diabetes [18]. These effects may translate to a decreased risk of cancer development and progression in TZD-treated patients.

While growing evidence suggests that TZDs may have anticancer effects, the clinical use of these drugs for cancer treatment is still in their early stages. Several small clinical trials have been conducted to investigate the safety and efficacy of TZDs in cancer patients. One study found that treatment with rosiglitazone improved progression-free survival in patients with metastatic breast cancer [19]. Another study found that pioglitazone reduced the risk of colorectal adenomas in patients with a history of colorectal cancer. More extensive clinical trials will be needed to further evaluate the potential of TZDs as anticancer agents. In conclusion, growing evidence suggests that TZDs may have anticancer effects through their direct effects on cancer cells and their indirect effects on insulin resistance and inflammation. While the clinical use of TZDs for cancer treatment is their early stages, these drugs may represent a promising avenue for cancer therapy and prevention [20].

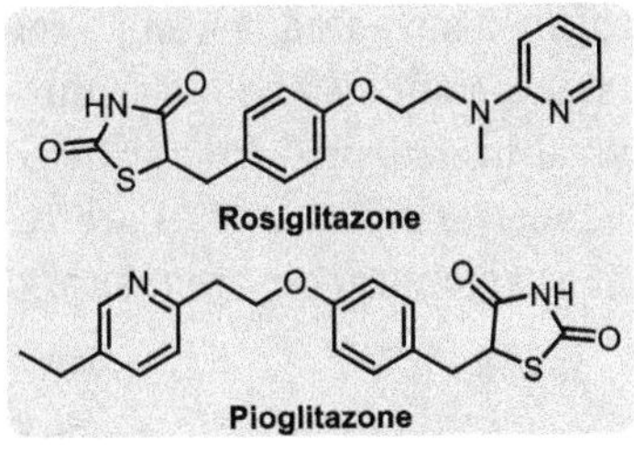

Rosiglitazone

Pioglitazone

10.2.3 Clarithromycin

Clarithromycin is a macrolide antibiotic repurposed for its potential anticancer effects. It has been found to inhibit the growth and proliferation of various cancer cells, including non-small cell lung cancer, gastric cancer, pancreatic cancer, and ovarian cancer. Clarithromycin's mode of action is believed to be due to its ability to induce apoptosis and inhibit the PI3K/Akt signaling pathway [21]. The PI3K/Akt signaling pathway is key to cancer cell survival, proliferation, and metastasis regulation. Clarithromycin has been shown to inhibit the phosphorylation of Akt, leading to decreased cancer cell survival and proliferation. It also induces apoptosis in cancer cells by activating caspase-dependent and -independent pathways. Moreover, clarithromycin has also been found to exert immunomodulatory effects by activating the T cells and the natural killer cells, which play a critical role in cancer immunosurveillance. Clarithromycin has been shown to enhance the antitumor effects of immune cells and increase the production of cytokines and chemokines that promote tumor cell death [22]. In addition, clarithromycin has exhibited antiangiogenic effects by inhibiting the production of vascular endothelial growth factor (VEGF) and decreasing angiogenesis in tumors. This effect may contribute to its anticancer activity by inhibiting the formation of new blood vessels, necessary for tumor growth and metastasis [23].

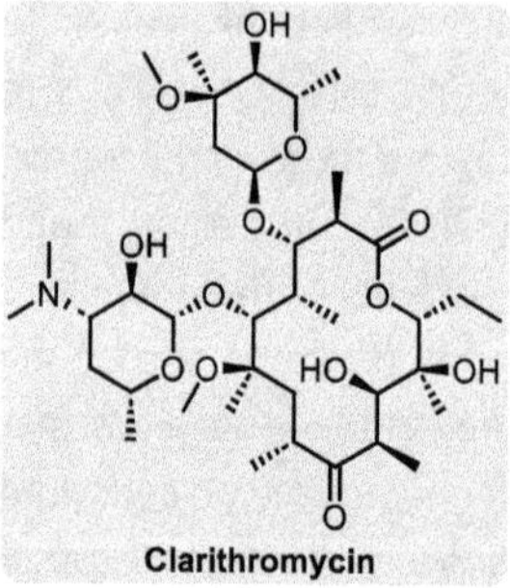

Clarithromycin

Several preclinical and clinical studies have evaluated the efficacy of clarithromycin as an anticancer agent. For instance, a randomized controlled trial in Japan showed that, combined with chemotherapy, clarithromycin is against advanced non-small cell lung cancer [24]. Another study reported that gemcitabine, combined with clarithromycin, improved the response rate and overall survival in patients with advanced pancreatic cancer [25]. Despite the promising results, more clinical studies are required to fully evaluate the potential of clarithromycin as an anticancer agent as also its long-term and potential adverse effects [26].

10.2.4 Doxycycline

Doxycycline is a tetracycline antibiotic that has been repurposed for anticancer therapy. Studies have shown that doxycycline has antitumor effects on various types of cancer cells by inducing cell cycle arrest, apoptosis, and inhibiting metastasis. Here are some of the recent findings on doxycycline's repurposing for anticancer therapy. In a study published in the journal, Oncotarget, Doxycycline was found to have antitumor effects in glioma stem cells (GSCs). Researchers found that Doxycycline treatment inhibited the growth and migration of GSCs and induced apoptosis *via* the mitochondrial pathway. They also demonstrated that Doxycycline decreased the expression of stem cell markers, suggesting that it may potentially target cancer stem cells, which are known to be responsible for cancer recurrence and treatment resistance [27]. Another study published in the journal Cancer Letters investigated the effect of Doxycycline on breast cancer cells. Researchers found that doxycycline inhibited the proliferation and migration of breast cancer cells by downregulating the expression of matrix metalloproteinases (MMPs), which are involved in cancer cell invasion and metastasis. The study also demonstrated that doxycycline induced apoptosis in breast cancer cells *via* the mitochondrial pathway [28].

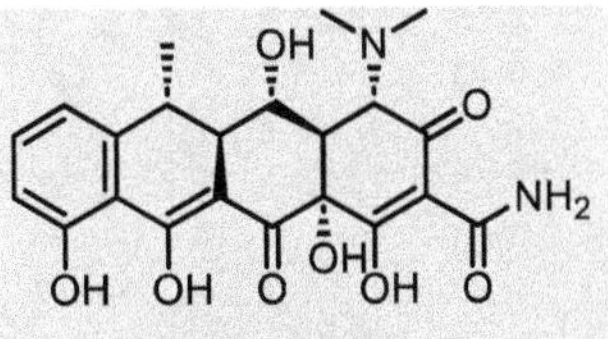

Doxycycline

In addition to its direct antitumor effects, Doxycycline has also been shown to enhance the efficacy of chemotherapy in cancer treatment. A study published in the journal Cell Death and Disease investigated the effect of doxycycline on pancreatic cancer cells treated with gemcitabine, a standard chemotherapy drug for pancreatic cancer. Researchers found that Doxycycline treatment increased the sensitivity of pancreatic cancer cells to gemcitabine and induced apoptosis *via* the mitochondrial pathway [29]. Overall, the repurposing of Doxycycline for anticancer therapy has shown promising results in preclinical studies. Further clinical studies are needed to evaluate the safety and efficacy of Doxycycline in cancer treatment.

10.2.5 Itraconazole

Janssen Pharmaceutical developed itraconazole for the first time in the early 1980s, and in 1992, the FDA approved it for use in people. The medication is a member of the triazole class of antifungal medications, along with fluconazole and voriconazole [30]. It blocks cell signaling and might cause a pause in autophagic development. Initial research on itraconazole centered on combating cancer's multi-drug resistance (MDR). It was demonstrated that it blocked the Hedgehog (Hh) signaling pathway and angiogenesis. It has been demonstrated in clinical trials that itraconazole is effective in treating leukemia, ovarian, breast, and pancreatic cancers [31]. It also improved the anticancer effects of additional medications through potential combinations. Suppressing the AKT-mTOR signaling pathway causes cholesterol redistribution, initiating autophagy and ultimately limiting cell proliferation. Following the PI3K/AKT/mTOR pathway, it counteracts the P-glycoprotein-induced chemoresistance by controlling the signal transduction pathways of Hedgehog and preventing cancer cells' lymph angiogenesis and angiogenesis [32].

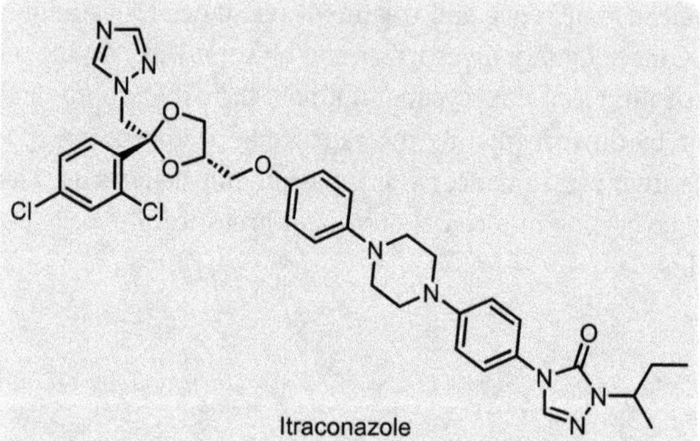

Itraconazole

10.2.6 Mebendazole

Mebendazole is a benzimidazole agent with a broad spectrum of anthelminthic activities. Janssen Pharmaceutical in Belgium developed Mebendazole, which was first used in 1971 [33]. Mebendazole decreased chemoresistance in melanoma and lung cancer cells, according to in vitro experiments. Mebendazole inhibits the growth of osteosarcoma, colon, breast, and ovarian carcinomas, and produced effective in vivo response in mice inoculated with H460 non-small cell lung cancer cells [34]. It also inhibits the growth of breast, ovary, colon carcinomas, adrenocortical, and osteosarcoma [35].

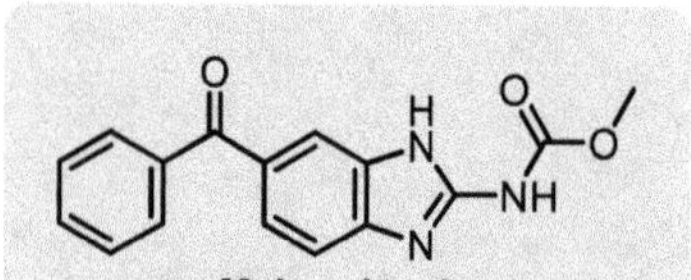

Mebendazole

10.2.7 Aspirin

Aspirin, also known as acetylsalicylic acid, is one of the most widely used NSAIDs globally. In addition to its anti-inflammatory action through the inhibition of COX-1,2, analgesic, and antiplatelet activity, its application in cancer therapy is reported against various cancers [31, 36]. Numerous mechanisms of action have been reported in the literature through which Aspirin exerts its anticancer effect. For example, Aspirin, through the inhibition of COX-2, inhibited angiogenesis in lymphoma and colon cancer. Similarly, in the ovarian and breast cancer cells, Aspirin exerts its anticancer effect through the inhibition of COX-2, targeting enzymes that are keenly involved in their proliferation [37]. In the context of metastasis, Aspirin inhibited metastasis by targeting anoikis resistance both in vitro and in vivo in breast cancer models. Aspirin and other drugs, such as decitabine, inhibited non-small cell lung cancer via downregulating the Beta Symbol should be placed-catenin/STAT3 signaling pathway [38]. The other mechanisms for anticancer activity include inhibiting the Wnt pathway [39], reducing the proliferation and glycolysis in hepatoma cells [40], increasing the α2 integrin levels, decreasing the migration in prostate cancer cells [41], and many more. Thus, Aspirin is one of the most explored drugs for anticancer activity and has demonstrated powerful inhibitory effects against cancer.

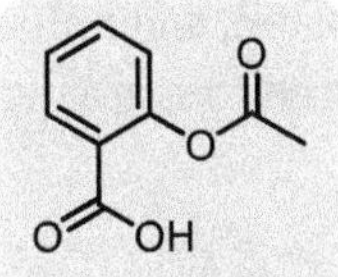

Aspirin

10.2.8 Sulindac

Sulindac, a substituted indene-3-acetic acid, is another NSAID with demonstrated anticancer activity [42]. In addition, the analogues of sulindac, such as sulindac sulfide amide, depicted higher anticancer activity than sulindac in the colon cancer model [43]. A study by Lim et al. showed the effectiveness of sulindac metabolites on the apoptosis and inhibition of growth in a prostate cancer model [41]. It was shown that sulindac caused a substantial decline in employing M2 macrophages and also in angiogenesis and cancer-related inflammation [44]. Other mechanisms of anticancer effects include the downregulation of β-catenin expression [45], production of ROS, mitochondrial dysfunction [46], etc. Thus, sulindac, its analogues, metabolites, and combination with other drugs *via* various mechanisms have shown promising results in attenuating multiple cancers.

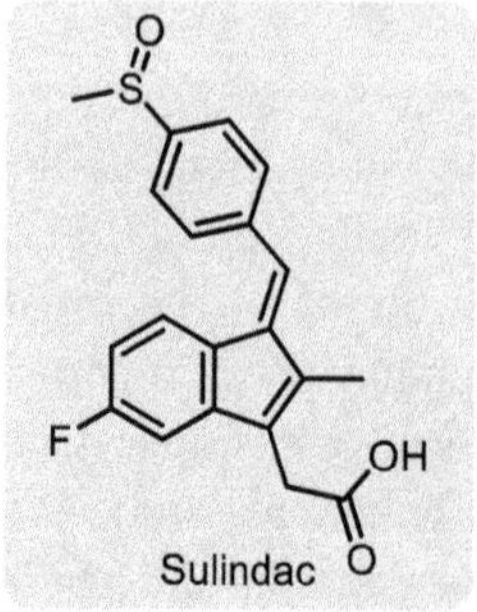

Sulindac

10.2.9 Acyclovir

Acyclovir, also known as 9-[(2-hydroxyethoxy) methyl] guanine, is one the frequently used antiviral drug for the treatment of herpes simplex virus disease [47]. This drug has also been explored for its anticancer activity. It was demonstrated that acyclovir could be explored as an adjuvant in breast cancer therapy [48]. It was shown that mitochondria would be the first target for the acyclovir cytotoxic effects, in the absence of HHVs (human herpes viruses) in non-small cell lung cancer cells [49]. In another study in the nasopharyngeal carcinoma cell line, the EBV (Epstein-Barr Virus)-infected C666-1 cell line's survival is substantially reduced when acyclovir, cisplatin, and radiation are combined [50]. However, the use of acyclovir can be explored to a greater extent since the literature is less, in our opinion.

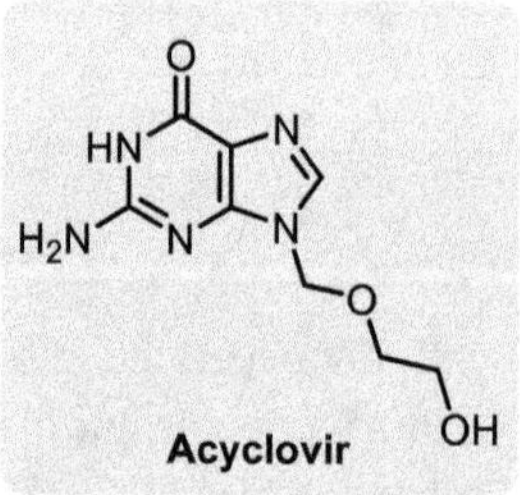

Acyclovir

10.2.10 Ritonavir

Ritonavir is a HIV (human immunodeficiency virus) protease inhibitor used for the therapy of acquired immunodeficiency virus (AIDS) and was explored for its action against cancer [51, 52]. The growth of breast cancer was inhibited by ritonavir through the Hsp90 substrates, including Akt [53]. Ritonavir induced G1 cell cycle arrest and reduced the phosphorylated AKT in accordance with the dose [54]. A combination of Ritonavir and lopinavir was tested against urological cancer cells. It was demonstrated that the co-delivery caused the induction of ER stress and led to the inhibition of renal and bladder cancer cells [55]. Similarly, ritonavir has been explored for pancreatic cancer [56], renal cancer [57], lung cancer [51], and other cancers too.

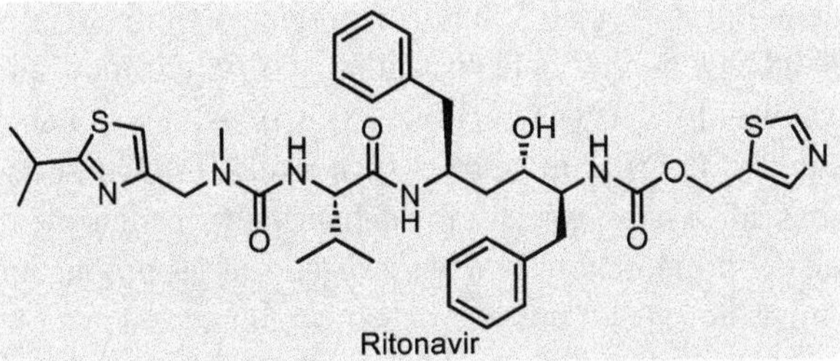

Ritonavir

10.2.11 Salinomycin

Salinomycin is a potent visionary drug isolated from *Streptomyces albus* and has been used as an anticoccidial agent. The anticancer activity of salinomycin was well discussed by Gupta et al. [58]. Various studies revealed the potential of salinomycin in targeting cancer stem cells. The possible mechanisms for the targeting ability of salinomycin include induction of cell death, apoptosis, blocking the oxidative phosphorylation, interfering with BAC transporters, etc. [59]. Furthermore, it was reported that salinomycin, through the induction of oxidative stress and decreasing the expression of oncogenes, stems cell fraction and inhibits growth and migration of prostate cancer cells [60]. In another study, it was demonstrated that Salinomycin, together with Gemcitabine both, differentiated CSCs also in pancreatic cells, both in vitro and in vivo [61]. Also, in breast cancer, this drug causes induction of cell differentiation affects, autophagic flux, etc. [62]. Various studies demonstrated Salinomycin's anticancer effectiveness, and further clinical studies need to be done to evaluate the effectiveness of salinomycin further.

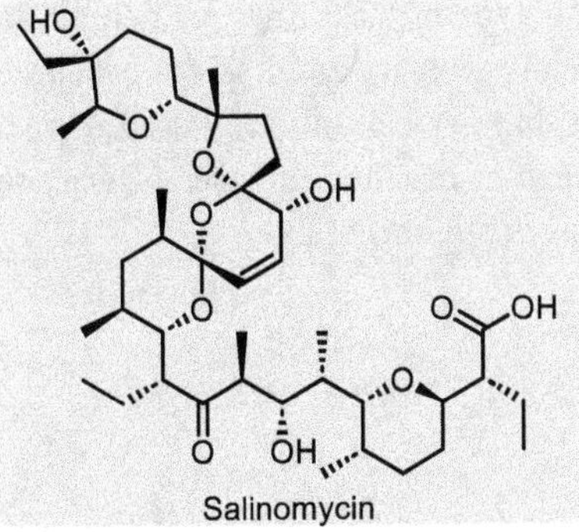

Salinomycin

10.2.12 Fenbendazole

Fenbendazole is commonly used to treat parasitic infections in animals and humans, and the literature evidence suggests anticancer activity of Fenbendazole [63]. A study demonstrated that the fundamental processes of FZ-induced preferred elimination of cancer cells in vitro and in vivo include microtubule disruption, p53 stabilization, and interference with glucose metabolism [64]. Its use in clinics is restricted due to issues owing to solubility and permeability. In order to address such issues, Esfahani et al. developed PEGylated mesoporous silica nanoparticles for delivering fenbendazole to prostate cancer cells, increasing the promotion of reactive oxygen species production and suppressing the PC-3 cell migration [65]. Thus, we believe, fenbendazole can be explored as an anticancer agent to a greater extent and various other nanocarriers can also be developed to deliver fenbendazole efficiently.

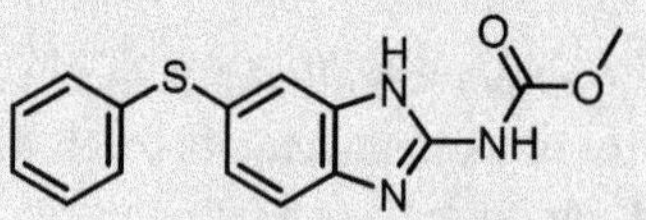

Fenbendazole

10.3 Conclusions

Drug repurposing means the use of existing or previously known drugs in newer domains. Developing a new drug takes years of hard work and billions of dollars. Therefore, repurposing aims to deliver effective molecules in a less time, and lower the cost to society. There are many successful instances of drug repurposing, based on evidence. Various factors favor drug repurposing, including preclinical and clinical phases. Target-based preclinical drug discovery mainly emphasize binding affinity to the primary target rather than to a secondary target. Hence, selectivity is one of the major concerns in this stage. However, due to this type of task, it leaves behind the other target and the pharmacokinetics behavior of the drug amid safety concerns. Many more blockbuster drugs are on the market that are continuously used in various diseases. In the future, pharma companies need to broaden the area of drug repurposing, which could solve many puzzles and saves millions of lives.

References

[1] Talevi A. Drug repurposing. Compr Pharmacol. 2022;813–824.
[2] Sanz MA, Fenaux P, Tallman MS, Estey EH, Löwenberg B, Naoe T, et al. Management of acute promyelocytic leukemia: Updated recommendations from an expert panel of the European LeukemiaNet. Blood. 2019;133(15):1630–1643.
[3] Abaza Y, Kantarjian H, Garcia-Manero G, Estey E, Borthakur G, Jabbour E, et al. Long-term outcome of acute promyelocytic leukemia treated with all-trans-retinoic acid, arsenic trioxide, and gemtuzumab. Blood. 2017;129(10):1275–1283.
[4] Dimopoulos MA, Kastritis E. Thalidomide for myeloma: Still here? Lancet Haematol. 2018;5(10):e439–40.
[5] Liu Q, Tong D, Liu G, Gao J, Wang LA, Xu J, et al. Metformin inhibits prostate cancer progression by targeting tumor-associated inflammatory infiltration. Clin Cancer Res. 2018;24(22):5622–5634.
[6] Quinn BJ, Kitagawa H, Memmott RM, Gills JJ, Dennis PA. Repositioning metformin for cancer prevention and treatment. Trends Endocrinol Metab. 2013;24(9):469–480.
[7] Morales DR, Morris AD. Metformin in cancer treatment and prevention. Annu Rev Med. 2015;66:17–29.
[8] Gallagher EJ, LeRoith D. Diabetes, cancer, and metformin: Connections of metabolism and cell proliferation. Ann N Y Acad Sci. 2011 Dec;1243(1):54–68.
[9] Franciosi M, Lucisano G, Lapice E, Strippoli GFM, Pellegrini F, Nicolucci A. Metformin therapy and risk of cancer in patients with type 2 diabetes: Systematic review. PLoS One. 2013;8:8.
[10] DeCensi A, Puntoni M, Goodwin P, Cazzaniga M, Gennari A, Bonanni B, et al. Metformin and cancer risk in diabetic patients: A systematic review and meta-analysis. Cancer Prev Res. 2010;3 (11):1451–1461.
[11] Margel D, Urbach DR, Lipscombe LL, Bell CM, Kulkarni G, Austin PC, et al. Metformin use and all-cause and prostate cancer-specific mortality among men with diabetes. J Clin Oncol. 2013;31 (25):3069–3075.
[12] Kulkarni AS, Gubbi S, Barzilai N. Benefits of metformin in attenuating the hallmarks of aging. Cell Metab. 2020;32(1):15–30.
[13] Lee MS, Hsu CC, Wahlqvist ML, Tsai HN, Chang YH, Huang YC. Type 2 diabetes increases and metformin reduces total, colorectal, liver and pancreatic cancer incidences in Taiwanese: A representative population prospective cohort study of 800,000 individuals. BMC Cancer. 2011;11.
[14] Hwang JT, Ha J, Ock JP. Combination of 5-fluorouracil and genistein induces apoptosis synergistically in chemo-resistant cancer cells through the modulation of AMPK and COX-2 signaling pathways. Biochem Biophys Res Commun. 2005;332(2):433–440.
[15] Elstner E, Müller C, Koshizuka K, Williamson EA, Park D, Asou H, et al. Ligands for peroxisome proliferator-activated receptory and retinoic acid receptor inhibit growth and induce apoptosis of human breast cancer cells in vitro and in BNX mice. Proc Natl Acad Sci U S A. 1998;95(15):8806–8811.
[16] Chang SS, Hu HY. Association of thiazolidinediones with gastric cancer in type 2 diabetes mellitus: A population-based case-control study. BMC Cancer. 2013;13.
[17] Ninomiya I, Yamazaki K, Oyama K, Hayashi H, Tajima H, Kitagawa H, et al. Pioglitazone inhibits the proliferation and metastasis of human pancreatic cancer cells. Oncol Lett. 2014;8(6):2709–2714.
[18] Ipsen EØ, Madsen KS, Chi Y, Pedersen-Bjergaard U, Richter B, Metzendorf MI, et al. Pioglitazone for prevention or delay of type 2 diabetes mellitus and its associated complications in people at risk for the development of type 2 diabetes mellitus. Cochrane Database Syst Rev. 2020;2020:11.
[19] Hatton JL, Yee LD. Clinical use of PPARγ ligands in cancer. PPAR Res. 2008.
[20] Liu Y, Jin PP, Sun XC, Hu TT. Thiazolidinediones and risk of colorectal cancer in patients with diabetes mellitus: A meta-analysis. Saudi J Gastroenterol. 2018;24(2):75–81.
[21] Sukhatme V, Bouche G, Meheus L, Sukhatme VP, Pantziarka P. Repurposing Drugs in Oncology (ReDO) – Nitroglycerin as an anti-cancer agent. Ecancermedicalscience. 2015;9.

[22] Petroni G, Bagni G, Iorio J, Duranti C, Lottini T, Stefanini M, et al. Clarithromycin inhibits autophagy in colorectal cancer by regulating the hERG1 potassium channel interaction with PI3K. Cell Death Dis. 2020;11:3.
[23] Zhou B, Xia M, Wang B, Thapa N, Gan L, Sun C, et al. Clarithromycin synergizes with cisplatin to inhibit ovarian cancer growth in vitro and in vivo. J Ovarian Res. 2019;12:1.
[24] Heudobler D, Schulz C, Fischer JR, Staib P, Wehler T, Südhoff T, et al. A randomized phase II trial comparing the efficacy and safety of pioglitazone, clarithromycin and metronomic low-dose chemotherapy with single-agent nivolumab therapy in patients with advanced non-small cell lung cancer treated in second or further line. Front Pharmacol. 2021 16;12.
[25] Sakai H, Yoneda S, Kobayashi K, Komagata H, Kosaihira S, Kazumoto T, et al. Phase II study of bi-weekly docetaxel and carboplatin with concurrent thoracic radiation therapy followed by consolidation chemotherapy with docetaxel plus carboplatin for stage III unresectable non-small cell lung cancer. Lung Cancer. 2004;43(2):195–201.
[26] Kiura K, Ueoka H, Segawa Y, Tabata M, Kamei H, Takigawa N, et al. Phase I/II study of docetaxel and cisplatin with concurrent thoracic radiation therapy for locally advanced non-small-cell lung cancer. Br J Cancer. 2003 Sep 26;89(5):795–802.
[27] Karp I, Lyakhovich A. Targeting cancer stem cells with antibiotics inducing mitochondrial dysfunction as an alternative anticancer therapy. Biochem Pharmacol. 2022;198.
[28] Qin Y, Zhang Q, Lee S, Zhong W-L, Liu Y-R, Liu H-J, et al. Doxycycline reverses epithelial-to-mesenchymal transition and suppresses the proliferation and metastasis of lung cancer cells. Oncotarget. 2015;6(38):40667–40679.
[29] Zhao Y, Wang X, Li L, Li C. Doxycycline inhibits proliferation and induces apoptosis of both human papillomavirus positive and negative cervical cancer cell lines. Can J Physiol Pharmacol. 2016;94 (5):526–533.
[30] Teixeira MM, Carvalho DT, Sousa E, Pinto E. New antifungal agents with azole moieties. Pharmaceuticals. 2022;15(11):1427.
[31] Olgen S, Kotra LP. Drug repurposing in the development of anticancer agents. Curr Med Chem. 2019;26(28):5410–5427.
[32] Liu R, Li J, Zhang T, Zou L, Chen Y, Wang K, Lei Y, Yuan K, Li Y, Lan J, Cheng L, Xie N, Xiang R, Nice EC, Huang C, Wei Y. Itraconazole suppresses the growth of glioblastoma through induction of autophagy. 2017.
[33] Conterno LO, Turchi MD, Corrêa I, Monteiro de Barros Almeida RA. Anthelmintic drugs for treating ascariasis. Cochrane Database Syst Rev. 2020;2020:4.
[34] Guerini AE, Triggiani L, Maddalo M, Bonù ML, Frassine F, Baiguini A, et al. Mebendazole as a candidate for drug repurposing in oncology: An extensive review of current literature. Cancers (Basel). 2019;11:9.
[35] Martarelli D, Pompei P, Baldi C, Mazzoni G. Mebendazole inhibits growth of human adrenocortical carcinoma cell lines implanted in nude mice. Cancer Chemother Pharmacol. 2008;61(5):809–817.
[36] Thun MJ, Jacobs EJ, Patrono C. The role of Aspirin in cancer prevention. Nat Rev Clin Oncol. 2012;9(5):259–267.
[37] Elwood P, Protty M, Morgan G, Pickering J, Delon C, Watkins J. Aspirin and cancer: Biological mechanisms and clinical outcomes. Open Biol. 2022;12(9):2–8.
[38] Xu M, Song B, Yang X, Li N. The combination of decitabine and Aspirin inhibits tumor growth and metastasis in non-small cell lung cancer. J Int Med Res. 2022;50:7.
[39] Feng Y, Tao L, Wang G, Li Z, Yang M, He W, et al. Aspirin inhibits prostaglandins to prevents colon tumor formation via down-regulating Wnt production. Eur J Pharmacol. 2021;906(April):174173.
[40] Yuan Y, Yuan HF, Geng Y, Zhao LN, Yun HL, Wang YF, et al. Aspirin modulates 2-hydroxyisobutyrylation of ENO1K281 to attenuate the glycolysis and proliferation of hepatoma cells. Biochem Biophys Res Commun. 2021;560:172–178.

[41] Lim JTE, Piazza GA, Han EKH, Delohery TM, Li H, Finn TS, et al. Sulindac derivatives inhibit growth and induce apoptosis in human prostate cancer cell lines. Biochem Pharmacol. 1999;58 (7):1097–1107.
[42] Piazza GA, Keeton AB, Tinsley HN, Whitt JD, Gary B, Mathew B, et al. NSAIDs: Old drugs reveal new anticancer targets. Pharmaceuticals. 2010;3(5):1652–1667.
[43] Mathew B, Hobrath JV, Connelly MC, Guy RK, Reynolds RC. Diverse amide analogs of sulindac for cancer treatment and prevention. Bioorganic Med Chem Lett. 2017;27(20):4614–4621.
[44] Yin T, Wang G, Ye T, Wang Y. Sulindac, a non-steroidal anti-inflammatory drug, mediates breast cancer inhibition as an immune modulator. Sci Rep. 2016;6:1–8.
[45] Han A, Song Z, Tong C, Hu D, Bi X, Augenlicht LH, et al. Sulindac suppresses β-catenin expression in human cancer cells. Eur J Pharmacol. 2008;583(1):26–31.
[46] Marchetti M, Resnick L, Gamliel E, Kesaraju S, Weissbach H, Binninger D. Sulindac enhances the killing of cancer cells exposed to oxidative stress. PLoS One. 2009;4:6.
[47] Yao J, Zhang Y, Ramishetti S, Wang Y, Huang L. Turning an antiviral into an anticancer drug: Nanoparticle delivery of acyclovir monophosphate. J Control Release. 2013;170(3):414–420.
[48] Siddiqui S, Deshmukh AJ, Mudaliar P, Nalawade AJ, Iyer D, Aich J. Drug repurposing: Re-inventing therapies for cancer without re-entering the development pipeline – A review. J Egypt Natl Canc Inst. 2022;34(1).
[49] Benedetti S, Catalani S, Canonico B, Nasoni MG, Luchetti F, Papa S, et al. The effects of Acyclovir administration to NCI-H1975 non-small cell lung cancer cells. Toxicol Vitr. 2022;79.
[50] Thandoni A. Acyclovir improves the efficacy of chemoradiation in nasopharyngeal cancer containing the epstein barr virus genome. 2020;614.
[51] Marima R, Hull R, Dlamini Z, Penny C. Efavirenz and Lopinavir/Ritonavir alter cell cycle regulation in lung cancer. Front Oncol. 2020;10(August):1–11.
[52] Moawad EY. Identifying the optimal dose of ritonavir in the treatment of malignancies. Metab Brain Dis. 2014;29(2):533–540.
[53] Srirangam A, Mitra R, Wang M, Gorski JC, Badve S, Baldridge LA, et al. Effects of HIV protease inhibitor ritonavir on Akt-regulated cell proliferation in breast cancer. Clin Cancer Res. 2006;12 (6):1883–1896.
[54] Kumar S, Bryant CS, Chamala S, Qazi A, Seward S, Pal J, et al. Ritonavir blocks AKT signaling, activates apoptosis and inhibits migration and invasion in ovarian cancer cells. Mol Cancer. 2009;8:1–12.
[55] Okubo K, Isono M, Asano T, Sato A. Lopinavir-ritonavir combination induces endoplasmic reticulum stress and kills urological cancer cells. Anticancer Res. 2019;39(11):5891–5901.
[56] Batchu RB, Gruzdyn OV, Bryant CS, Qazi AM, Kumar S, Chamala S, et al. Ritonavir-mediated induction of apoptosis in pancreatic cancer occurs via the RB/E2F-1 and AKT pathways. Pharmaceuticals. 2014;7(1):46–57.
[57] Isono M, Sato A, Asano T, Okubo K, Asano T. Delanzomib interacts with ritonavir synergistically to cause endoplasmic reticulum stress in renal cancer cells. Anticancer Res. 2018;38(6):3493–3500.
[58] Dewangan J, Srivastava S, Rath SK. Salinomycin: A new paradigm in cancer therapy. Tumor Biol. 2017;39:3.
[59] Naujokat C, Steinhart R. Salinomycin as a drug for targeting human cancer stem cells. J Biomed Biotechnol. 2012;2012.
[60] Ketola K, Hilvo M, Hyötyläinen T, Vuoristo A, Ruskeepää AL, Orešič M, et al. Salinomycin inhibits prostate cancer growth and migration via induction of oxidative stress. Br J Cancer. 2012;106 (1):99–106.
[61] Zhang GN, Liang Y, Zhou LJ, Chen SP, Chen G, Zhang TP, et al. Combination of salinomycin and gemcitabine eliminates pancreatic cancer cells. Cancer Lett. 2011;313(2):137–144.

[62] Wang H, Zhang H, Zhu Y, Wu Z, Cui C, Cai F. Anticancer mechanisms of salinomycin in breast cancer and its clinical applications. Front Oncol. 2021;11.

[63] Duan Q, Liu Y, Rockwell S. Fenbendazole as a potential anticancer drug. Anticancer Res. 2013;33(2):355–362.

[64] Dogra N, Kumar A, Mukhopadhyay T. Fenbendazole acts as a moderate microtubule destabilizing agent and causes cancer cell death by modulating multiple cellular pathways. Sci Rep. 2018;8(1):1–15.

[65] Esfahani MKM-L-MMSNAPC for the TD of F into PCC, Alavi SE, Cabot PJ, Islam N, Izake EL. $β$-lactoglobulin-modified mesoporous silica nanoparticles: a promising carrier for the targeted delivery of fenbendazole into prostate cancer cells. Pharmaceutics. 2022;14(4).

Debajyoti Roy, Naresh Kumar Rangra, Bhupinder Kumar

11 Drug repurposing of drugs from natural and marine origins

Abstract: Natural and marine sources are a wide pool of pharmacologically active molecules with diverse pharmacological activities and therapeutic applications. From 1981 to 2019, two-third of the approved drugs was small molecules and mostly inspired by natural resources. Natural products and marine organisms have been found to produce a range of compounds with potential anti-infective activity, including artemisinin, which is used to treat malaria, and tetrodotoxin, which has potent activity against a range of bacterial and viral infections. The use of natural sources for a healthy lifestyle is increasing day by day in modern life. The chemical defence mechanism of natural and marine products can be more advantageous for their efficiency over synthetic molecules. Repurposing of the clinically approved natural and marine-derived molecules provide a great possibility of finding molecules with high effectiveness against emerging infectious diseases and highly prevalent disease states that are uncurable. Drug repurposing of natural and marine products can provide an effective therapy with lower to minimal side effects in a low cost and at lesser time. In this chapter, we have enlisted the various natural and marine products that are under clinical investigation for repurposing against various disease states. This chapter will also provide brief information about the various drug repurposing strategies and their applications.

Keywords: Drug repurposing, drug repositioning, natural products, marine products, phytochemicals

11.1 Introduction

Natural sources such as herbs, plants, trees, etc. and their parts have been utilized from ancient times in human history for the treatment of various ailments and diseases. However, in today's era, these sources have gained more interest in the discovery of new small therapeutic molecules for particular ailments [1, 2]. From 1981 to 2019, two-third of the approved drugs was small molecules and mostly inspired by natural resources. Nature-derived products have long been a basis for drug design and are the inspiration from development to drug discovery, and many of the drugs we use today were originally derived from natural sources. For example, the anticancer drug, paclitaxel, was originally derived from the Pacific yew tree, and the antibiotic penicillin was first isolated from the Penicillium mould. Similarly, marine organisms offer pharmacologically active molecules, possessing potential therapeutic uses [3]. The importance of nat-

https://doi.org/10.1515/9783110791150-011

ural products can further be understood from the fact that WC Campbell, S Omura, and Y Tu received the Nobel Prize for discovering two natural compounds, avermectin, and artemisinin, which are used to treat parasitic disorders [4, 5]. Natural and marine source-based/derived drugs have gained significant importance in our daily lives as they provide a natural alternative to conventional drugs [6]. These therapeutic molecules obtained from plants, animals, marine sources, and microorganisms, present them as more human-body compatible and potentially safer molecules than synthetic drugs [7]. They have been used for centuries as traditional medicines and are currently gaining more attention for their medicinal benefits. One of the primary reasons why natural and marine drugs are better than conventional drugs is that they often have fewer side effects while synthetic drug molecules are associated with severe adverse reactions and even toxicities, which can be harmful to our bodies [8]. Natural and marine drugs are generally more compatible with our body's natural chemistry and have been used for generations without causing significant harm [9, 10]. The impact of natural and marine drugs in the market is causing alarm in modern life as more of the human population is turning to use natural resources for improving healthy lifestyles as well as for treatment options for various ailments and diseases. This trend is driven by a growing awareness of the benefits of natural and marine drugs, including their potential to provide targeted, effective treatment with fewer side effects [11]. With the ongoing research and development, these natural remedies have the potential to revolutionize the healthcare industry and provide a healthy and quality lifestyle for people around the world.

In the last two decades, the trend of utilizing natural and marine sources for discovery and development of drugs has increased to large extent. The repurposing of already discovered natural products for targets other than the approved has also garnered further interest from scientists [3]. The search for new bioactive molecules from natural and marine sources has been fuelled by the need for new and effective treatments for a wide range of diseases, including cancer, infectious diseases, and neurological disorders [12]. Natural products and marine organisms have evolved complex chemical defence mechanisms, which make them a valuable source of novel drug candidates with potential therapeutic properties. The use of these sources for drug discovery has led to the development of a wide spectrum of bioactive molecules, some of which have already been approved for clinical use, while others are currently undergoing preclinical or clinical trials [13].

Drug repurposing is one of the techniques to identify new therapeutic indications for old existing drug molecules. Recently, drug repurposing has gained so much interest that one-third of recently approved drugs are repurposed for some therapeutic indication, in addition to the previously identified targets [14, 15]. In a most recent trend, a large number of drug discovery groups have worked on drug repurposing for the identification of medications against COVID-19 from synthetic as well as natural resources [16]. One of the most promising areas of drug repurposing from natural and marine sources is the development of anticancer drugs [17]. Cancer is a major public health problem, with millions of people worldwide being diagnosed with the disease each year. Although there are many treatments available for cancer, including chemotherapy, radiation, and surgery, there is a need for more effective new treatments. Natural habitants and marine organisms have been found to produce a range of compounds with potential anticancer activity, including paclitaxel, vinblastine, and vincristine, which have been approved for clinical use [18, 19]. Another area of drug repurposing from natural and marine sources is the development of anti-infective drugs. Infectious diseases remain a major global health problem, with millions of people dying each year from infections such as malaria, tuberculosis, and HIV/AIDS. Natural products and marine organisms have been found to produce a range of compounds with potential anti-infective activity, including artemisinin, which is used to treat malaria, and tetrodotoxin, which has potent activity against a range of bacterial and viral infections [20, 21]. The use of natural products and marine organisms for drug repurposing is supported by advances in technology and analytical techniques. For example, high-throughput screening methods have been developed to rapidly screen large numbers of compounds for potential therapeutic activity. In addition, advances in genomics and metabolomics have enabled scientists to identify and characterize novel compounds produced by natural products and marine organisms [22].

11.2 Repurposed drugs from plants and herbs

Plants and herbs have been used for centuries to treat various ailments and diseases. Many of these natural compounds have been studied extensively and have been found to have therapeutic potential. Some of these compounds have been repurposed as drugs to treat different diseases. The process of drug repurposing involves taking a compound originally derived from a plant or herb and developing it into a drug to treat a different disease. This approach can be more efficient and cost-effective than developing new drugs from scratch. Moreover, natural compounds often have fewer side effects and are better tolerated by patients than conventional drugs. There are numerous examples of drugs that were repurposed from plants and herbs, such as paclitaxel, artemisinin, resveratrol, curcumin, aspirin, quinine, morphine, taxol, artesunate, etoposide, aconitine, reserpine, salinomycin, and psilocybin as described in Table 11.1 and Figure 11.1. These drugs traditionally have been approved for their clinical application against various ailments, such as cancer, malaria, cardiovascular disease, and pain.

In one research study, the anticancer drug, paclitaxel, displayed neuronal rescue against tau-induced toxicity [23] and β-amyloid toxicity [24], which are potential hallmarks of Alzheimer's disease (AD). To enhance the blood-brain barrier permeability, an intranasal formulation of paclitaxel was also reported [25]. Artemisinin is another plant-derived medicine used for the treatment of malaria. This front-line antimalarial drug is widely explored for its anticancer properties and for the treatment of hepatitis too [26]. An anti-inflammatory drug, resveratrol, is being explored for its pharmacological potential in the treatment of cancer and cardiovascular diseases [27]. Similarly, there are a large number of molecules being explored for drug repurposing as indicated in Table 11.1.

Table 11.1: List of plants-derived phytochemicals that have potential use for drug repurposing.

S. no.	Drug name	Name of source	Pharmacological use	Drug repurposing use	References
1.	Paclitaxel	*Taxus brevifolia*	Chemotherapy drug for treating breast, lung, and ovarian cancer	Anti-Alzheimer	[23]
2.	Artemisinin	*Artemisia annua*	Antimalarial drug	Anticancer, hepatitis	[26]
3.	Resveratrol	*Vitis vinifera*	Anti-inflammatory and antioxidant effects	Anticancer, cardiovascular disease	[27]
4.	Curcumin	*Curcuma longa*	Management of oxidative and inflammatory conditions, metabolic syndrome, arthritis, anxiety, and hyperlipidaemia	Anticancer, Anti-Alzheimer's disease	[28]
5.	Aspirin	*Salix alba*	Anti-inflammatory, analgesic, antipyretic	Anticancer and cardiovascular disease	[29]
6.	Morphine	*Papaver somniferum*	Painkiller	Antidepressant, Antipsychotic	[30]
7.	Taxol	*Taxus brevifolia*	Chemotherapy drug that is used to treat breast, lung, and ovarian cancer	Alzheimer's disease and other neurological disorders	[31]
8.	Artesunate	*Artemisia annua*	Antimalarial drug	Anticancer	[26]
9.	Etoposide	*Podophyllum peltatum*	Anticancer	Alzheimer's disease and other neurological disorders	[32]
10.	Aconitine	*Aconitum napellus*	Fear, anxiety, and restlessness, acute sudden fever, influenza, pulsating headaches	COVID-19	[33]
11.	Reserpine	*Rauvolfia serpentina*	Antihypertensive	Refractory hypertension	[34]

Table 11.1 (continued)

S. no.	Drug name	Name of source	Pharmacological use	Drug repurposing use	References
12.	Salinomycin	*Streptomyces albus*	Antibiotic	Antiviral	[35]
13.	Psilocybin	*Psilocybe cubensis*	Anxiety, migraines, depression, PTSD	Anti-inflammatory	[36]
14.	Bromelain	*Ananas comosus*	Osteoarthritis, anticancer, muscle soreness, and digestive problem	Anti-inflammation, painkiller	[37]

As the demand for effective and safe drugs continues to grow, drug repurposing from natural compounds is becoming increasingly important. With the advancement of technology and research methods, more natural compounds are being discovered and studied for their therapeutic potential. This approach not only benefits patients but also has the potential to impact the pharmaceutical industry by reducing the time and cost associated with drug development [22].

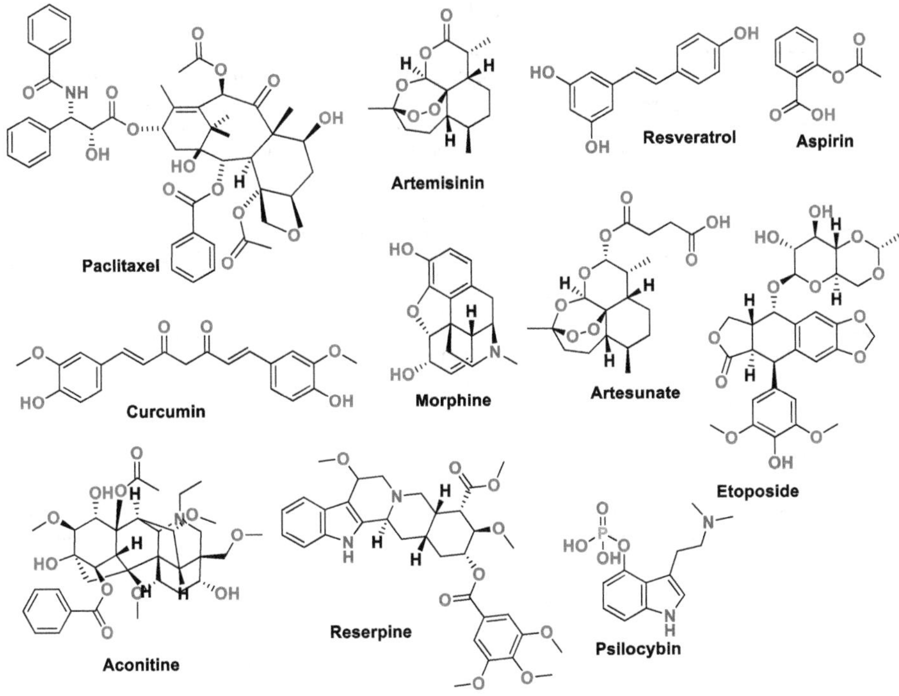

Paclitaxel Artemisinin Resveratrol Aspirin Curcumin Morphine Artesunate Etoposide Aconitine Reserpine Psilocybin

Figure 11.1: Structure of some natural-derived medicinal compounds being explored for drug repurposing.

11.3 Repurposed drugs from marine sources

The ocean is a vast and largely unexplored resource that is rich in biodiversity. Many of the organisms that live in the ocean produce compounds with potential therapeutic value [38]. As a result, the interest in repurposing of these marine-derived compounds as drugs to treat various diseases has grown significantly [39]. Drug repurposing from marine-derived compounds can provide multiple benefits, such as lower timeline and cost of development, as compared traditional drug discovery and development methods. There are numerous examples of drugs that were repurposed from marine organisms, such as trabectedin, ziconotide, bryostatin, and discodermolide. These drugs have been approved clinically for the treatment of a range of diseases like cancer, chronic pain, and Alzheimer's disease. As the demand for new and effective drugs continues to increase, there is a growing interest in the potential therapeutic value of marine-derived compounds [40]. Advances in technology and research methods are making it possible to explore and identify new compounds from marine organisms [41]. There are many natural and marine drugs that have been repurposed for new therapeutic uses and some of examples are included in Table 11.2 and their structures are given in Figure 11.2.

Table 11.2: List of marine-based drugs with potential drug repurposing and use.

S. no.	Drug name	Name of source	Pharmacological use	Drug repurposing use	References
1.	Tetrodotoxin	*Tetraodontidae*	To relieve headache associated with heroin withdrawal	Analgesic and Anticancer	[42]
2.	Cephalosporins	*Acremonium falciforme*	Treat a variety of bacterial infections, including pneumonia and meningitis	Anti-Alzheimer	[43, 44]
3.	Lovastatin	*Aspergillus terreus*	Lower cholesterol levels and reduce the risk of heart disease	Anti-Alzheimer	[45]
4.	Fucoidan	*Sarghassum Wightii*	Immune-modulatory and anti-inflammatory	Anticancer	[46]
5.	Doxorubicin	*Streptomyces peucetius*	Antibiotic	Anticancer	[47, 48]
6.	Omega-3 fatty acids	*Salmo salar Sardina pilchardus*	Prevent heart disease and stroke	Treatment of cardiovascular disease, depression	[49, 50]

Table 11.2 (continued)

S. no.	Drug name	Name of source	Pharmacological use	Drug repurposing use	References
7.	Chitosan	*Crustacea*	Reduce fat and cholesterol absorption, Wound healing	Treatment of obesity	[51, 52]
8.	Marine peptides	Various Marine Organism	Antimicrobial, antiviral, antitumor, antioxidative, antihypertensive, anti-atherosclerotic, anticoagulant, immunomodulatory, analgesic, anxiolytic, antidiabetic	Anticancer, anti-inflammation	[53]

Marine-derived sources can provide a large range of chemical constituents that can be further useful in a wide range of pharmacological indications. The earlier reported clinical agents, derived from marine sources such as tetrodotoxin and cephalosporins, are best examples that can be further explored for their anticancer potential along with the approved pharmacological indications.

Tetrodotoxin

Cephalosporins

Lovastatin

Fucoidan

Doxorubicin

Chitosan

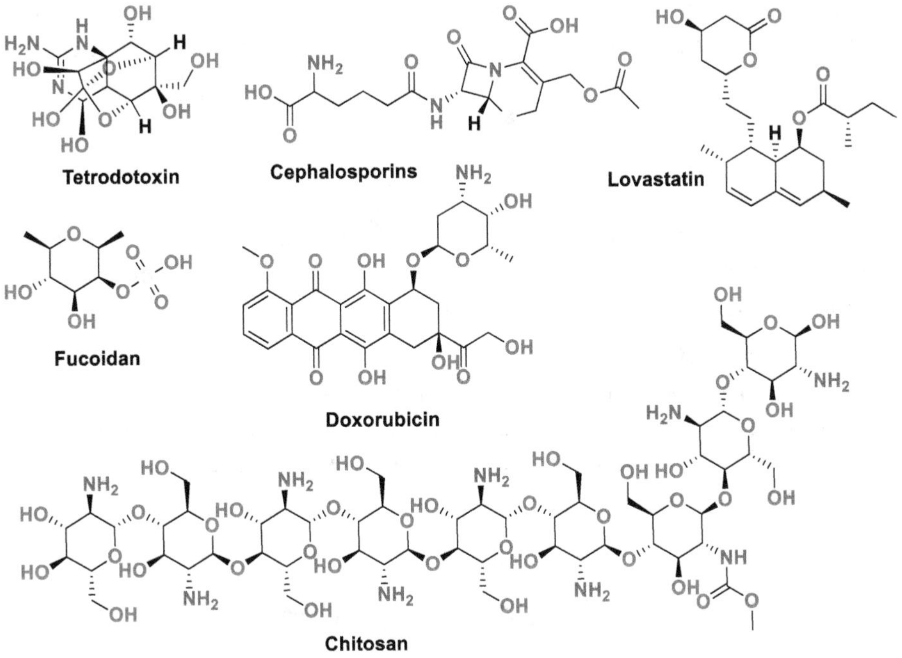

Figure 11.2: Structure of some marine-derived medicinal compounds being explored for drug-repurposing.

11.4 Methods used in drug repurposing

Drug repurposing is a method to identify a new pharmacological application or target for already clinically approved drug molecules. There are multiples strategies employed in drug repurposing. Some of the most common methods used are as following:

I. Phenotypic screening
As the name indicates, the pool of clinically approved drug molecules are screened in vitro (Cell-based studies) or in vivo (animal models) for their desired phenotype. This strategy is helpful in the discovery of unexpected and novel pharmacological effects with new potential pharmacological targets for new therapeutic applications.

II. Target-based screening
This strategy is based for the identification of interactions of drug molecules with some new specific targets or their ability to target a molecular pathway associated with the progression of a disease. Target-based screening is frequently employed when the target is clearly identified or the disease pathway/mechanism is well established.

III. Computational approaches
This strategy involves the application of mathematical algorithms and data analysis techniques for the identification of new drug-target interaction with the best suitable combination. These computational approaches help to identify new drug targets with help of interactions shown by drug (small molecules) with macromolecules (proteins or receptors).

IV. Drug repositioning databases
This technique involves the collection of all the known information regarding physical and chemical properties of approved drug molecules, their application, target, and their mechanism. Based on the collected information, data is analysed for identification of a new target for the existing molecules.

V. Clinical observation and serendipity
It involves the identification of new applications of existing drugs via continuous clinical observation or serendipity. For example, sildenafil was originally developed as a treatment for hypertension but was later found to be effective in treating erectile dysfunction [54].

11.5 Application of drug repurposing

Drug repurposing presents wide applications in the field of medicine, and its potential benefits are becoming increasingly recognized. Some of the most notable applications of drug repurposing include:

I. Treatment of rare diseases

Drug repurposing offers a way to potentially treat rare diseases that may not have many available treatment options. For example, the drug thalidomide was originally developed as a sedative and anti-nausea medication, but it was discovered later that it is effective in the treatment of leprosy and multiple myeloma, a rare type of blood cancer [55].

II. Speeding up drug development

Repurposing existing drugs can save time and resources compared to developing entirely new drugs from scratch. This process becomes more important in the case of treatment of some emerging infectious diseases where immediate treatment development is critical. For example, during the COVID-19 pandemic, researchers quickly repurposed several existing drugs for potential use in treating the virus [56].

III. Development of personalized medicine

Repurposed drugs can be used to develop personalized treatments that are tailored to an individual's specific needs. This is particularly important in the treatment of cancer, where personalized medicine can help to identify the most effective treatments for individual patients. For example, the drug imatinib was originally developed as a treatment for chronic myeloid leukaemia but has since been repurposed for use in several other types of cancer [57].

IV. Treatment of multiple conditions

Many repurposed drugs are effective in the treatment of multiple conditions, offering a way to potentially treat several different conditions with a single medication. For example, the drug metformin, originally developed as a treatment for type 2 diabetes, has been repurposed for use in the treatment of cancer, polycystic ovary syndrome, and other conditions [58].

11.6 Conclusion

Drug repurposing from drugs of natural and marine origin is an exciting area of drug discovery that has the potential to accelerate the drug development process and reduce costs. Natural and marine organisms have long been a source of inspiration for drug discovery, and their complex chemical defence mechanisms make them as a valuable source of novel drug candidates with potential therapeutic properties. However, the successful repurposing of drugs from natural and marine sources requires robust preclinical and clinical testing to ensure their safety and efficacy, as well as the overcoming of intellectual property rights challenges. With the right approach, drug repurposing from drugs of natural and marine origin has the potential to revolutionize. This is just a small selection of the many natural and marine drugs that are used in drug repurposing. As research in this field continues, it is likely that more drugs will be discovered and developed from these sources.

References

[1] Singh S, et al. Paradigms and success stories of natural products in drug discovery against. Curr Neuropharmacol. 2023.
[2] Harvey AL. Natural products in drug discovery. Drug Discov Today. 2008;13(19–20):894–901.
[3] Rudrapal M, Khairnar SJ, Jadhav AG. Drug repurpos Emerg approach Drug Discov. Drug Repurposing-hypothesis, Molecular Aspects and Therapeutic Applications. 2020;10.
[4] Efferth T, et al. Nobel Prize for artemisinin brings phytotherapy into the spotlight. Phytomedicine. 2015;22(13):A1–A3.
[5] Shen B. A new golden age of natural products drug discovery. Cell. 2015;163(6):1297–1300.
[6] Chin Y-W, et al. Drug discovery from natural sources. AAPS J. 2006;8:E239–E253.
[7] Newman DJ. Natural products as leads to potential drugs: An old process or the new hope for drug discovery?. J Med Chem. 2008;51(9):2589–2599.
[8] Krettli AU, Adebayo JO, Krettli LG. Testing of natural products and synthetic molecules aiming at new antimalarials. Curr Drug Targets. 2009;10(3):261–270.
[9] Stonik V. Marine natural products: A way to new drugs. Acta Nat. 2009;1(2):15–25.
[10] Schmidt BM, et al. Revisiting the ancient concept of botanical therapeutics. Nat Chem Biol. 2007;3(7):360–366.
[11] Research GV. Natural and Marine Derived Products Market Size, Share & Trends Analysis Report by Source (Plant, Animal, Microbial), by Product Type, by Application, by Region, and Segment Forecasts, 2022–2030. 2022.
[12] Lahlou M. The Success of Natural Products in Drug Discovery. 2013.
[13] Martins M, et al. Marine natural products, multitarget therapy and repurposed agents in Alzheimer's disease. Pharmaceuticals. 2020;13(9):242.
[14] Pantziarka P, Pirmohamed M, Mirza N. New Uses for Old Drugs. British Medical Journal Publishing Group, 2018.
[15] Sachs RE, Ginsburg PB, Goldman DP. Encouraging new uses for old drugs. Jama. 2017;318(24):2421–2422.
[16] Khan SA, Al-Balushi K. Combating COVID-19: The role of drug repurposing and medicinal plants. J Infect Public Health. 2021;14(4):495–503.
[17] Rastelli G, et al. Repositioning natural products in drug discovery. MDPI. 2020:1154.
[18] El-Sherbeni AA, El-Kadi AO. Repurposing resveratrol and fluconazole to modulate human cytochrome P450-mediated arachidonic acid metabolism. Mol Pharm. 2016;13(4):1278–1288.
[19] Maruca A, et al. Natural products extracted from fungal species as new potential anti-cancer drugs: A structure-based drug repurposing approach targeting HDAC7. Molecules. 2020;25(23):5524.
[20] Tu Y. The discovery of artemisinin (qinghaosu) and gifts from Chinese medicine. Nat Med. 2011;17(10):1217–1220.
[21] Narahashi T. Tetrodotoxin – A brief history –. Proc Jpn Acad Ser B. 2008;84(5):147–154.
[22] Cragg GM, Grothaus PG, Newman DJ. Impact of natural products on developing new anti-cancer agents. Chem Rev. 2009;109(7):3012–3043.
[23] Shemesh OA, Spira ME. Rescue of neurons from undergoing hallmark tau-induced Alzheimer's disease cell pathologies by the antimitotic drug paclitaxel. Neurobiol Dis. 2011;43(1):163–175.
[24] Michaelis M, et al. Protection against β-amyloid toxicity in primary neurons by paclitaxel (taxol). J Neurochem. 1998;70(4):1623–1627.
[25] Cross DJ, et al. Intranasal paclitaxel alters Alzheimer's disease phenotypic features in 3xTg-AD mice. J Alzheimers Dis. 2021;83(1):379–394.
[26] Meng Y, et al. Recent pharmacological advances in the repurposing of artemisinin drugs. Med Res Rev. 2021;41(6):3156–3181.

[27] Zhang M, Chen X, Radacsi N. New tricks of old drugs: Repurposing non-chemo drugs and dietary phytochemicals as adjuvants in anti-tumor therapies. J Control Release. 2021;329:96–120.
[28] Roy R, et al. Computational repurposing model of curcumin as a drug. Int J Pharm Sci Rev Res. 2021;68(1).
[29] Mohapatra TK, Subudhi BB. Repurposing of aspirin: Opportunities and challenges. Res J Pharm Technol. 2019;12(4):2037–2044.
[30] Sehrawat H, et al. Unraveling the interaction of an opium poppy alkaloid noscapine ionic liquid with human hemoglobin: Biophysical and computational studies. J Mol Liq. 2021;338:116710.
[31] Calcul L, et al. Natural products as a rich source of tau-targeting drugs for Alzheimer's disease. Future Med Chem. 2012;4(13):1751–1761.
[32] Ganesan V, et al. Repurposing the antibacterial activity of etoposide— a chemotherapeutic drug in combination with eggshell-derived hydroxyapatite. ACS Biomater Sci Eng. 2022;8(2):682–693.
[33] Lakhera S, et al. In-Silico Investigation of Inhibiting Property of Phytoconstituents of Medicinal Herb 'Aconitum Heterophyllum' Against Omicron Variant of SARS-CoV-2. 2022.
[34] Weir MR. Reserpine: A new consideration of an old drug for refractory hypertension. Am J Hypertens. 2020;33(8):708–710.
[35] Pindiprolu SKS, et al. Pulmonary delivery of nanostructured lipid carriers for effective repurposing of salinomycin as an antiviral agent. Med Hypotheses. 2020;143:109858.
[36] Smedfors G, et al. Psilocybin Combines Rapid Synaptogenic and Anti-Inflammatory Effects in Vitro. 2022.
[37] Hong J-H, et al. Anti-inflammatory and mineralization effects of bromelain on lipopolysaccharide-induced inflammation of human dental pulp cells. Medicina. 2021;57(6):591.
[38] Santos JD, et al. From ocean to medicine: Pharmaceutical applications of metabolites from marine bacteria. Antibiotics. 2020;9(8):455.
[39] Jaspars M, et al. The marine biodiscovery pipeline and ocean medicines of tomorrow. J Mar Biol Assoc UK. 2016;96(1):151–158.
[40] Püsküllüoğlu M, Michalak I. An ocean of possibilities: A review of marine organisms as sources of nanoparticles for cancer care. Nanomedicine. 2022.
[41] Fayed MA, et al. Structure-and ligand-based in silico studies towards the repurposing of marine bioactive compounds to target SARS-CoV-2. Arab J Chem. 2021;14(4):103092.
[42] Permana KR, Septian DD. Revolutionary therapy for breast cancer (BCa) by using Tetrodotoxin (TTX) extracted from the masked puffer fish arothron diadematus as best investment in community. In Proceedings of the 1st Annual International Scholars Conference in Taiwan. 2013.
[43] Kumari S, Deshmukh R. β-lactam antibiotics to tame down molecular pathways of Alzheimer's disease. Eur J Pharmacol. 2021;895:173877.
[44] Tikka T, et al. Tetracycline derivatives and ceftriaxone, a cephalosporin antibiotic, protect neurons against apoptosis induced by ionizing radiation. J Neurochem. 2001;78(6):1409–1414.
[45] Peyclit L, et al. Drug repurposing in medical mycology: Identification of compounds as potential antifungals to overcome the emergence of multidrug-resistant fungi. Pharmaceuticals. 2021;14(5):488.
[46] Mustafa S, Pawar JS, Ghosh I. Fucoidan induces ROS-dependent epigenetic modulation in cervical cancer HeLa cell. Int J Biol Macromol. 2021;181:180–192.
[47] Mandell JB, et al. Combination therapy with disulfiram, copper, and doxorubicin for osteosarcoma: In vitro support for a novel drug repurposing strategy. Sarcoma. 2019;2019.
[48] Yousefnezhad M, et al. PCL-based nanoparticles for doxorubicin-ezetimibe co-delivery: A combination therapy for prostate cancer using a drug repurposing strategy. BioImpacts. 2023.
[49] Chang JP-C, et al. Omega-3 polyunsaturated fatty acids in cardiovascular diseases comorbid major depressive disorder–Results from a randomized controlled trial. Brain Behav Immun. 2020;85:14–20.

[50] Carney RM, et al. A randomized placebo-controlled trial of omega-3 and sertraline in depressed patients with or at risk for coronary heart disease. J Clin Psychiatry. 2019;80(4):13302.
[51] Lütjohann D, et al. Influence of chitosan treatment on surrogate serum markers of cholesterol metabolism in obese subjects. Nutrients. 2018;10(1):72.
[52] Huang H, et al. The effects of chitosan supplementation on body weight and body composition: A systematic review and meta-analysis of randomized controlled trials. Crit Rev Food Sci Nutr. 2020;60(11):1815–1825.
[53] Kumar LV, Shakila RJ, Jeyasekaran G. In vitro anti-cancer, anti-diabetic, anti-inflammation and wound healing properties of collagen peptides derived from unicorn leatherjacket (Aluterus monoceros) at different hydrolysis. Turkish J Fish Aquat Sci. 2019;19(7):551–560.
[54] Pushpakom S, et al. Drug repurposing: Progress, challenges and recommendations. Nat Rev Drug Discov. 2019;18(1):41–58.
[55] Jourdan J-P, et al. Drug repositioning: A brief overview. J Pharm Pharmacol. 2020;72(9):1145–1151.
[56] Touret F, et al. In vitro screening of a FDA approved chemical library reveals potential inhibitors of SARS-CoV-2 replication. Sci Rep. 2020;10(1):13093.
[57] Druker BJ, et al. Efficacy and safety of a specific inhibitor of the BCR-ABL tyrosine kinase in chronic myeloid leukemia. N Engl J Med. 2001;344(14):1031–1037.
[58] Kort E, Jovinge S. Drug repurposing: Claiming the full benefit from drug development. Curr Cardiol Rep. 2021;23(6):62.

Nirjhar Saha, Soumili Biswas, Asim Kumar, Asit K. Chakraborti

12 Drug repurposing of selective COX-2 inhibitors in the pursuit of new therapeutic avenues

Abstract: Drug repurposing is an alternative and effective strategy in drug discovery. Compared to the de novo drug discovery approach, drug repurposing is a rapid process to identify new uses of existing therapeutic entities for other newly emerging and existing diseases. Drug repurposing approach often relies on some commonalities in the biochemical pathways involved in the pathogenesis of the disease against which a particular drug has been approved and the pathogenesis of the new disease against which the repurposing of the existing drug is planned for. It has been recently observed that COX-2 expression, popularly known to be associated with the pathogenesis of inflammatory disorder causing arthritis, is increased in other pathophysiological conditions such as cancer, CNS disorders, respiratory disorders, diabetic peripheral neuropathy, and various microbial infections. These findings form the basis for considering the repurposing of COX-2 selective agents/drugs against these diseases. Thus, there have been reports on the uses of various selective COX-2 inhibitors for drug repurposing in the combination therapy regimen for the treatment of cancer, neuronal disorders, and microbial infections. Some clinical trials for the repurposing of COX-2 inhibitors are still in progress. This chapter deliberates on the recent developments in drug repurposing of selective COX-2 inhibitors in the pursuit of new therapeutic avenues.

Keywords: Cyclooxygenase, drug Repurposing, selectivity, arthritis, inflammation

12.1 Introduction

Inflammation, a natural and crucial defense mechanism in biological systems, is the protective response by the host immune system against foreign bodies/stimuli (microorganisms, pollen, allergen, etc.) or injuries (chemical and physical) to restore the homeostatic condition of the human body. Inflammation is carried out by various mediators such as NF-kB [1], granulocyte colony-stimulating factor (G-CSF) [2], granulocyte-macrophage colony-stimulating factor (GM-CSF) [2, 3], cholinergic system [4–6], neutrophil [7–10], protein kinase signaling [11], signaling platforms [12], extracellular DNA traps [13], p38 MAP kinase [14–16], Reactive Oxygen Species (ROS) [17], and metalloproteinases [18]. Among these inflammatory targets/species, the enzyme cyclooxygenase (COX) is targeted widely to discover therapeutic agents for the treatment of anti-inflammatory disorder. Determination of the crystal structures of cyclooxygenase-1 (COX1) in 1994 [19] and cyclooxygenase-2 (COX2) in 1996 [20] opened up a new

https://doi.org/10.1515/9783110791150-012

and vast horizon to discover drugs for anti-inflammatory therapy. The adverse effects (gastric ulceration and delayed platelet aggregation) of aspirin-like drugs that act on both of the COX1 and COX2 enzymes, popularly referred to as nonsteroidal anti-inflammatory drugs (NSAIDs), brought down the level of enthusiasm in the search for newer non-selective NSAIDs. It was till the findings of the second isoform of the cyclo-oxygenase enzyme COX2 that the search for selective NSAIDs ultimately ushered in an exciting field of drug discovery, introducing an entirely new generation of anti-inflammatories [21]. Apart from inhibiting the inflammatory mediators, agonistic action on the anti-inflammatory mediators can also lead to the resolution of the inflammatory process. Such anti-inflammatory mediators are heme oxygenase-1 and carbon monoxide [22], cannabinoid [23], and lipid mediators [24]. Development of new chemical entities by the academia and the API industries face many obstacles, including the cost-benefit challenge, requirement of longer time and large manpower, adverse drug reaction challenges, toxicity challenges, dosage-formulation obstacles, and difficulty in large-scale production protocol setup. Another option to deliver effective treatments to patients is through the findings on new therapeutic applications of already known drugs, originally discovered for other disease indications, termed as the drug repurposing approach. The repurposing approach uses the FDA-approved tools (e.g., drugs, devices, and nutraceuticals) to apply them against diseases other than the actual indication. Therapies with repurposed drugs can be delivered to patients in a shorter time period and are less costly than new drug development [25, 26]. The perpetual interest of our research group in the search for novel anti-inflammatory compounds, particularly COX-2 inhibitors [27–33], led us to also focus our interest to the repurposing of selective COX 2 inhibitors in various diseases, other than inflammation and pain.

12.2 Drug repurposing

Drug repurposing (or drug repositioning or drug reprofiling or drug re-tasking) is an alternative and effective strategy in drug discovery and development program for identifying new therapeutic indications of the approved/investigational/abandoned drugs, other than its original medical indications [34, 35]. In the older days, drug repurposing was random and clinical/preclinical identification of on-target or off-target drug was considered for its commercial production. This practice leads to the repurposing of various unsuccessful drugs. Nowadays, more systematic approaches (e.g., in vitro and in vivo experimental investigation, bioinformatics, and cheminformatics) to identify repurposable drugs for another biological target have resulted in the successful invention of a number of repurposed drugs for the treatment of both common and rare diseases. However, the technical and regulatory problems are the crucial challenges in drug repurposing till now [35].

12.2.1 Advantages of drug repurposing

Drug repurposing is a faster and safer technology to identify the uses of existing therapeutic entities for new/existing diseases for which new/better medical treatment is required. It provides certain advantages compared to the de novo drug discovery approach.

The increasing cost and longer time period required to launch a new drug in the market deters the interest of investors to invest in the pharmaceutical industry. In contrast, the lesser investment and the shorter time that are needed to identify the repurposed drug encourage both academia and industry to pursue research on drug repurposing. However, the requirement of time span and cost vary from one drug candidate to another, depending on the type of protocol required for repurposing a particular candidate drug. The drug repurposing approach offers cost savings in the preclinical as well as clinical (phase I and phase II) studies as these have been already documented during the investigations of this drug for the original therapeutic use. However, in the regulatory and phase III clinical trials, the costs associated with the repurposed drug are similar to that for a new drug for a particular therapeutic indication [35]. The development of a new chemical scaffold costs approximately $2–3 billion with a time span of 13–15 years for its approval while the expenses for drug repurposing lies between $40 million and $80 million with a required time span of 3–12 years [36]. The success rate of the drug repurposing approach (~30% of new FDA-approved drugs) [34] is also higher than that of the new drug discovery. This expenditure reduction, time saving, and higher success rate draw the attention of the pharmaceutical companies to invest in the drug repurposing program. In addition, various phases of the drug discovery pipeline, such as formulation development, and clinical and pre-clinical trials (safety and efficacy assessment) can be bypassed in the case of the repurposed drug candidates as the data for the investigations of these steps/stages were previously documented at the time of investigation against the original therapeutic indication(s) [34]. In the case of drug repurposing, the risk of clinical failure is lower than that of a new chemical entity discovery because the repurposed drug was approved for its original therapeutic use after evaluating its therapeutic safety parameters in clinical trials. Further, the understanding of the mechanism of action of the repurposed drugs can be of help to invent new drug targets and mechanistic pathways for future research purposes that might broaden the scope for therapeutic applications.

Thus, the aforementioned advantages make the drug repurposing process less risky and more beneficial with respect to the time span and cost needed for drug development, and it generates scope for rapid return of profit for the investors on repurposed drugs [35].

12.2.2 Approaches used for drug repurposing

Drug repurposing protocol follows three fundamental steps to declare the candidate drug as "repurposed drug": a) hypothesis generation for the drug candidate molecule against a biological target utilizing the modern drug design techniques/approaches; b) investigation of the mode of action and effect of the candidate drug in preclinical trials; and c) efficacy assessment in phase II clinical trials. Two common approaches of drug repositioning are: wet-lab experimental and computational approaches [35]. The wet-lab experimental approaches for drug repurposing involve in vitro and in vivo assays to find out the efficacy and binding interactions of the drug molecules to the desired target and to find out the lead repurposed drug candidate from the selection of drugs. Computational approaches utilize cheminformatics, bioinformatics, artificial intelligence, network biology, and systems biology tools, and gear up the drug repurposing process. Both of these approaches consider already explored/approved drugs, drug targets, and disease pathways to reduce the time period of the drug repurposing protocol [34]. Repurposing of the approved drugs has become an attractive/additional strategy for the treatment of various inflammatory diseases (e.g., ankylosing spondylitis, psoriatic arthritis, rheumatoid arthritis, etc.), primary Sjögren syndrome, and systemic lupus erythematosus [36].

12.2.3 Examples of drug repurposing

Historically, thalidomide, a sedative drug used in the first trimester of pregnancies was unexpectedly found to be active against erythema nodosum leprosum, ENL1, in 1964 and against multiple myeloma in 1999. Sildenafil, an antihypertensive drug, was repurposed and marketed for erectile dysfunction, with the brand name of Viagra in 2012 [35]. Miltefosine, the phospholipid analogue, was originally invented for breast cancer and solid tumor therapy in 1980s [37, 38]. It enhances the cancer cell apoptosis activity through the inhibition of the enzyme phosphocholine cytidylyltransferase, and subsequently, the downregulation of phosphatidylcholine biosynthesis, which is required for cell survival [39, 40]. However, miltefosine could not get through the phase II clinical trial and was not approved by FDA for anticancer therapy [41]. In the next attempt, it was repurposed for the therapy of visceral leishmaniosis in 2002 [42, 43]. In recent days, miltefosine is the only available oral medication for the antileishmanial therapy. Apart from the repositioning against leishmanial activity, miltefosine was also repurposed as the vaccine adjuvant for influenza [44]. The antibiotic, cycloserine, was earlier indicated for UTI [45–47] and antitubercular therapy [48, 49]. Due its partial agonistic property towards the NMDA receptor [50–52], it is now repurposed for the management of various neuropsychiatric and neurological conditions, such as depressive disorders [53] and substrate reduction therapy in Krabbe disease [54]. Likewise, later on, various approved or abandoned drugs were successfully repurposed for both common and rare diseases. A few repurposed drugs are found to be effective against multiple diseases and a particular single disease can also be treated with several repurposed therapeutic agents. Representative examples of repurposed drugs are compiled in Table 12.1, with reference to the original and repurposed therapeutic indications of the particular drug.

12.3 Cyclooxygenase (COX) inhibitors

Cyclooxygenase (COX) is a heme-containing membrane-bound protein that catalyzes the conversion of arachidonic acid (AA) to PGG_2 and prostaglandin H_2 (PGH_2). It has three isoforms COX-1, COX-2, and COX-3. The detailed description of these enzymes and the names of their inhibitors are enlisted in Table 12.2. Prostaglandin-endoperoxidesynthase-2, also known as cyclooxygenase-2 or COX-2, is an enzyme that is encoded in humans by the *PTGS2* (COX-2) gene. It is involved in the conversion of AA to PGH_2, an important precursor of prostanoids. The PGHSs are targets for NSAIDs- and COX-2-specific inhibitors (COXibs). Each monomer of the enzyme has a peroxidase and cyclooxygenase active site. These enzymes catalyze the conversion of AA to prostaglandins in two steps. First, hydrogen is abstracted from C13 of AA, and then two molecules of oxygen are added by the COX-2, giving PGG_2, which is reduced to PGH_2 in the peroxidase active site. The synthesized PGH_2 is converted to prostaglandins (PGD_2, PGE_2, and $PGF_{2\alpha}$), prostacyclin (PGI_2), and thromboxane A_2 by tissue-specific isomerases [85]. COX-3 is an enzyme, discovered in 2002, encoded by the COX-1 gene, but is not functional in humans. COX-3 isoform is the third and most recently discovered COX enzyme isoform [86, 87]. It can be selectively inhibited by paracetamol, phenacetin, antipyrine, and dipyrone.

The NSAIDs were discovered for the treatment of inflammation and pain. The structures of some traditional (non-selective) NSAIDs are provided in Figure 12.1. In 1970s, it was revealed that aspirin and aspirin-like drugs (i.e., traditional NSAIDs) exhibit their anti-inflammatory and analgesic activities through inhibiting the cyclooxygenase enzyme and prostaglandin production [88, 89]. Later on, the discovery of COX-2 enzyme and the determination of COX-2 crystal structure opened up the field of COX-2 inhibitors.

Table 12.1: Examples of some repurposed drugs.

Drug [ref]	Original indication	Repurposed indication	Drug [ref]	Original indication	Repurposed indication
Repurposed for neuronal disorders					
Tetrabenazine [55]	Movement Disorders	Tourette syndrome	Latrepirdine [35]	Antihistamine	Huntington disease
Bumetanide [56]	Edema, High Blood Pressure	Autism	Ceftriaxone [35]	Antibiotic	Amyotrophic lateral sclerosis
Fenfluramine [57]	Weight loss	Dravet Syndrome	Fingolimod [59]	Transplant rejection	MS [35]
Omega Fatty Acids [58]	Dietary Supplement	Autism and ADHD	Zardaverine [25]	Selective PDE3/4 inhibitor	FXS [60]
Guanfacine [25]	ADHD	Tourette Syndrome	Metoprolol, Penbutolol [25]	beta blocker	FXS [61]
Clioquinol [25]	antifungal	Alzheimer's disease	Sulindac [25]	NSAIDs	FXS, PPAR-g agonist [62]
Atomoxetine [25]	Parkinson's disease	ADHD	Topiramate [25]	Anti-epileptic drug	FXS
Atomoxetine [35]	Parkinson disease	ADHD	Phenformin [25]	Antidiabetic drug	FXS [63]
Saracatinib [35]	Experimental anticancer drug	Alzheimer disease Cancer-induced bone pain Lymphangioleiomyomatosis Psychosis	Quercetin [25]	Antioxidant	'pre-FXS' [64]
LY500307 [35]	Benign prostatic hyperplasia	Schizophrenia	Disulfiram [60]	Chronic alcoholism	FXS, MMP9 inhibitor [65], NF-kB inhibitor [66]

Repurposed for *Mycobacterium abscessus* pulmonary infection					
Rifabutin [67]	anti-Mycobacterium avium infection	*Mycobacterium abscessus* pulmonary infection	Avibactam [67]	UTI, in combination with ceftazidim	*Mycobacterium abscessus* pulmonary infection
Tedizolid [67]	acute bacterial skin infections	*Mycobacterium abscessus* pulmonary infection	Omadacycline [67]	Pneumonia and skin infection	*Mycobacterium abscessus* pulmonary infection
Bedaquiline [67]	MDR TB	*Mycobacterium abscessus* pulmonary infection	Disulfiram [67, 68]	Chronic alcoholism	*Mycobacterium abscessus* pulmonary infection
Orlistat [67, 69]	Anti-obesity		D-cycloserine [67, 70]	Anti-Mycobacterium tubercular infection	*Mycobacterium abscessus* pulmonary infection
Repurposed for anti-cancer and anti-tumor activity					
Raloxifene [35]	Osteoporosis	Breast cancer	Brivudine [71]	antiviral	Anticancer
Aspirin [35]	Analgesia	Colorectal cancer	Celecoxib [72]	Anti-inflammatory	Against colorectal carcinoma
Ribavirin [35]	Antiviral	Acute myeloid leukemia and breast cancer	Metformin [73] Doxorubicin [73]	Antidiabetic anticancer	Against tumor growth and metastasis
Nelfinavir [35]	HIV	Various cancers			
Repurposed for anti-aging					
Metformin [74]	antidiabetic	Anti-aging (topical application)	Rapamycin [75]	Antibiotics	Anti-aging
Repurposed for miscellaneous disease conditions					
Duloxetine [35]	Depression	SUI	Rituximab [35]	Various cancers	Rheumatoid arthritis
Zidovudine [35]	Cancer	HIV/AIDS	Dapoxetine [35]	Analgesia and depression	Premature ejaculation
Minoxidil [35]	Hypertension	Hair loss	Topiramate [35]	Epilepsy	Obesity

(continued)

Table 12.1 (continued)

Drug [ref]	Original indication	Repurposed indication	Drug [ref]	Original indication	Repurposed indication
Sildenafil [35]	Angina	Erectile dysfunction	Ketoprofen [76]	Anti-inflammatory	Treatment of glioblastoma multiforme
Thalidomide [35]	Morning sickness	Erythema nodosum leprosum and multiple myeloma	Miltefosine [77]	Anticancer	Treatment of schistosomiasis mansoni
Celecoxib [35]	Pain and inflammation	Familial adenomatous polyps	Disulfiram [78]	Alcoholic syndrome	Treatment of glaucoma
Topiramate [35]	Epilepsy	Inflammatory bowel disease	Nilvadipine [79]	Antihypertensive	Treatment of retinopathy
PF-05190457 [35]	Ghrelin receptor inverse agonist	Alcoholism	Atorvastatin [80]	Antihyperlipidemic	Edema
Denosumab [35]	Osteoporosis	Crohn's disease	Simvastatin [81,82]	Antihyperlipidemic	Bone regeneration
Ketoconazole [35]	Fungal infections	Cushing syndrome	Fasudil [83] Superoxide Dismutase [83]	Aubarachnoid hemorrhage Antioxidant	Treatment of pulmonary Arterial hypertension
AZD4017 [35]	11β-HSD1 inhibitor	Idiopathic intracranial hypertension	Ebselen [84]	anti-inflammatory	Antifungal (topical)
AZD4901 (MLE4901) [35]	Neurokinin 3 receptor antagonist	Menopausal hot flushes			

Table 12.2: Classification of COX enzymes [21, 90].

Gene	Enzyme	Tissue expression	Functions	Inhibitors	Comments
COX1	COX-1	Constitutively expressed in most tissues	Platelet aggregation, GI protection, production of vascular prostacyclin	Most 'classical' NSAIDs; some selective inhibitors	First identified COX
COX1	COX-3	Brain, heart and aorta; constitutive?	Pain perception?	Paracetamol, diclofenac ibuprofen, dipyrone phenacetin, antipyrine	Still to be explored
COX2	COX-2	Induced by many stimuli, including growth factors, cytokines, oxidative stress, brain hypoxia or seizures, and other forms of injury or stress; constitutively present in the brain, kidney, and elsewhere	Inflammation, fever, some pain, parturition, and renal function	Many NSAIDs, COX2-selective drugs, such as the coxibs and others	

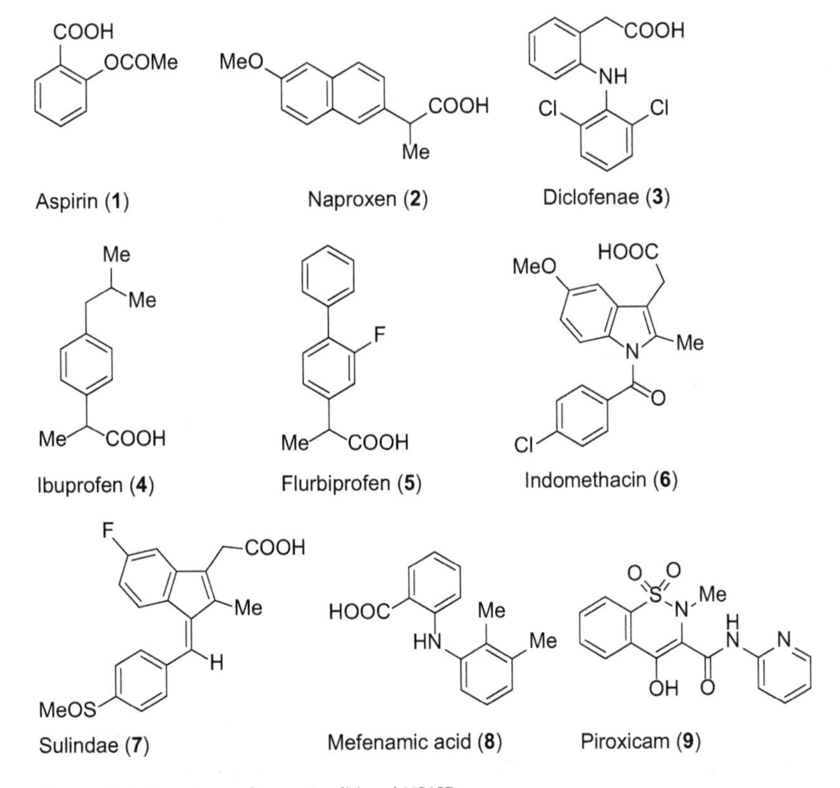

Figure 12.1: Structures of some traditional NSAIDs.

12.3.1 Adverse Drug Reaction (ADR) of nonselective COX inhibitors

The mostly observed adverse drug reactions (ADRs) associated with nonselective NSAIDs are ulceration and bleeding in the gastrointestinal (GI) tract. The acidic functionality of the nonselective NSAIDs causes direct irritation in the gastric mucosa and also inhibits the COX-1-mediated production of mucosa-protective prostaglandin PGE_1. Decrease in the level of PGE_1 in the GI tract elevates the gastric acid level, reduces the bicarbonate and mucus secretion, and attenuates the trophic effects on the epithelial mucosa. All these physiological outputs lead to the erosion of the gastric mucosa and to ulceration and bleeding in future. Clinically observed symptoms of the gastrointestinal side effects are nausea/vomiting, dyspepsia, gastric ulceration/bleeding, and diarrhea. Ulcer formation due to the intake of the nonselective NSAIDs is independent of the route of administration of the drug (e.g., oral, rectal, or parenteral) and is relatable to the mechanism of action of the drug. The possibility and extent of ulceration depends on the drug molecule, duration of the therapy, and the dosage amount. For example, indomethacin, ketoprofen, and piroxicam cause more gastric side effects than ibuprofen and diclofenac. Gastric ADRs can be minimized by the simultaneous intake of a proton pump inhibitor or misoprostol (prostaglandin analogue) [92]. Another clinically observed adverse drug reaction associated with nonselective NSAIDs is photosensitivity reactions. This side effect is prevalent with the use of 2-arylpropionic acids, diclofenac, and benoxaprofen. Decarboxylation of the nonselective COX inhibitors is the cause of photosensitivity reactions. The absorbance type of the molecular chromophore is directly correlated with the extent of photosensitivity reactions by the drugs. For example, weak absorption by ibuprofen makes it a lesser photosensitizing agent [93]. Nonselective COX inhibitors-related renal dysfunction is the observed nephrotoxicity caused through the inhibition of renal vasodilatory prostaglandins. In some cases, ingestion of nonselective NSAIDs may cause renal failure, depending on the age of the patients, cardiovascular conditions, previous occurrence of renal failure, comorbidities, and plasma-volume contraction. These potential drawbacks of nonselective COX inhibitors drove the drug discovery program toward the invention of selective COX-2 inhibitors that are devoid of the abovementioned side effects.

12.3.2 Selective COX-2 inhibitor

Selective COX-2 inhibitors (also known as COXibs) are classified according to their chemical classes such as methanesulfoanilides, di-aryl heterocyclics, acetic acid derivatives, and analogues, derived from nonselective inhibitors (Figure 12.2).

Depending on the extent of selectivity to inhibit the COX isoforms, the NSAIDs are classified into four classes (Table 12.3). This classification includes both nonselective and selective COX inhibitors.

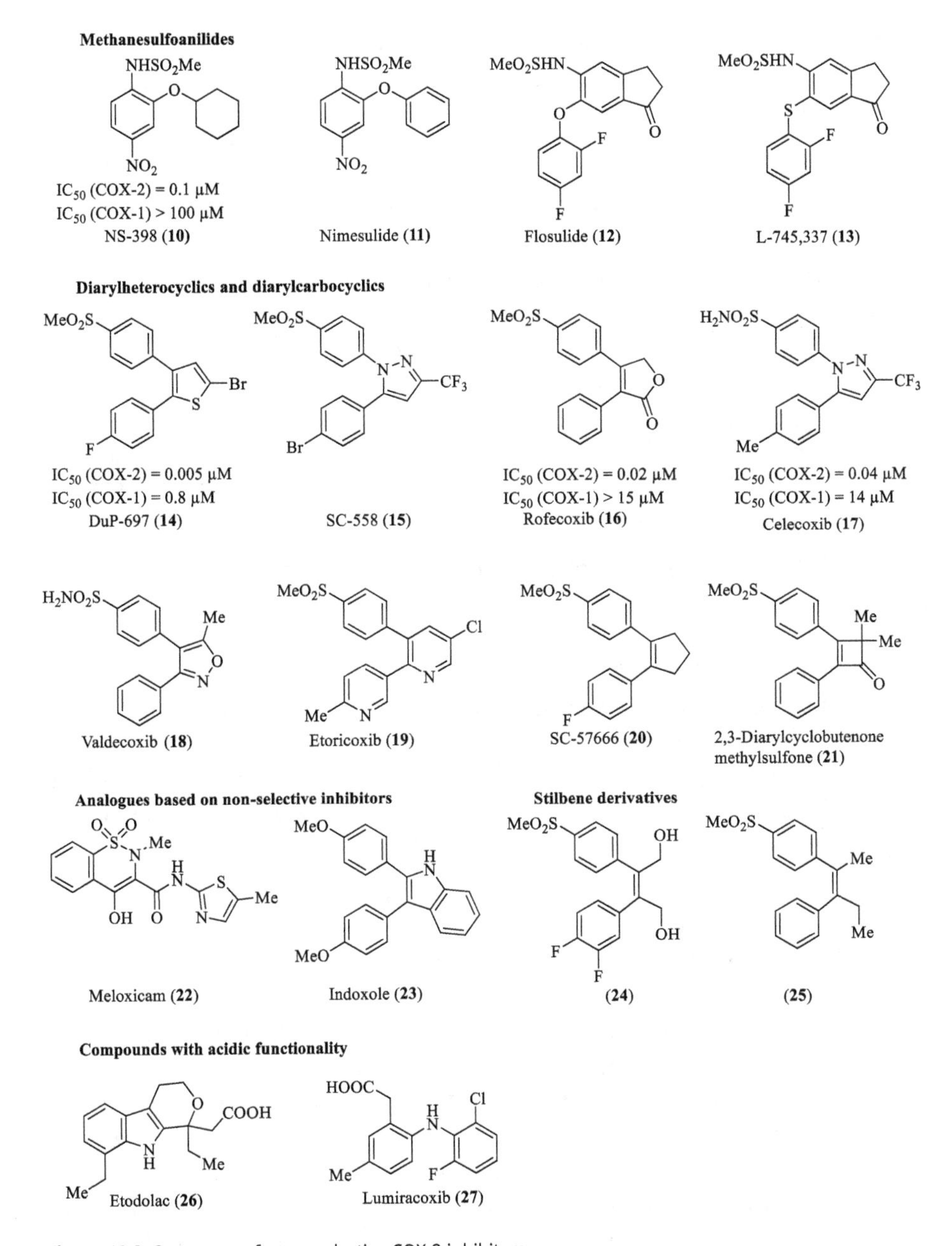

Figure 12.2: Structures of some selective COX-2 inhibitors.

Table 12.3: Classification of NSAIDs based on COX inhibition selectivity [21, 91].

Class	Properties	Examples
Group 1	NSAIDs that can completely inhibit both COX1 and COX2 with little selectivity	Aspirin, diclofenac, fenoprofen, flurbiprofen, indomethacin, ibuprofen, ketoprofen, mefenamic acid, naproxen, piroxicam, and sulindac sulfide
Group 2	NSAIDs that inhibit COX2 with a 5–50-fold selectivity	Celecoxib, etodolac, and meloxicam
Group 3	NSAIDs that inhibit COX2 with a > 50-fold selectivity	Rofecoxib
Group 4	NSAIDs that are weak inhibitors of both isoforms	5-amino salicylic acid, diflunisal, sodium salicylate, nabumetone, and sulphasalazine

12.3.3 Potential benefits of COXibs as anti-inflammatory and analgesic agents

The discovery of COX-2 selective agents (COXibs) provided certain benefits to the anti-inflammatory and analgesic treatment regimen over the nonselective COX inhibitors. These include better analgesia, reduced occurrence of gastrointestinal side effects, long duration of drug action, preoperative application for pain management, opioid sparing effect, synergistic effects with opioids, and no platelet inhibition. The current status of the selective COX-2 inhibitors is provided in Table 12.4.

12.4 Relevance of COX-2 in various diseases

Nowadays, researchers have found that COX-2 expression is increased in different pathophysiological conditions (Figure 12.3), such as cancer (colorectal cancer [98, 99], pancreatic cancer [100, 101], ovarian cancer [102], hepatic cancer [103], prostate cancer [104], breast cancer [105, 106], endometrial cancer [107], lung cancer [108], oral squamous cell carcinoma [109], esophageal cancer [110], urinary bladder cancer [111]), CNS disorders (Alzheimer's disease [112], parkinsonism [113], schizophrenia [114], depression [115], epilepsy [116] etc.), respiratory disorders (Chronic obstructive pulmonary disorder [117, 118], asthma [119]), arthritis (rheumatoid [120] and osteoarthritis [121]), diabetic peripheral neuropathy [122, 123], various viral infections [124–133], and tuberculosis [134].

Table 12.4: Current status of COXibs in the treatment of inflammatory disorder [94–97].

Drug (Brand name)	Structural class of the drug	Manufacturer (Time span in the market)	IC_{50} (COX-1) (A)	IC_{50} (COX-2) (B)	COX-2 selectivity ratio (A/B)	Toxicity	Reason of Toxicity
Celecoxib (Celebrex)	Pyrazole	Pfizer (1999-till now)	6.7 ± 0.9	0.87 ± 0.18	7.6	Cardiotoxicity & Interstitial nephritis	Blockade of COX-2 results in increased level of TXA_2 and hence vasoconstriction and platelet aggregation.
Valdecoxib (Bextra)	Isoxazole	G.D Searle & Company (2001–2005)	26.1 ± 4.3	0.87 ± 0.11	30	Cardiotoxicity	Blockade of COX-2, resulting in increased level of vasoconstricting agents in blood.
Rofecoxib (Vioxx)	Furan-2-one	Merck & Co. (1999–2004)	18.8 ± 0.9	0.53 ± 0.02	35	Cardiotoxicity	Blockade of COX-2, resulting in increased level of TXA_2.
Etoricoxib (Arcoxia)	Bi-pyridine	Merck & Co. (Not approved due to high risk-benefit ratio but clinical researches are still going on)	116 ± 18	1.1 ± 0.1	106	Cardiotoxicity & Epidermal necrosis	Blockade of COX-2 causes cardiotoxicity, and excessive activation of the effector memory, T-cell resident, causes epidermal necrosis.
Lumiracoxib (Prexib)	Phenyl acetic acid	Novartis 2006–2007	--------	--------	515	Hepatotoxicity	Formation of an adduct of Quinone-imine with N-acetyl cysteine after bioactivation of lumiracoxib.

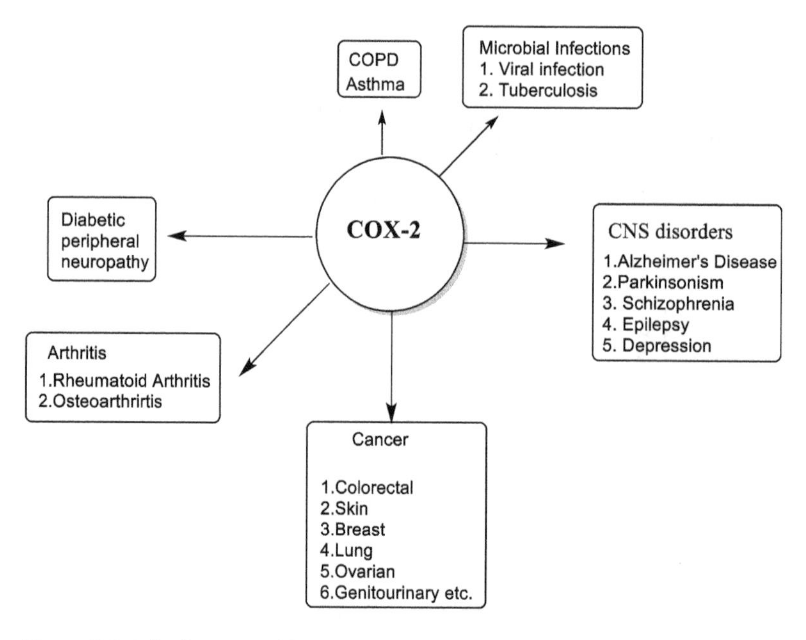

Figure 12.3: Role of COX-2 in different pathophysiological conditions.

12.4.1 Role of COX-2 in cancer

The COX enzyme in involved in the production of prostaglandin that has a role in the progress of colon cancer [135]. Later on, the role of prostaglandin was investigated and it was found that overproduction of COX-2 enzyme is responsible for the development of cancer cells [136]. Many experiments have revealed that COX-2 enzyme plays an important role in angiogenesis [101, 137], increased cell proliferation, and reduced apoptosis [138].

12.4.2 Role of COX-2 in the pathogenesis of neurodegenerative diseases

Up till now it was thought that COX-2 is involved only in inflammatory related responses but recently, the role of COX-2 has been established in the development of various pathophysiological conditions, other than inflammatory pain [139], such as schizophrenia [114], depression [115], epilepsy [116], Parkinsonism [140], ischemic brain injury [141], and diabetic peripheral neuropathy [142]. Clinical studies indicate that COX-2 is involved in the development and progression of various neurodegenerative diseases such as Alzheimer's disease, schizophrenia, Parkinsonism, amyotrophic lateral sclerosis [143], and multiple sclerosis [144]. Additionally, it was observed that

COX-2 expression was upregulated in the brain neurons in Alzheimer's disease (AD) [145]. COX-2 retards the clearance of amyloid beta peptides, which is involved in the brain neuro-degeneration cascade [146]. Amyloid-β-peptide aggregates, synthesized from amyloid precursor protein (APP), trigger neurotoxic microglia through toll-like receptors and RAGE (receptor for advanced glycation end products). These result in the further activation of the transcription factors (e.g., NF-κB and AP-1) to produce reactive oxygen species (ROS) and cytokines. These inflammatory mediators exhibit neurotoxic effects on cholinergic neurons via amplification of pro-inflammatory signals and stimulating astrocytes (Figure 12.4) [147–149].

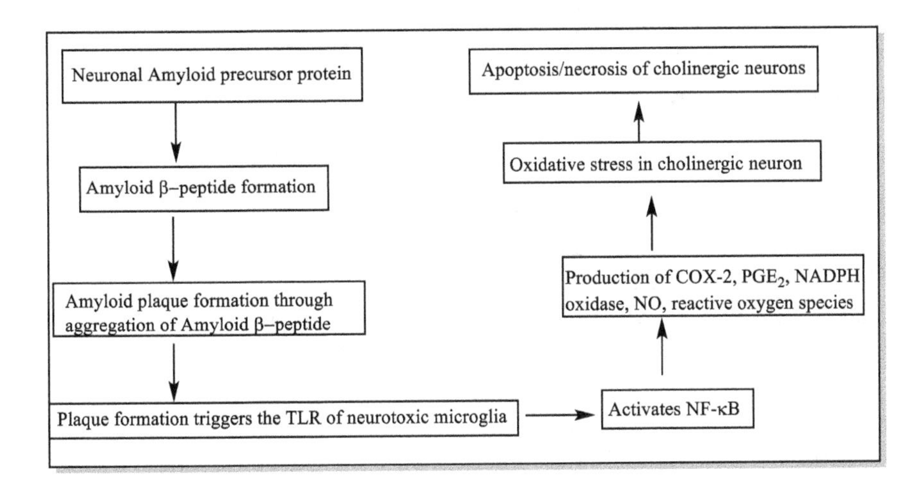

Figure 12.4: Role of COX-2 in Alzheimer's disease.

Neuroinflammatory process results in neurodegeneration and the subsequent loss of cognitive functions. COXibs are being explored for the treatment of neurodegenerative disorders [150, 151]. COX-2 and PGE_2 expressions were also found upregulated under some of the other psychiatric conditions like schizophrenia and depression. The COX-2 enzyme is found to be involved in the pathogenesis of schizophrenia, which has been proven by various clinical studies. During the last decade, several clinical studies were performed that showed the efficacy of celecoxib in reducing the progression of schizophrenia [152]. Epilepsy is a neuronal disorder in brain, characterized by repetitive seizure attacks. It has been reported that transcriptional activation of the genes encoding COX-2 was upregulated in patients with epileptic condition [116]. The increased level of COX-2 leads to neurodegeneration; therefore, a selective COX-2 inhibitor could act as a novel and alternative therapy to treat epilepsy. In fact, etoricoxib, a selective COX-2 inhibitor has shown promising results in the animal model [153].

12.4.3 Role of COX-2 in COPD

Chronic obstructive pulmonary disease (COPD) is an obstructive lung disease associated with chronically low air flow. The COX-2 enzyme was found to be over expressed in lung fibroblasts of patients with COPD, suggesting a role of COX-2 in the pathophysiology of COPD. The enzyme COX-2 catalyzes the conversion of arachidonic acid to thromboxanes (TXA_2) and various prostaglandins (PG), such as PGE_2, which cause the constriction of smooth airways, suggesting therapeutic benefits of COX-2 inhibitors against COPD. As a matter of fact, celecoxib, a COX-2 inhibitor in clinical use, is effective in treating COPD models [154].

12.4.4 Role of COX-2 in diabetes

Diabetes, a metabolic disorder, often leads to peripheral neuropathy, which is a severe peripheral nerve dysfunction, affecting 30–50% of diabetic patients [155]. Convincing evidences are emerging from various clinical studies that establish the role of COX-2 in causing nerve conduction velocity deficit and oxidative stress [156]. Selective inhibition of pro-inflammatory COX-2 prevents the progression and occurrence of diabetic peripheral neuropathy [157].

12.4.5 Role of COX-2 in viral infections

The level of COX-2 enzyme is elevated due to infection caused by the dengue virus (DENV) [124], enterovirus (EV) [125], hepatitis C virus (HCV) [126], hepatitis B virus (HBV) [127], cytomegalovirus [128], HIV [129], herpes simplex virus type 1 [130], bovine leukemia virus [131], human T-lymphotropic virus type 1 [132], and H5N1 virus [133]. Various signaling pathways, including NF-kB, MAPK, and JNK [125, 158, 159] are activated due to the viral infection. DENV and EV71 promote the phosphorylation of NF-kB and JNK. The binding of phosphorylated NF-kB and CEBPβ with the COX-2 promoter enhances the COX-2/PGE_2 production (Figure 12.5) [124, 125]. PGE_2 enhances the activity of NS5 polymerase, which is responsible for DENV replication [124]. In the case of EV71, PGE_2 activates the cAMP-PKA signaling pathway to promote viral replication (Figure 12.5) [125].

HBx (hepatits B viral protein) protein directly interacts with CEBPβ and helps in the binding of CEBPβ with the COX-2 promoter to increase the production of COX-2/ PGE_2 [127]. Entry of HCV into hepatocytes causes the activation of the NF-kB and MAPK signaling pathway, which leads to an increased level of COX-2 and PGE_2 [126]. PGE_2 also induces hepatocyte proliferation [160]. Thus, PGE_2 prepares a platform for the hepatitis (B and C) virus to replicate (Figure 12.6).

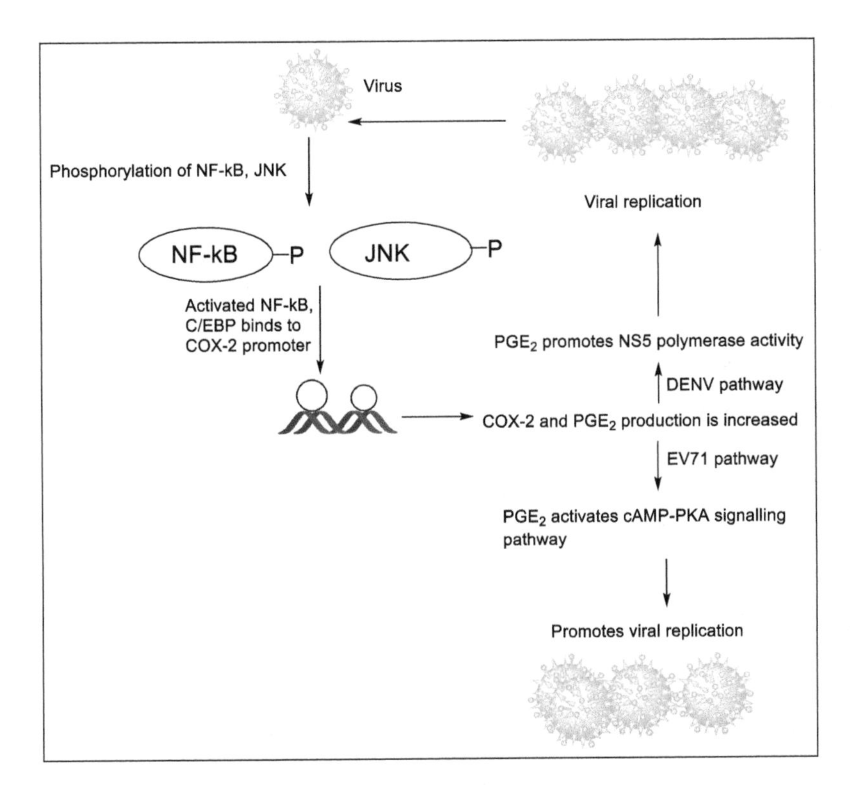

Figure 12.5: Role of COX-2 in DENV and EV replication.

12.4.6 Role of COX-2 in tuberculosis

Studies revealed that the COX-2 enzyme may also play a role in the pathogenesis of tuberculosis. ESAT-6 (antigenic protein) of *mycobacterium tuberculosis* acts on the toll-like receptor (TLR2) and induces the COX-2 expression through MAPK signaling pathway [161, 162]. *M. tuberculosis* can also increase the level of PGES in the host. Increased levels of COX-2 and PGES elevate the production of PGE_2. High concentration of PGE_2 reduces the host immunity [163] by a decrement of the Th1 cytokine level [164–167], inhibiting lymphocyte proliferation, causing intrinsic intracellular killing of the bacteria [168,169], and activating macrophage [170]. Hence, PGE_2 helps in immunopathogenesis and progression of tuberculosis.

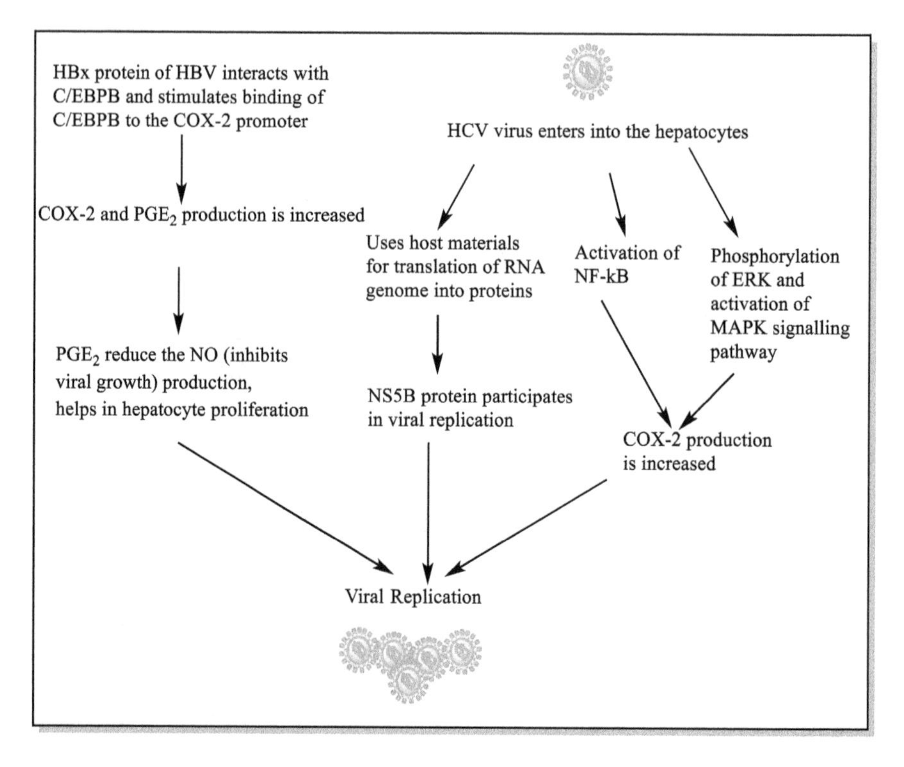

Figure 12.6: Role of COX-2 in HBV and HCV replication.

12.5 Drug repurposing of COX-2 inhibitors

The involvement of COX-2 in the development and progression of various pathophysiological conditions leads to the logical application of selective COX-2 inhibitors in the treatment of various diseases, other than arthritis. For a particular therapeutic target, a few selective COX-2 inhibitors are also repurposed, depending on the structural motif similarities between the already marketed standard drugs and COXib. In a few examples of drug repurposing, selective COX-2 inhibitors were repositioned only on an empirical basis. Such examples of repurposing of selective COX-2 inhibitors are discussed in the following sections.

12.5.1 COX-2 inhibitors in cancer

12.5.1.1 Celecoxib in breast cancer

Celecoxib, alone or in combination with other chemotherapeutic agents, was utilized in the treatment of breast cancers. Combination therapy is more advantageous in terms of higher therapeutic efficacy, less possibility of cardiovascular risks, and pharmacoresistance. The synergistic effect in combination therapy of celecoxib and curcumin (a bioactive natural triterpenoid) augments the antitumor effect of celecoxib on breast cancer via COX-2 inhibition, and induction of apoptosis [171]. Endoplasmic reticulum (ER) stress response system reduces the apoptotic process and in turn facilitates the tumor cell survival and development of chemoresistance against therapeutic agents. In contrast, the severe aggravation of ER stress augments the proapoptotic module. A combination of two marketed drugs, viracept (nelfinavir, HIV protease inhibitor) and Celebrex (celecoxib, COX-2 inhibitor), on the enhancement of ER stress via different mechanisms was explored for the death of breast cancer cells. Celecoxib enhances the antitumor effect of viracept in a COX-2-independent manner via the escalation of ER stress, and this synergistic effect can be explored in future in the treatment of advanced and drug-resistant breast cancers [172].

Successful delivery of celecoxib to the desired therapeutic site is still the challenging factor for drug discovery scientists [173–175]. However, the development of various celecoxib nanoformulations mitigated the problems associated with celecoxib delivery. Nanoscale drug carriers were employed for efficacious drug delivery in the treatment of various types of cancer [176, 177]. Physically stable PEGylated solid lipid nanoparticles, nanostructured lipid carriers, and nanoemulsion of celecoxib were formulated utilizing the hot high pressure homogenization technique for use in the treatment of breast cancer and acute promyelocytic leukemia. Lipid nanoformulations of celecoxib contributed to the controlled drug delivery to provide an efficient therapy. A mechanistic study of their activity on cancer cells suggests that the mechanism of action of these nanoformulations proceeds via the MK pathway, combined with other pathways such as drug-efflux proteins and COX-2 pathways [178]. In another attempt, oily core-based nanocarriers loaded with letrozole and celecoxib show enhanced antitumor activity against breast cancer cells [179].

12.5.1.2 Celecoxib in cervix cancer

Clinical strategies for targeting cervical carcinomas can be improved by adjuvant therapy of celecoxib and its analogue, dimethylcelecoxib (DMC), in combination with canonical anticancer drugs. This combination therapy exerts an effective inhibition of human cervix HeLa multicellular tumor spheroids (MCTS). These may provide more effective alternatives for cervix cancer treatment [180].

12.5.1.3 Celecoxib in oral carcinoma

Oral cancers with high morbidity and mortality are still a challenge in public health management [181]. High concentration of intracellular proline and its metabolic factors are involved in the progression of oral cancer. Celecoxib attenuates proline metabolism, induces apoptosis cascade through elevation of pro-apoptotic factors (e.g., PRODH/POX and PPARγ), and reduces HIF-1α. All these suggest the potential of celecoxib in the prohibition of tumor growth and survival [182]. Celecoxib prevents the occurrence of oral cancer in dose- and time-dependent manner. Gradual increment of the daily dose and medication time of celecoxib improves oral cancer prevention. Celecoxib inhibits EMT and cell mobility through blocking of various transcription factors, cytoplasmic mediators, cell adhesion molecules, and surface receptors. These findings offer an alternative therapeutic strategy for oral cancer [183].

12.5.1.4 Celecoxib in glioma

Glioma, a brain tumor, starts in the glial cells of the central nervous system. Celecoxib is capable of improving the health conditions of glioma patient while used as combination therapy or as single therapy. Celecoxib and simvastatin were tethered to the dendrimer to target the cancer cells through biotin and R-glycidol. In vitro anticancer activity studies against human glioblastoma cells indicated that dendrimer conjugate, having celecoxib and simvastatin together, was more active than the single therapy of each drug, or other combination therapies [184]. Another approach of treating cancers is to target programmed cell death or its ligand. High concentration of PD-L1 affects the proliferation of the cancer cell and its invasiveness. Combination therapy of celecoxib with anti-PD-1 antibody can be an alternative therapeutic strategy of glioblastoma. Celecoxib downregulates the PD-L1 expression, via FKBP5, to exhibit its antitumor effects [185].

12.5.1.5 Celecoxib in urothelial carcinoma

Celecoxib and MLN4924 synergistically induce cytotoxicity and inhibit EMT in urothelial carcinoma (UC). This combination therapy suppresses growth and invasiveness of urothelial cancer, and increases apoptosis. Celecoxib also inhibits the MLN4924-activated AKT-ERK pathways. These experimental evidences suggest that celecoxib can be utilized as a combination therapy for urothelial cancer treatment [186].

12.5.1.6 Celecoxib in ovarian and prostate cancer

Celecoxib inhibited ovarian cancer cell proliferation through different modes of actions after 72 h of administration. Celecoxib arrests the G1 phase of the cell cycle, induces apoptosis, inhibits cellular adhesion-invasiveness properties, and reduces the expression of various proteins (e.g., hTERT mRNA, MMP9 protein, and COX2 protein) in the ovarian cancer cells. Celecoxib downregulates the VEGF level and in turn reduces the blood vessel density around the ovarian cancer microenvironment to cut off the angiogenesis process [187].

COX-2 and E-cadherin are expressed in the cancer cells in an opposite manner. If COX-2 is downregulated, E-cadherin is increased in the ovarian cancer cells. This phenomenon was exploited for anticancer therapy by celecoxib. Celecoxib inhibited the snail nuclear translocation through COX-2 prohibition and simultaneously elevated the level of E-cadherin in EOC, resulting in the decline of invasiveness of the EOC cells. Therefore, blocking the COX-2-snail signaling pathway can be beneficial in the treatment of invasive and metastatic epithelial ovarian cancer [188]. Clinical concentrations of celecoxib inhibit the growth of prostate cancer cells by a COX-2-independent mechanism [189].

12.5.1.7 Celecoxib in CNS lymphomas

The number of clinical patients (older in age) associated with central nervous system lymphomas is gradually increasing day by day. Despite treatment with an effective therapeutic protocol (i.e., methotrexate with whole brain radiation therapy), this type of lymphoma relapses in a short period of time and becomes fatal in some cases. As most of the older patients are usually immunocompromised, a combination of chemotherapy with radiation therapy is not suitable for them. To overcome this problem, celecoxib (which can inhibit lymphoma cell growth) can be added to the treatment regimen to prevent the relapse and improve the treatment protocol. Mechanistically, celecoxib augments the apoptic process through declination of bcl-2 and surviving the expression in the cancer cells [190].

12.5.1.8 Celecoxib in colon cancer

Piperine, a natural chemo-preventive agent, acts as drug bioavailability enhancer. A combination therapy of celecoxib with piperine inhibited colon adenocarcinoma via downregulation of COX-2 activity. Piperine synergizes celecoxib-induced inhibition of cancer cell growth and promotes apoptosis. The combination therapy works through β-catenin degradation, Axin stabilization, downregulating the Wnt/β-catenin signaling molecules, and subsequently arresting the cell cycle at the G1 phase [191].

12.5.1.9 Celecoxib in colorectal cancer

The prevalence of colorectal cancer is one of the burdens on public health. Overexpression of COX-2 regulates the colorectal cancer stem cell (CSC). Within the NSAIDs, celecoxib is most potent to inhibit c-Met in a COX-2 independent manner for colorectal cancer prevention. Hence, c-Met plays a critical role in the development and progression of colorectal cancer [192]. Celecoxib inhibited colorectal cancer cell growth *via* blocking of miRNAs, which are the potential targets of colorectal cancer progression [193]. The combination of celecoxib with cepharanthine at different doses inhibits the colorectal cancer synergistically through different mechanisms, such as (a) cell cycle inhibition at the G1 phase, (b) aggravation of p21 mRNA levels, (c) decline in cyclin-A2 mRNA levels, (d) initiation of apoptosis cascade through escalating the mRNA expression, BAX and reducing the mRNA expression for Bcl-xL. CLX-CEP combinations had synergistic cytotoxic and apoptotic effects, suggesting that their combination is useful for reducing their toxicity and side effects when treating colorectal cancer in humans. This result suggests the application of celecoxib as an adjuvant therapy without enhancing the side effect of each of the drugs [194].

12.5.1.10 Celecoxib in hepatocellular carcinoma

Celecoxib inhibits hepatocellular carcinoma cell growth via a PNO1-dependent mechanism. PNO1 knockdown significantly reduced tumor proliferation and metastasis [195]. In addition, celecoxib also inhibits the cancer cell proliferation through modification of the PTEN/NF-kB/PRL-3 pathway. The mode of action of celecoxib proceeds through the upregulation of PTEN and downregulation of NF-kB and PRL-3 in the liver of the HCC model [196].

The effects of metronomic celecoxib treatment on cell proliferation, metastatic ability, and rate of tumor growth were explored. Metronomic celecoxib therapy approach was selected to reduce the cardiovascular risk and to enhance the effective anticarcinogenesis of COXibs in advanced HCC patients. This novel therapeutic approach attenuated cancer cell invasion via the NF-kB/MMP9 pathway in COX-2 independent manner. Hence, metronomic celecoxib can be further explored as chemotherapeutic agents in advanced HCC patients [197].

Hepatitis B virus X protein (HBx) and COX-2 are potential therapeutic targets involved in the progression of HCC. Celecoxib exhibits its anti-HCC activity through inhibiting cell growth and triggering cell apoptosis by releasing cytochrome c and activating caspase enzymes in HBx-positive cells. These experimental evidences enlighten the novel mode of action of celecoxib against HBx-positive HCC [198].

12.5.1.11 Celecoxib in lung cancer

With the highest mortality rate and lack of effective therapies, lung cancer is still a challenging area for medical personnel. Due to the glucophilic characteristics of cancer cells, metformin exhibits its inhibitory effect on oxidative phosphorylation. Hence, the combination therapy of low dose celecoxib and metformin was applied to reduce the side effects of these drugs and to avoid the possibility of drug resistance. This combination therapy impedes tumor growth by attenuating cell proliferation and metastasis, arresting the cell cycle, and inducing apoptosis. Further exploration on the mode of action of the combination therapy revealed that it breaks the DNA double-strand through ROS aggregation in NSCLC cells, and also increases the expression of p53 to arrest the cell cycle. Metformin also synergizes the celecoxib-induced apoptosis by inducing caspase enzymes, increasing the expressions of proapoptotic proteins, Bad and Bax, and reducing the antiapoptotic proteins, Bcl-xl and Bcl-2 [199]. In another experiment, celecoxib induced the susceptibility of lung cancer cells to NK cell via targeting COX-2 [200]. In another combination therapy approach, celecoxib with chemotherapy or TKIs increased the overall response rate [201]. The radiosensitizing effect of celecoxib in the presence of radiotherapy in non-small-cell lung cancer cells promotes the apoptosis process via the Akt/mTOR signaling pathway without interfering the cell cycle distribution [202].

12.5.1.12 Celecoxib in bladder cancer

Celecoxib inhibits migration, invasion, and epithelial-to-mesenchymal transition via the miRNA-145/TGFBR2/Smad3 axis in bladder cancer cells. A combination therapy of celecoxib and miR-145 results in a satisfactory antitumor effect via downregulation of TGF-β signaling cascade [203].

12.5.1.13 Celecoxib in gastric cancer

Celecoxib elevates the expression of the tumor suppressor, miR-29c, resulting in the induction of apoptosis in the gastric cancer cells [204]. Celecoxib, in combination with first-line chemotherapy, provides additional clinical benefits for COX-2-dependent advanced (metastatic or postoperative recurrent) gastric cancer patients [205]. Selective inhibition of COX-2 by celecoxib attenuates the cell proliferation of 5-FU-resistant gastric cancer cells and simultaneously induces cell death [206]. In a nutshell, modification of COX-2 and its biocatalytic activity can be an alternative therapeutic strategy to overcome gastric cancer.

12.5.1.14 Celecoxib in pancreatic cancer

Elevation of L1CAM expression in pancreatic cancer tissue promotes the STAT/NF-κB signaling pathway. Celecoxib is capable of inhibiting L1CAM, STAT3, and the NF-κB signaling pathway for arresting cell growth and invasion [207].

12.5.1.15 Celecoxib in anticancer immune response activity

Chidamide, in combination with celecoxib (CC-01), augments the antitumor effect by modulating the immune system in tumor-infiltrating lymphocytes (TILs) and is also associated with the immune cell activation. CC-01 is also capable of downregulating the number and function of several immunosuppressive cells [208].

Apart from the pharmacodynamic explorations of celecoxib in cancer, attempts to improve the pharmacokinetics of celecoxib were also noticed. Various approaches for the successful delivery of celecoxib and to escalate the overall therapeutic efficacy were adopted, such as (a) celecoxib encapsulated in EGFR-targeted immunoliposomes [209], (b) liposomes co-loaded with celecoxib and doxorubicin [210], (c) solid lipid nanoparticles impregnated with celecoxib for colon delivery system [211], (d) Spherical nanoparticles of guar gum encapsulating celecoxib for colon cancer delivery [212].

12.5.1.16 Celecoxib in multi-drug resistance cancer

Treatment of multidrug resistance (MDR) cancer is a major challenge. The causative biomolecule of MDR is P-glycoprotein (Pgp), which facilitates in developing pharmacoresistance by extruding out the anticancer drugs from the cancer cells. Celecoxib prevented the occurrence of drug-induced MDR in lymphoma cells. For example, celecoxib attenuates MDR development by inhibiting the doxorubicin-induced overexpression of efflux protein, Pgp [213].

The status of celecoxib in the treatment of cancer is summarized in Table 12.5 [214].

Table 12.5: Status of celecoxib in the treatment of cancer [214].

entry	name of the drug(s)	clinical stage	disease for treatment
1	celecoxib	phase 2	endometrium cancer
2	celecoxib	phase 3	breast cancer
3	chidamide + celecoxib	phase 1	metastatic colorectal cancer
4	folferi + celecoxib	phase 4	metastatic colorectal cancer
5	celecoxib + gemcitabine + cisplatin	phase 1	bladder cancer
6	celecoxib + gefitinib	phase 2	nasopharyngeal carcinoma
7	celecoxib + docetaxel	phase 2	lung cancer
8	celecoxib	phase 2	rectal cancer

Table 12.5 (continued)

entry	name of the drug(s)	clinical stage	disease for treatment
9	celecoxib	phase 2	head and neck cancer
10	erlotinib hydrochloride + celecoxib	phase 2	lung cancer
11	celecoxib	phase 3	colorectal cancer
12	celecoxib + gemcitabine hydrochloride	phase 2	pancreatic cancer
13	exemestane + celecoxib	phase 2	breast cancer
14	celecoxib	phase 2	bladder cancer
15	celecoxib	phase 2	thyroid cancer
16	celecoxib	phase 2	breast cancer
17	celecoxib + cholecalciferol	phase 2	breast cancer
18	celecoxib	phase 2	head and neck cancer
19	celecoxib + zd1839	phase 2	lung cancer
20	celecoxib	phase 2	lung cancer
21	celecoxib	phase 3	prostate cancer
22	celecoxib	phase 3	bladder cancer
23	celecoxib	phase 2	prostate cancer
24	erlotinib + celecoxib	phase 2	head and neck cancer
25	celecoxib + docetaxel	phase 2	lung cancer
26	celecoxib + sorafenib	phase 4	hepatocellular carcinoma
27	celecoxib + carboplatin + paclitaxel + radiation therapy	phase 2	head and neck cancer
28	celecoxib + cisplatin + cpt-11 + radiation therapy	phase 1	lung cancer
29	celecoxib	phase 2	colorectal carcinoma
30	celecoxib	phase 2	cervical carcinoma
31	celecoxib	phase 2	metastatic breast cancer
32	celecoxib	phase 2	uterine cancer
33	celecoxib + 5-fluorouracil + oxaliplatin + leucovorin	phase 3	colorectal cancer
34	celecoxib + vinorelbine detartrate	phase 1	breast cancer

Table 12.5 (continued)

entry	name of the drug(s)	clinical stage	disease for treatment
35	celecoxib + gemcitabine	phase 3	malignant pleural mesothelioma
36	carboplatin + celecoxib + paclitaxel	phase 2	lung cancer
37	gemcitabine + cisplatin + celecoxib	phase 2	pancreatic cancer
38	cisplatin + chemokine modulatory + celecoxib + dc vaccine	phase 2	ovarian cancer
39	dfmo + celecoxib + cyclophosphamide + topotecan	phase 1	neuroblastoma
40	atorvastatin calcium + celecoxib (2)	phase 2	prostate cancer
41	celecoxib + rintatolimod + interferon -2b	phase 2	liver cancer, colorectal cancer
42	nivolumab + ipilimumab + celecoxib	phase 2	colon carcinoma
43	erlotinib + celecoxib	phase 1	head and neck cancer
44	cyclophosphamide + celecoxib	phase 2	fallopian tube cancer, peritoneal cavity cancer, ovarian epithelial cancer

12.5.1.17 Etoricoxib in colon cancer and hepatocarcinogenesis

Etoricoxib prevents the early stage of colon cancer by inducing the mitochondria-promoted apoptosis. Etoricoxib downregulates the Bcl-2 expression to elevate the apoptotic susceptibility of the cancer cells to exhibit its anticancer effect [215]. Hepatocellular carcinoma is one of the global burdens in cancer area with an increasing trend. Etoricoxib reduces ROS and tumor biomarkers and inhibits the inflammatory cascade generated by NF-κB/COX-2/PGE2 signaling in hepatocellular carcinoma [216].

12.5.1.18 Etoricoxib in lung cancer

Etoricoxib, loaded into nano-emulsion (ETO-NE), exhibited substantial anticancer activity against lung cancer cells. The ETO-NE induces cancer cell apoptosis/necrotic cell death, S-phase cell cycle arrests, and reduces the inflammatory mediators involved in cancer progression such as IL-12, IL-6, TNF, COX-2, and NF-kB [217]. The role of angiogenesis in lung cancer and its chemoprevention by etoricoxib was also investigated. Etoricoxib helps in the downregulation of the protein expression of MCP-1, MMP-2, MMP-9, MIP-1b, and VEGF. Hence, etoricoxib is capable of chemoprevention of lung cancer through the inhibition of angiogenesis [218]. Etoricoxib also induces cell apoptosis and reduces the expressions of IL-1β, TNF-α, and IFN-γ in lung cancer patients, exhibiting anticancer activity [219].

12.5.1.19 Parecoxib in colorectal cancer and esophageal squamous cell carcinoma

Parecoxib synergizes radiation sensitivity in colorectal cancer cells through direct and indirect effects on the tumor cells [220]. Experiments indicated that parecoxib induces cell cycle arrest in the G2 phase, and inhibits cell proliferation, invasion, and migration in vitro in esophageal cancer. Parecoxib may also attenuate the phosphorylation levels of AKT and PDK1, the expression of p53, cyclin B1, and CDK1.

12.5.1.20 Rofecoxib in lung cancer

In the combination therapy approach, gefitinib, combined with rofecoxib, improves the therapeutic efficacy, and is well tolerated [221]. A clinical study was carried out, anticipating that the COX-2 inhibitors, in combination with prolonged constant infusion (PCI) of gemcitabine, may improve the clinical condition of non-small-cell lung cancer patients. The study outcome indicated that PCI gemcitabine or rofecoxib failed to increase the survival rate. However, rofecoxib increased the overall response rate, several parameters of quality of life, and reduced pain-related symptoms [222].

12.5.1.21 Rofecoxib in colon cancer and bladder cancer

Rofecoxib escalates the sensitivity of radiation therapy by inhibiting endothelial cell function. This result opens up an alternative strategy to target angiogenesis in cancer patients [223]. Rofecoxib attenuated the growth of human bladder cancer by inducing cell apoptotic mechanisms [224].

12.5.1.22 Rofecoxib in gastric and pancreatic cancer

In a different role, rofecoxib acts as a chemotherapeutic sensitizer to the standard anticancer agents in gastric cancer cell by altering the multidrug resistance genes, MRP1 and GST-p [225]. Rofecoxib modulates the cell cycle arresting genes expression to downregulate the human pancreatic cancer development and progression [226].

12.5.1.23 Rofecoxib in esophagus cancer

Rofecoxib exhibits antiproliferative activity in adenocarcinoma and squamous cell carcinoma of the oesophagus through augmenting the apoptotic activity and COX-2 inhibition [227].

The clinical applications of the rofecoxib in various types of cancers are enlisted in Table 12.6 [214].

Table 12.6: Status of the rofecoxib in the treatment of cancer [214].

entry	name of the drug	clinical stage	Disease for treatment
1	rofecoxib	phase 3	prostate cancer
2	rofecoxib	phase 3	colorectal cancer

12.5.2 Alzheimer's disease (AD) and inflammation

Alzheimer's disease (AD), a neurodegenerative disorder, can be reversed by promoting the neurogenesis and attenuating the neuronal cell apoptosis. Inflammatory processes accumulate the β-Amyloid (Ab) peptide in the form of amyloid plaques in AD and, subsequently, decrease the number of cholinergic, serotonergic, and dopaminergic neurons. Hence, anti-inflammatory drugs can counteract the Aβ-induced neuronal decline by targeting key enzymes such as lipoxygenase and cyclooxygenase [228].

12.5.2.1 Celecoxib in AD

Earlier literature reports revealed that celecoxib may provide neuroprotection through inhibition of COX-2. Celecoxib, loaded erythrocyte membranes, prolonged the release of the drug and increased the bioavailability of the drug in the brain. This approach improves the clearance of the aggregated β-amyloid proteins from brain neurons. The larger accumulation of celecoxib in neurons increases the migratory activity of neurons, and improves the cognitive function. Therefore, this strategy can be a rational approach to cure AD by influencing the self-healing property of the brain [229].

12.5.2.2 Valdecoxib in AD

A combination therapy of zafirlukast and valdecoxib provides better neuroprotective effect via dual pharmacological action in AD subjects through the inhibition of the release of LOX metabolites and downregulation of mito-oxidative stress, causing a decline in the oxido-nitrosative burden [230].

12.5.3 COX inhibitors in Parkinsonism

12.5.3.1 Celecoxib in Parkinsonism

Parkinson's disease (PD) is the second most prevalent age-dependent neurodegenerative disease after Alzheimer's disease. Clinical approaches to treat Parkinsonism include symptomatic relief but there are no therapeutic drugs. Celecoxib recovered the dopaminergic neuronal cells through elevating the levels of APOD and the transcription factor, TFEB [231].

12.5.3.2 Parecoxib in Parkinsonism

Parecoxib provides a neuroprotective effect on motor neurons, tyrosine hydroxylase expression, and cognitive functions in the early phase of Parkinsonism [232]. In recent studies, it was found that mitochondrial dysfunction is the primary manifestation observed in the substantia nigra of Parkinson disease. Parecoxib alleviates the mitochondrial dysfunction by inducing antioxidative activity, activating mitochondrial biogenesis, and improving the dysregulated mitochondrial dynamics situation. Parecoxib also ameliorates the inflammatory dysfunction at the nigrostriatal dopaminergic neurons during the aging process [233].

12.5.3.3 Rofecoxib in Parkinsonism

Although the individual administration of rofecoxib and creatine inhibited the striatal dopamine neurons and substantia nigra immunoreactive neurons, the combination therapy of rofecoxib with creatine synergizes the neuroprotective effects of dopaminergic neurons via attenuating the oxidative damage process [234]. This suggests the possible benefits of rofecoxib in the treatment of this disease.

12.5.3.4 Valdecoxib in Parkinsonism

COX-2 is the key enzyme found in the inflammatory cascade of a PD brain. Therefore, COX-2 inhibition results in the reduction in microglial-derived proinflammatory markers for its neuroprotective action [235]. Valdecoxib inhibited microglial activation and the loss of tyrosine hydroxylase-containing cells by reducing the level of COX-2 and, subsequently, microglial activation [236].

12.5.4 COXibs in Amyotrophic Lateral Sclerosis (ALS)

Combination therapy with different mode of action may be useful in the treatment of ALS. For example, the celecoxib-creatine combination was preferred over minocycline-creatine combination for ALS treatment [237]. In the mono drug therapy, ciprofloxacin provides mild effect, and celecoxib is ineffective. In the combination therapy of ciprofloxacin and celecoxib, there is improvement in the locomotor and cellular deficits of ALS subjects [238]. Clinical application of this study is mentioned in Table 12.7 [214].

Table 12.7: Status of celecoxib for the treatment of ALS [214].

entry	name of the drug(s)	clinical stage	Disease for treatment
1	ciprofloxacin and celecoxib (fixed dose combination)	phase 2	ALS

12.5.5 COXibs in Diabetic Peripheral Neuropathy (DPN)

Among the selective COX-2 inhibitors, only celecoxib is found to be applied in the treatment of DPN, characterized by peripheral neuropathic pain due to diabetes. Increase in glucose level results in the decline in dorsal root ganglia (DRG) neuron cell viability, induced apoptosis, elevated levels of ROS, and downregulated SOD activity. Application of celecoxib in DPN guards the DRG neurons against high glucose-induced damage. The mode of action of celecoxib includes the prohibition of the miR-155/COX-2 axis in the DPN model [239]. The combinatorial action of proglumide and celecoxib in DPN proceeds through the blocking of CCK receptors, sensitization of opioid receptors, and inhibition of COX-2 [240]. The use of COX-2 or NOS inhibitors in combination with CB1 and/or CB2 receptor agonists alleviates diabetic neuropathic pain by escalating the analgesic activity of CB1 and CB2 receptor agonists [241].

12.5.6 COXibs in epilepsy

Epilepsy is an abnormal neuronal disorder of excessive electrical discharge in brain neurons. Despite detailed investigations on the pathophysiology of epilepsy and continuous effort in inventing a clinical remedy for epilepsy, very few approaches are available for epilepsy treatment. Involvement of neuroinflammation in epilepsy encouraged drug discovery scientists to design novel anti-neuroinflammatory agents for epilepsy treatment. COX-2, the key enzyme involved in seizures, is a potential therapeutic target for epilepsy [242]. In the brain, the elevated level of proinflammatory mediators (PGs) reduces the threshold of seizures and escalates the frequency-intensity of epilepsy. Low-dose celecoxib reverses these phenomena through its anti-inflammatory property [243]. COX-2 inhibition downregulates the P-glycoprotein induction and subsequently attenuates epileptic seizures [244]. Abnormal neurogenesis, astrogliosis, microglial activation, and COX-2 induction are key factors in the development of epilepsy. Celecoxib attenuated the development of spontaneous recurrent seizures, neuronal death, and microglia activation [245].

Spontaneous epileptic seizures are generated from synaptic reorganization and various neurotransmitter (GABA and glutamate) abnormalities. Synaptic activity promotes COX-2 transcription in the pathogenesis of epilepsy. COX-2 creates oxidative stress and its metabolic product, prostaglandins, furnishes neuronal injuries. Celecoxib improves the epileptic condition by reducing the neuronal excitability, inhibiting astrogliosis, and causing ectopic neurogenesis. Celecoxib induces the $GABA_A$ receptors' expression in the brain to mitigate the neurotransmitter abnormalities. Hence, celecoxib may be utilized in combination therapy with other antiepileptic drugs [246].

Celecoxib is a promising agent for intervention therapy for autosomal dominant lateral temporal epilepsy (ADLTE) as it is a partly but significantly suppressed epileptogenic activity, and extended the lifetime. The efficient therapeutic activity in ameliorating seizure susceptibility makes celecoxib a promising long-term treatment to reduce the risk of seizures in family members carrying an LGI1 mutation [247]. Combination therapy of celecoxib with VPA effectively reduces the pro-inflammatory cytokines in the brain, oxidative stress, and facilitates HMGB1 translocation into peripheral circulation. Therefore, COX-2 inhibition can be an adjuvant therapy for epilepsy treatment in the future [248].

The combination therapy with celecoxib was explored clinically for the treatment of neuronal disorders (Table 12.8) [214].

Table 12.8: Status of celecoxib in the treatment of depression [214].

entry	name of the drug(s)	clinical stage	Disease for treatment
1	minocycline + celecoxib	phase 3	depression, bipolar disorder, bipolar depression, mood disorders
2	escitalopram + celecoxib	phase 4	depression

12.5.7 COX inhibitors in diabetes

12.5.7.1 Celecoxib in diabetes

Elevated inflammatory adipocytokines and decreased levels of adiponectin aggravate insulin resistance in obese diabetic patients, which may be restored by celecoxib treatment by targeting the inflammatory cascade. Celecoxib, in combination with glimepiride, ameliorates insulin resistance and also reduces the hyperglycemic condition and the inflammatory process in obese type 2 diabetic subjects [249]. Increased lipid levels that induce pancreatic β-cell apoptosis through the activation of COX-2/PGE_2/EP3 receptor axis are another source of type 2 diabetes occurrence. These phenomena can be reversed by treatment with either EP3 antagonist or selective COX-2 inhibitor. Celecoxib attenuated β-cell apoptosis by downregulating the COX-2/PGE_2/EP3 pathway [250]. The clinical application of celecoxib in type 2 diabetes is enlisted in Table 12.9 [214].

Table 12.9: Status of celecoxib in the treatment of type 2 diabetes [214].

entry	name of the drug(s)	clinical stage	disease for treatment
1	valsartan, celecoxib, and metformin	phase 2	type 2 diabetes

12.5.7.2 Rofecoxib in diabetes

Infants with congenital nephrogenic diabetes insipidus (disorder of water metabolism) suffer from frequent hypernatremic dehydration, central nervous system depletion, and reduced growth rate. This problem can be restored by altering the daily lifestyle, such as by reducing water loss, increasing intake of fluids, and maintaining a low solute diet. In medicine, diuretics and prostaglandin (PG) synthesis inhibitors are useful to reduce the urine volume. The combination of hydrochlorothiazide with rofecoxib has the potential for treating congenital nephrogenic diabetes insipidus infants [251].

12.5.7.3 Valdecoxib in diabetes

Valdecoxib improves the insulin resistance by inhibiting inflammation, and ER stress. This approach can be useful for the treatment of type 2 diabetes through drug repurposing rather than searching for a novel drug candidate [252].

12.5.8 Carbonic anhydrase (CA) and cyclooxygenase inhibitors

Carbonic anhydrases (CA), the zinc metalloenzymes, are present in kidney for acid-base catalysis. Several research outcomes have proposed that few COX-2 inhibitors can inhibit the carbonic anhydrase without blocking the COX enzymes. Celecoxib and valdecoxib are structurally similar to many CA inhibitors. Among the selective COX-2 inhibitors, celecoxib potently inhibits the human carbonic anhydrase II (hCAII) activity. Valdecoxib is less potent and rofecoxib is inactive due to the absence of the sulfonamide moiety in the scaffold [253].

12.5.8.1 Celecoxib and carbonic anhydrase inhibition

Selective drug action is preferred to reduce the toxic side effects and promote the desired therapeutic effects. The trifluoromethyl group of celecoxib occupies the hydrophobic part of the CAII-active site, whereas the p-tolyl group is docked to the hydrophilic portion [254]. The nanomolar activity of celecoxib and valdecoxib toward CA I, II, IV, and IX, was observed due to the interaction of the sulfonamide group with the Zn^{2+} ion of CA. Celecoxib and valdecoxib are also found to be Zn-Cam inhibitors [255].

12.5.8.2 Valdecoxib and carbonic anhydrase inhibition

The sulfonamide moiety of valdecoxib interacts with the Zn(II) ion of the carbonic anhydrase, and the phenyl-isoxazole moiety interacts with Gln92, Val121, Leu198, Thr200, and Pro202 amino acid residues. The three phenyl groups of valdecoxib are docked to a hydrophobic pocket of the carbonic anhydrase enzyme and create van der Waals interactions with the aliphatic side chains. Valdecoxib also forms π-stacking interaction with the aromatic ring of Phe131. Valdecoxib has an unexpectedly low affinity for CA XIII, found in the salivary glands, kidney, brain, lung, and the gut. The clinically used compounds acetazolamide, methazolamide, dichlorophenamide, and dorzolamide inhibited CA XIII more potently than sulfanilamide, halogenated sulfanilamides, homosulfanilamide, 4-aminoethyl-benzenesulfonamide, and orthanilamides [256]. The inhibitory activities of celecoxib and valdecoxib on different isoforms of CA are mentioned in Table 12.10.

Table 12.10: Inhibition of various CA isoforms by valdecoxib and celecoxib [254, 256, 257].

Inhibitor	K_I^* (nM)								
	hCA I	hCA II α	bCA IV	hCA IX	hCA XII	mCA XIII	LdcCA β	PgiCA γ	TweCA δ
valdecoxib	54,000	43	340	16	13		338	755	378
celecoxib	50,000	21	290	27	18	425	705	169	265

12.5.9 COXibs in COVID-19

The worldwide spread of COVID-19 is an additional burden to the world of infectious diseases. Severe illness in COVID-19 patients generally arises from the cytokine storm that is generated by the highly active immune system. Hence, the need of the hour is to come up with drugs that can inhibit the cytokine storm. As new drug discovery will need a large amount of money and will require long time, it is beneficial to find use of the existing anti-inflammatory drugs for the treatment of COVID-19. Recent research activities established that COX-2, p38 MAPK, IL-1b, IL-6, and TGF-β have crucial roles in inhibiting the COVID-19 cytokine storm, pulmonary interstitial fibrosis, and cell death. Due to the lack of an effective treatment and the necessity of immunomodulation in COVID-19, COX-2 inhibition can be an adjuvant treatment strategy [258]. The outcome of various research activities toward this highlighted that etoricoxib may effectively inhibit the cytokine storm in the treatment of COVID-19. However, further scientific investigations and systematic clinical trials are needed to entitle etoricoxib as a "repurposed drug for COVID-19" [259]. The combination effect of celecoxib with famotidine (anti-histaminic) was evaluated in clinical trials against COVID-19 (Table 12.11) [214].

Table 12.11: Status of celecoxib in the treatment of COVID-19 [214].

entry	name of the drug(s)	clinical stage	Disease for treatment
1	famotidine + celecoxib	phase 2	COVID-19

12.5.10 COXibs in HIV

Chronic HIV infection results in T cell dysfunction and cyclic AMP (cAMP) level upregulation to inhibit T cell activation. Lipopolysaccharide induced upregulation of COX-2 and PGE2 in monocytes, elevating the cAMP level in chronic HIV patients. Application of selective COX-2 inhibitors, alone or in combination with standard anti-HIV therapy, reduces the CD38 densities, IgA levels, expression of PD-1 on CD8 cells, and also ameliorates T cell responsiveness [260]. The acute therapy of HIV patients with rofecoxib provides improved T-cell function [261]. Rofecoxib-celecoxib, together with anti-retroviral treatment, alleviates the chronic immune activation in viraemia patients [262].

12.5.11 COXibs in tuberculosis

Tuberculosis (TB) is the world's second leading cause of fatality due to infectious illness, next to SARS-CoV-2 but is above HIV/AIDS [263]. The standard clinical practice to treat TB includes the use of first-line drugs, isoniazid (INH), rifampicin (R), pyrazinamide (Z), and ethambutol (E) for six months under the directly observed treatment short strategy (DOTS). The availability of medicines to combat TB remained limited due to the scarcity of new and more effective anti-TB drugs. The incidence of drug-resistance adds further hurdle to treat TB. The drug-resistant TB falls into five types e.g., isoniazid-resistant, rifampicin-resistant, multi drug resistant (MDR), pre-extensively drug-resistant (Pre-XDR), and extremely drug resistant (XDR). It is only recently that bedaquiline and delamanid have been added to the armory of anti-TB drugs after receiving FDA approval for the treatment of multi-drug resistant tuberculosis (MDR-TB) [264]. Pre-XDR-TB is resistant to rifampicin and any fluoroquinolone, and XDR-TB is resistant to rifampicin, any fluoroquinolone, and at least one of the drugs such as bedaquiline and linezolid [263]. Therefore, the lack of adequate number of effective anti-TB drugs makes the treatment of TB a fitting case for consideration of drug repurposing approach as an alternative way to relocate the existing marketed drugs. Eicosanoids and their biosynthetic enzymes are potential therapeutic targets in the treatment of TB. Proinflammatory mediators and proinflammatory cytokines were involved in the related inflammatory process of TB patients, and this provides rationale for considering the control of inflammatory pathways as a viable approach to manage TB. Selective COX-2 inhibitors inhibit the inflammatory cascade in TB patients via alteration of the phosphorylation process and signal transduction in monocytes [265]. Celecoxib was combined with antitubercular drugs for clinical treatment of tuberculosis (Table 12.12) [214].

Table 12.12: Status of celecoxib in treatment of TB [214].

entry	name of the drug(s)	clinical stage	disease for treatment
1	celecoxib + rifampicin + pyrazinamide	phase 1	tuberculosis

12.6 Conclusions

In the drug discovery process, drug repurposing is an approach to identify new therapeutic indications of approved/investigational/abandoned drugs. Drug repurposing provides a faster and cost-effective way to utilize the therapeutic benefits of existing therapeutic agents to treat new/existing diseases for which, new/better medical treatment is required. It provides certain advantages compared to the de novo drug discovery approach. Less investment and shorter time to identify the repurposed drug encourages both the academia and the industry to pursue research on drug repurposing. Various drugs have been already repurposed for neuronal disorders, *Mtb* infection, anticancer therapy, rheumatoid arthritis, erectile dysfunction etc. COX-2 inhibitors were initially invented for the treatment of inflammation, arthritis (rheumatoid arthritis and osteoarthritis), and gout. Later on, the understanding of involvement of COX-2 enzyme in the development and progression of other diseases (e.g., cancer, neuronal disorders, microbial diseases, etc.) widened the therapeutic avenue of selective COX-2 inhibitors towards these diseases (i.e., the repurposing of selective COX-2 inhibitors). Mostly, the marketed selective COX-2 inhibitor, celecoxib, is being repurposed for various diseases due to the less risk of cardiovascular side effects with celecoxib. Selective COX-2 inhibitors are also repurposed in the combination therapy regimen for the treatment of cancer, neuronal disorders, and microbial infections. Some clinical trials for the repurposing of COX-2 inhibitors are still in progress. However, more in vitro-in vivo preclinical experiments, clinical trials, and experimental evidences are still required for the logical/empirical repurposing of selective COX-2 inhibitors for the treatment of various pathophysiological conditions.

Acknowledgments: AKC and NS thank the Department of Atomic Energy, Mumbai, India, for the award of Raja Ramanna Fellowship and Research Associateship, respectively.

References

[1] Taniguchi K, Karin M. NF-κB, inflammation, immunity and cancer: Coming of age. Nat Rev Immunol. 2018;18:309–324.

[2] Cornish AL, Campbell IK, McKenzie BS, Chatfield S, Wicks IP. G-CSF and GM-CSF as therapeutic targets in rheumatoid arthritis. Nat Rev Rheumatol. 2009;5:554–559.

[3] Choi Y, Arron JR, Townsend MJ. Promising bone-related therapeutic targets for rheumatoid arthritis. Nat Rev Rheumatol. 2009;5:543–548.

[4] Van Maanen MA, Vervoordeldonk MJ, Tak PP. The cholinergic anti-inflammatory pathway: Towards innovative treatment of rheumatoid arthritis. Nat Rev Rheumatol. 2009;5:229–232.

[5] Vezzani A, Balosso S, Ravizza T. Neuroinflammatory pathways as treatment targets and biomarkers in epilepsy. Nat Rev Neurol. 2019;15:459–472.

[6] Declèves AE, Sharma K. Novel targets of antifibrotic and anti-inflammatory treatment in CKD. Nat Rev Nephrol. 2014;10:257–267.
[7] Németh T, Sperandio M, Mócsai A. Neutrophils as emerging therapeutic targets. Nat Rev Drug Discov. 2020;19:253–275.
[8] Ji RR, Xu ZZ, Gao YJ. Emerging targets in neuroinflammation-driven chronic pain. Nat Rev Drug Discov. 2014;13:533–548.
[9] Barnes PJ. New anti-inflammatory targets for chronic obstructive pulmonary disease. Nat Rev Drug Discov. 2013;12:543–559.
[10] Charo IF, Taub R. Anti-inflammatory therapeutics for the treatment of atherosclerosis. Nat Rev Drug Discov. 2011;10:365–376.
[11] Gaestel M, Kotlyarov A, Kracht M. Targeting innate immunity protein kinase signalling in inflammation. Nat Rev Drug Discov. 2009;8:480–499.
[12] McCulloch CA, Downey GP, El-Gabalawy H. Signalling platforms that modulate the inflammatory response: New targets for drug development. Nat Rev Drug Discov. 2006;5:864–876.
[13] Daniel C, Leppkes M, Muñoz LE, Schley G, Schett G, Herrmann M. Extracellular DNA traps in inflammation, injury and healing. Nat Rev Nephrol. 2019;15:559–575.
[14] Kumar S, Boehm J, Lee JC. p38 MAP kinases: Key signalling molecules as therapeutic targets for inflammatory diseases. Nat Rev Drug Discov. 2003;2:717–726.
[15] Afonina IS, Zhong Z, Karin M, Beyaert R. Limiting inflammation – The negative regulation of NF-κB and the NLRP3 inflammasome. Nat Immunol. 2017;1:861–869.
[16] Han J, Ulevitch RJ. Emerging targets for anti-inflammatory therapy. Nat Cell Biol. 1999;1:39–40.
[17] Virgilio FD. New pathways for reactive oxygen species generation in inflammation and potential novel pharmacological targets. Curr Pharm Des. 2004;10:1647–1652.
[18] Leppert D, Lindberg RL, Kappos L, Leib SL. Matrix metalloproteinases: Multifunctional effectors of inflammation in multiple sclerosis and bacterial meningitis. Brain Res Rev. 2001;36:249–257.
[19] Picot D, Loll PJ, Garavito RM. The X-ray crystal structure of the membrane protein prostaglandin H 2 synthase-1. Nature. 1994;367:243–249.
[20] Kurumbail RG, Stevens AM, Gierse JK, McDonald JJ, Stegeman RA, Pak JY, Gildehaus D, Iyashiro JM, Penning TD, Seibert K, Isakson PC. Structural basis for selective inhibition of cyclooxygenase-2 by anti-inflammatory agents. Nature. 1996;384:644–648.
[21] Flower RJ. The development of COX-2 inhibitors. Nat Rev Drug Discov. 2003;2:179–191.
[22] Pae HO, Lee YC, Chung HT. Heme oxygenase-1 and carbon monoxide: Emerging therapeutic targets in inflammation and allergy. Recent Pat Inflamm Allergy Drug Discov. 2008;2:159–165.
[23] Klein TW. Cannabinoid-based drugs as anti-inflammatory therapeutics. Nat Rev Immunol. 2005;5:400–411.
[24] Lawrence T, Willoughby DA, Gilroy DW. Anti-inflammatory lipid mediators and insights into the resolution of inflammation. Nat Rev Immunol. 2002;2:787–795.
[25] Tranfaglia MR, Thibodeaux C, Mason DJ, Brown D, Roberts I, Smith R, Guilliams T, Cogram P. Repurposing available drugs for neurodevelopmental disorders: The fragile X experience. Neuropharmacol. 2019;147:74–86.
[26] Roin BN. Solving the problem of new uses. Available at SSRN 2337821. 2013.
[27] Kumar R, Saha N, Purohit P, Garg SK, Seth K, Meena VS, Dubey S, Dave K, Goyal R, Sharma SS, Banerjee UC. Cyclic enaminone as new chemotype for selective cyclooxygenase-2 inhibitory, anti-inflammatory, and analgesic activities. Eur J Med Chem. 2019;182:111601.
[28] Seth K, Garg SK, Kumar R, Purohit P, Meena VS, Goyal R, Banerjee UC, Chakraborti AK. 2-(2-Arylphenyl) benzoxazole as a novel anti-inflammatory scaffold: Synthesis and biological evaluation. ACS Med Chem Lett. 2014;5:512–516.
[29] Chakraborti AK, Garg SK, Kumar R, Motiwala HF, Jadhavar PS. Progress in COX-2 inhibitors: A journey so far. Curr Med Chem. 2010;17:1563–1593.

[30] Chakraborti AK, Thilagavathi R. Computer-aided design of non sulphonyl COX-2 inhibitors: An improved comparative molecular field analysis incorporating additional descriptors and comparative molecular similarity indices analysis of 1, 3-diarylisoindole derivatives. Bioorg Med Chem. 2003;11:3989–3996.

[31] Selvam C, Jachak SM, Thilagavathi R, Chakraborti AK. Design, synthesis, biological evaluation and molecular docking of curcumin analogues as antioxidant, cyclooxygenase inhibitory and anti-inflammatory agents. Bioorg Med Chem Lett. 2005;15:1793–1797.

[32] Chawla G, Gupta P, Thilagavathi R, Chakraborti AK, Bansal AK. Characterization of solid-state forms of celecoxib. Eur J Pharm Sci. 2003;20:305–317.

[33] Selvam C, Jachak SM, Oli RG, Thilagavathi R, Chakraborti AK, Bhutani KK. A new cyclooxygenase (COX) inhibitory pterocarpan from Indigofera aspalathoides: Structure elucidation and determination of binding orientations in the active sites of the enzyme by molecular docking. Tetrahedron Lett. 2004;45:4311–4314.

[34] Pillaiyar T, Meenakshisundaram S, Manickam M, Sankaranarayanan M. A Med Chem perspective of drug repositioning: Recent advances and challenges in drug discovery. Eur J Med Chem 2020. 195:112275.

[35] Pushpakom S, Iorio F, Eyers PA, Escott KJ, Hopper S, Wells A, Doig A, Guilliams T, Latimer J, McNamee C, Norris A. Drug repurposing: Progress, challenges and recommendations. Nat Rev Drug Discov. 2019;18:41–58.

[36] Kingsmore KM, Grammer AC, Lipsky PE. Drug repurposing to improve treatment of rheumatic autoimmune inflammatory diseases. Nat Rev Rheumatol. 2020;16:32–52.

[37] Braga SS. Multi-target drugs active against leishmaniasis: A paradigm of drug repurposing. Eur J Med Chem. 2019;183:111660.

[38] Ferreira LG, Andricopulo AD. Drug repositioning approaches to parasitic diseases: A medicinal chemistry perspective. Drug Discov Today. 2016;21:1699–1710.

[39] Geilen CC, Wieder T, Reutter W. Hexadecylphosphocholine inhibits translocation of CTP: Choline-phosphate cytidylyl transferase in Madin-Darby canine kidney cells. J Biol Chem. 1992;267:6719–6724.

[40] Cui Z, Houweling M, Chen MH, Record M, Chap H, Vance DE, Tercé F. A genetic defect in phosphatidylcholine biosynthesis triggers apoptosis in Chinese hamster ovary cells. J Biol Chem. 1996;271:14668–14671.

[41] Chong CR, Sullivan DJ Jr. New uses for old drugs. Nature. 2007;448:645–646.

[42] Sundar S, Jha TK, Thakur CP, Engel J, Sindermann H, Fischer C, Junge K, Bryceson A, Berman J. Oral miltefosine for Indian visceral leishmaniasis. N Engl J Med. 2002;347:1739–1746.

[43] Sindermann H, Croft SL, Engel KR, Bommer W, Eibl HJ, Unger C, Engel J. Miltefosine (Impavido): The first oral treatment against leishmaniasis. Med Microbiol Immunol. 2004;193:173–180.

[44] Lu L, Fong CH, Zhang AJ, Wu WL, Li IC, Lee AC, Dissanayake TK, Chen L, Hung IF, Chan KH, Chu H. Repurposing of Miltefosine as an adjuvant for influenza vaccine. Vaccines. 2020;8:754.

[45] Murdoch JM, Sleigh JD, Frazer SC. Cycloserine in treatment of infection of urinary tract. Brit Med J. 1959;2:1055.

[46] Hughes J, Coppridge WM, Roberts LC. Cycloserine in chronic urinary infections. J Urol. 1958;80:75–79.

[47] Syme J, Sleigh JD, Richardson JE, Murdoch JM. Cycloserine in the treatment of infection of the urinary tract. Brit J Urol. 1961;33:261–266.

[48] Epstein IG, Nair KG, Boyd LJ. Cycloserine, a new antibiotic, in the treatment of human pulmonary tuberculosis: A preliminary report. Antibiotic Med. 1955;1:80–93.

[49] Patnode RA, Hudgins PC, Cummings MM. Further observations on the effect of cycloserine on tuberculosis in guinea pigs. Am Rev Tuberc Pulmonary Dis. 1955;72:856–858.

[50] Watson GB, Bolanowski MA, Baganoff MP, Deppeler CL, Lanthorn TH. D-Cycloserine acts as a partial agonist at the glycine modulatory site of the NMDA receptor expressed in Xenopus oocytes. Brain Res. 1990;510:158–160.
[51] Hood WF, Compton RP, Monahan JB. D-cycloserine: A ligand for the N-methyl-D-aspartate coupled glycine receptor has partial agonist characteristics. Neurosci Lett. 1989;98:91–95.
[52] Sheinin A, Shavit S, Benveniste M. Subunit specificity and mechanism of action of NMDA partial agonist D-cycloserine. Neuropharmacol. 2001;41:151–158.
[53] Mohammad Sadeghi H, Adeli I, Mousavi T, Daniali M, Nikfar S, Abdollahi M. Drug repurposing for the management of depression: Where do we stand currently? Life. 2021;11:774.
[54] LeVine SM, Tsau S. Substrate reduction therapy for Krabbe disease: Exploring the repurposing of the antibiotic D-Cycloserine. Front Pediatr. 2022;9:1662.
[55] Ondo WG, Jong D, Davis A. Comparison of weight gain in treatments for Tourette syndrome: Tetrabenazine versus neuroleptic drugs. J Child Neurol. 2008;23:435–437.
[56] Lemonnier É, Degrez C, Phelep M, Tyzio R, Josse F, Grandgeorge M, Hadjikhani N, Ben-Ari Y. A randomised controlled trial of bumetanide in the treatment of autism in children. Transl Psychiatry. 2012;2:202.
[57] Ceulemans B, Schoonjans AS, Marchau F, Paelinck BP, Lagae L. Five-year extended follow-up status of 10 patients with Dravet syndrome treated with fenfluramine. Epilepsia. 2016;57:129–134.
[58] Boone KM, Gracious B, Klebanoff MA, Rogers LK, Rausch J, Coury DL, Keim SA. Omega-3 and-6 fatty acid supplementation and sensory processing in toddlers with ASD symptomology born preterm: A randomized controlled trial. Early Hum Dev. 2017;115:64–70.
[59] Pushpakom S, Iorio F, Eyers PA, Escott KJ, Hopper S, Wells A, Doig A, Guilliams T, Latimer J, McNamee C, Norris A. Drug repurposing: Progress, challenges and recommendations. Nat Rev Drug Discov. 2019;18:41–58.
[60] Choi CH, Schoenfeld BP, Bell AJ, Hinchey J, Rosenfelt C, Gertner MJ, Campbell SR, Emerson D, Hinchey P, Kollaros M, Ferrick NJ. Multiple drug treatments that increase cAMP signaling restore long-term memory and aberrant signaling in fragile X syndrome models. Front Behav Neurosci. 2016;10:136.
[61] Cohen IL, Tsiouris JA, Pfadt A. Effects of long-acting propranolol on agonistic and stereotyped behaviors in a man with pervasive developmental disorder and fragile X syndrome: A double-blind, placebo-controlled study. J Clin Psychopharmacol. 1991;11:398–399.
[62] Puhl AC, Milton FA, Cvoro A, Sieglaff DH, Campos JC, Bernardes A, Filgueira CS, Lindemann JL, Deng T, Neves FA, Polikarpov I. Mechanisms of peroxisome proliferator activated receptor γ regulation by non-steroidal anti-inflammatory drugs. Nucl Recept Signal. 2015;13:13004.
[63] Monyak RE, Emerson D, Schoenfeld BP, Zheng X, Chambers DB, Rosenfelt C, Langer S, Hinchey P, Choi CH, McDonald TV, Bolduc FV. Insulin signaling misregulation underlies circadian and cognitive deficits in a Drosophila fragile X model. Mol Psychiatry. 2017;22:1140–1148.
[64] Song G, Napoli E, Wong S, Hagerman R, Liu S, Tassone F, Giulivi C. Altered redox mitochondrial biology in the neurodegenerative disorder fragile X-tremor/ataxia syndrome: Use of antioxidants in precision medicine. Mol Med. 2016;22:548–559.
[65] Kast RE, Halatsch ME. Matrix Metalloproteinase-2 and-9 in glioblastoma: A trio of old drugs – Captopril, disulfiram and nelfinavir – Are inhibitors with potential as adjunctive treatments in glioblastoma. Arch Med Res. 2012;43:243–247.
[66] De la Hoz B, Soria C, Fraj J, Losada E, Ledo A. Fixed drug eruption due to piroxicam. Int J Dermatol. 1990;29:672–673.
[67] Egorova A, Jackson M, Gavrilyuk V, Makarov V. Pipeline of anti-Mycobacterium abscessus small molecules: Repurposable drugs and promising novel chemical entities. Med Res Rev. 2021;41:2350–2387.

[68] Das S, Garg T, Chopra S, Dasgupta A. Repurposing disulfiram to target infections caused by non-tuberculous mycobacteria. J Antimicrob Chemother. 2019;74:1317-1322.

[69] Santucci P, Dedaki C, Athanasoulis A, et al. Synthesis of long-chain β-lactones and their antibacterial activities against pathogenic mycobacteria. ChemMedChem. 2019;14:349-358.

[70] Khosravi AD, Mirsaeidi M, Farahani A, et al. Prevalence of nontuberculous mycobacteria and high efficacy of d-cycloserine and its synergistic effect with clarithromycin against Mycobacterium fortuitum and Mycobacterium abscessus. Infect Drug Resist. 2018;11:2521-2532.

[71] Gupta MN, Alam A, Hasnain SE. Protein promiscuity in drug discovery, drug-repurposing and antibiotic resistance. Biochimie. 2020;175:50–57.

[72] Riahi MM, Sahebkar A, Sadri K, Nikoofal-Sahlabadi S, Jaafari MR. Stable and sustained release liposomal formulations of celecoxib: In vitro and in vivo anti-tumor evaluation. Int J Pharm. 2018;540:89–97.

[73] Lu Z, Long Y, Cun X, Wang X, Li J, Mei L, Yang Y, Li M, Zhang Z, He Q. A size-shrinkable nanoparticle-based combined anti-tumor and anti-inflammatory strategy for enhanced cancer therapy. Nanoscale. 2018;10:9957–9970.

[74] Rostamkalaei SS, Akbari J, Saeedi M, Morteza-Semnani K, Nokhodchi A. Topical gel of Metformin solid lipid nanoparticles: A hopeful promise as a dermal delivery system. Colloids Surf. 2019;175:150–157.

[75] Thapa RK, Nguyen HT, Jeong JH, Kim JR, Choi HG, Yong CS, Kim JO. Progressive slowdown/prevention of cellular senescence by CD9-targeted delivery of rapamycin using lactose-wrapped calcium carbonate nanoparticles. Sci Rep. 2017;7:43299.

[76] Da Silveira EF, Chassot JM, Teixeira FC, Azambuja JH, Debom G, Beira FT, Del Pino FA, Lourenço A, Horn AP, Cruz L, Spanevello RM. Ketoprofen-loaded polymeric nanocapsules selectively inhibit cancer cell growth in vitro and in preclinical model of glioblastoma multiforme. Invest New Drugs. 2013;31:1424–1435.

[77] El-Moslemany RM, Eissa MM, Ramadan AA, El-Khordagui LK, El-Azzouni MZ. Miltefosine lipid nanocapsules: Intersection of drug repurposing and nanotechnology for single dose oral treatment of pre-patent schistosomiasis mansoni. Acta Trop. 2016;159:142–148.

[78] Nagai N, Yoshioka C, Mano Y, Tnabe W, Ito Y, Okamoto N, Shimomura Y. A nanoparticle formulation of disulfiram prolongs corneal residence time of the drug and reduces intraocular pressure. Exp Eye Res. 2015;132:115–123.

[79] Deguchi S, Otake H, Nakazawa Y, Hiramatsu N, Yamamoto N, Nagai N. Ophthalmic formulation containing nilvadipine nanoparticles prevents retinal dysfunction in rats injected with streptozotocin. Int J Mol Sci. 2017;18:2720.

[80] Andalib S, Molhemazar P, Danafar H. In vitro and in vivo delivery of atorvastatin: A comparative study of anti-inflammatory activity of atorvastatin loaded copolymeric micelles. J Biomater Appl. 2018;32:1127–1138.

[81] Eskinazi-Budge A, Manickavasagam D, Czech T, Novak K, Kunzler J, Oyewumi MO. Preparation of emulsifying wax/glyceryl monooleate nanoparticles and evaluation as a delivery system for repurposing simvastatin in bone regeneration. Drug Dev Ind Pharm. 2018;44:1583–1590.

[82] Nath SD, Linh NT, Sadiasa A, Lee BT. Encapsulation of simvastatin in PLGA microspheres loaded into hydrogel loaded BCP porous spongy scaffold as a controlled drug delivery system for bone tissue regeneration. J Biomater Appl. 2014;28:1151–1163.

[83] Gupta N, Al-Saikhan FI, Patel B, Rashid J, Ahsan F. Fasudil and SOD packaged in peptide-studded-liposomes: Properties, pharmacokinetics and ex-vivo targeting to isolated perfused rat lungs. Int J Pharm. 2015;488:33–43.

[84] Jaromin A, Zarnowski R, Piętka-Ottlik M, Andes DR, Gubernator J. Topical delivery of ebselen encapsulated in biopolymeric nanocapsules: Drug repurposing enhanced antifungal activity. Nanomedicine. 2018;13:1139–1155.

[85] O'Banion MK. Cyclooxygenase-2: Molecular biology, pharmacology, and neurobiology. Crit Rev Neurobiol. 1999;13:45–82.
[86] Aronoff DM, Neilson EG. Antipyretics: Mechanisms of action and clinical use in fever suppression. Am J Med. 2001;111:304–315.
[87] Brown NG, Costanzo MC. Interactions among three proteins that specifically activate translation of the mitochondrial COX-3 mRNA in Saccharomyces cerevisiae. Mol Cell Biol. 1994;14:1045–1053.
[88] Smith MJ, Dawkins PD. Salicylate and enzymes. J Pharm Pharmacol. 1971;23:729–744.
[89] Collier HOJ. A pharmacological analysis of aspirin. Adv Pharmacol Chemother. 1969;7:333–405.
[90] Bazan NG, Flower RJ. Lipid signals in pain control. Nature. 2002;420:135–138.
[91] Warner TD, Giuliano F, Vojnovic I, Bukasa A, Mitchell JA, Vane JR. Nonsteroid drug selectivities for cyclo-oxygenase-1 rather than cyclo-oxygenase-2 are associated with human gastrointestinal toxicity: A full in vitro analysis. Proc Natl Acad Sci. 1999;96:7563–7568.
[92] Crofford LJ, Lipsky PE, Brooks P, Abramson SB. Basic biology and clinical application of specific cyclooxygenase-2 inhibitors. Arthritis Rheumatol. 2000;43:4–13.
[93] Moore DE. Drug-induced cutaneous photosensitivity. Drug Saf. 2002;25:345–372.
[94] Sun SX, Lee KY, Bertram CT, Goldstein JL. Withdrawal of COX-2 selective inhibitors rofecoxib and valdecoxib: Impact on NSAID and gastroprotective drug prescribing and utilization. Curr Med Res Op. 2007;(23):1859–1866.
[95] Sibbald B. Rofecoxib (Vioxx) voluntarily withdrawn from market. Can Med Assoc J. 2004;171:1027–1028.
[96] Riendeau D, Percival MD, Brideau C, Charleson S, Dube D, Ethier D, Falgueyret JP, Friesen RW, Gordon R, Greig G, Guay J. Etoricoxib (MK-0663): Preclinical profile and comparison with other agents that selectively inhibit cyclooxygenase-2. J Pharmacol Exp Ther. 2001;296:558–566.
[97] FDA drug approval letter. Available at https://www.accessdata.fda.gov/drugsatfda_docs/nda/98/20998.cfm
[98] Huang WW, Hsieh KP, Huang RY, Yang YH. Role of cyclooxygenase-2 inhibitors in the survival outcome of colorectal cancer patients: A population-based cohort study. Kaohsiung J Med Sci. 2017;33:308–314.
[99] Shao J, Jung C, Liu C, Sheng H. Prostaglandin E2 stimulates the β-catenin/T cell factor-dependent transcription in colon cancer. J Biol Chem. 2005;280:26565–26572.
[100] Chiblak S, Steinbauer B, Pohl-Arnold A, Kucher D, Abdollahi A, Schwager C, Müller-Decker K. K-Ras and cyclooxygenase-2 coactivation augments intraductal papillary mucinous neoplasm and Notch1 mimicking human pancreas lesions. Sci Rep. 2016;6:1–13.
[101] Hu H, Han T, Zhuo M, Wu LL, Yuan C, Wu L, Wang LW. Elevated COX-2 expression promotes angiogenesis through EGFR/p38-MAPK/Sp1-dependent signalling in pancreatic cancer. Sci Rep. 2017;7:1–10.
[102] Thill M, Fischer D, Kelling K, Hoellen F, Dittmer C, Hornemann A, Becker S. Expression of vitamin D receptor (VDR), cyclooxygenase-2 (COX-2) and 15-hydroxyprostaglandin dehydrogenase (15-PGDH) in benign and malignant ovarian tissue and 25-hydroxycholecalciferol (25 (OH 2) D 3) and prostaglandin E 2 (PGE 2) serum level in ovarian cancer patients. J Steroid Biochem Mol Biol. 2010;121:387–390.
[103] Chen H, Cai W, Chu ESH, Tang J, Wong CC, Wong SH, Yu J. Hepatic cyclooxygenase-2 overexpression induced spontaneous hepatocellular carcinoma formation in mice. Oncogene. 2017;36:4415.
[104] Ko CJ, Lan SW, Lu YC, Cheng TS, Lai PF, Tsai CH, Lin HH. Inhibition of cyclooxygenase-2-mediated matriptase activation contributes to the suppression of prostate cancer cell motility and metastasis. Oncogene. 2017: 1–13.
[105] Krishnamachary B, Stasinopoulos I, Kakkad S, Penet MF, Jacob D, Wildes F, Bhujwalla ZM. Breast cancer cell cyclooxygenase-2 expression alters extracellular matrix structure and function and numbers of cancer associated fibroblasts. Oncotarget. 2017;8:17981.

[106] Tian J, Hachim MY, Hachim IY, Dai M, Lo C, Al Raffa F, Lebrun JJ. Cyclooxygenase-2 regulates TGFβ-induced cancer stemness in triple-negative breast cancer. Sci Rep. 2017;7:1–16.
[107] Pan J, Cheng L, Bi X, Zhang X, Liu S, Bai X, Zhao AZ. Elevation of ω-3 polyunsaturated fatty acids attenuates PTEN-deficiency induced endometrial cancer development through regulation of COX-2 and PGE2 production. Sci Rep. 2015;5:1–12.
[108] Campillo N, Torres M, Vilaseca A, Nonaka PN, Gozal D, Roca-Ferrer J, Almendros I. Role of Cyclooxygenase-2 on Intermittent Hypoxia-Induced Lung Tumor Malignancy in a Mouse Model of Sleep Apnea. Sci Rep. 2017;7:1–11.
[109] Janakiraman H, House RP, Talwar S, Courtney SM, Hazard ES, Hardiman G, Palanisamy V. Repression of caspase-3 and RNA-binding protein HuR cleavage by cyclooxygenase-2 promotes drug resistance in oral squamous cell carcinoma. Oncogene. 2017;36:3137.
[110] Shao Y, Li P, Zhu ST, Yue JP, Ji XJ, Ma D, Zhang ST. MiR-26a and miR-144 inhibit proliferation and metastasis of esophageal squamous cell cancer by inhibiting cyclooxygenase-2. Oncotarget. 2016;7:15173.
[111] D'Arca D, LeNoir J, Wildemore B, Gottardo F, Bragantini E, Shupp-Byrne D, Baffa R. Prevention of urinary bladder cancer in the FHIT knock-out mouse with Rofecoxib, a Cox-2 inhibitor. Urol Oncol. 2010;28:189–194.
[112] Ho L, Purohit D, Haroutunian V, Luterman JD, Willis F, Naslund J, Pasinetti GM. Neuronal cyclooxygenase-2 expression in the hippocampal formation as a function of the clinical progression of Alzheimer disease. Arch Neurol. 2001;58:487–492.
[113] Teismann P, Tieu K, Choi DK, Wu DC, Naini A, Hunot S, Przedborski S. Cyclooxygenase-2 is instrumental in Parkinson's disease neurodegeneration. Proc Natl Acad Sci. 2003;100:5473–5478.
[114] Keller WR, Kum LM, Wehring HJ, Koola MM, Buchanan RW, Kelly DL. A review of anti-inflammatory agents for symptoms of schizophrenia. J Psychopharmacol. 2013;27:337–342.
[115] Müller N, Riedel M, Scheppach C, Brandstätter B, Sokullu S, Krampe K, Ulmschneider M, Engel RR, Möller HJ, Schwarz MJ. Beneficial antipsychotic effects of celecoxib add-on therapy compared to risperidone alone in schizophrenia. Am J Psychiatry. 2002;159:1029–1034.
[116] Katyal J, Kumar H, Gupta YK. Anticonvulsant activity of the cyclooxygenase-2 (COX-2) inhibitor etoricoxib in pentylenetetrazole-kindled rats is associated with memory impairment. Epilepsy Behav. 2015;44:98–103.
[117] Chen Y, Chen P, Hanaoka M, Droma Y, Kubo K. Enhanced levels of prostaglandin E2 and matrix metalloproteinase-2 correlate with the severity of airflow limitation in stable COPD. Respirology. 2008;13:1014–1021.
[118] Xaubet A, Roca-Ferrer J, Pujols L, Ramirez J, Mullol J, Marin-Arguedas A, Picado C. Cyclooxygenase-2 is up-regulated in lung parenchyma of chronic obstructive pulmonary disease and down-regulated in idiopathic pulmonary fibrosis. Sarcoidosis Vasc Diffuse Lung Dis. 2004;21:35–42.
[119] Sousa A, Pfister R, Christie PE, Lane SJ, Nasser SM, Schmitz-Schumann M, Lee TH. Enhanced expression of cyclo-oxygenase isoenzyme 2 (COX-2) in asthmatic airways and its cellular distribution in aspirin-sensitive asthma. Thorax. 1997;52:940–945.
[120] Woods JM, Mogollon A, Amin MA, Martinez RJ, Koch AE. The role of COX-2 in angiogenesis and rheumatoid arthritis. Exp Mol Pathol. 2003;74:282–290.
[121] De Boer TN, Huisman AM, Polak AA, Niehoff AG, Van Rinsum AC, Saris D, Mastbergen SC. The chondroprotective effect of selective COX-2 inhibition in osteoarthritis: Ex vivo evaluation of human cartilage tissue after in vivo treatment. Osteoarthr Cartil. 2009;17:482–488.
[122] Kellogg AP, Pop-Busui R. Peripheral nerve dysfunction in experimental diabetes is mediated by cyclooxygenase-2 and oxidative stress. Antioxid Redox Signal. 2005;7:1521–1529.
[123] Pop-Busui R, Marinescu V, Van Huysen C, Li F, Sullivan K, Greene DA, Stevens MJ. Dissection of metabolic, vascular, and nerve conduction interrelationships in experimental diabetic neuropathy by cyclooxygenase inhibition and acetyl-L-carnitine administration. Diabetes. 2002;51:2619–2628.

[124] Lin CK, Tseng CK, Wu YH, Liaw CC, Lin CY, Huang CH, Lee JC. Cyclooxygenase-2 facilitates dengue virus replication and serves as a potential target for developing antiviral agents. Sci Rep. 2017;7:1–15.
[125] Wang H, Zhang D, Ge M, Li Z, Jiang J, Li Y. Formononetin inhibits enterovirus 71 replication by regulating COX-2/PGE 2 expression. Virol J. 2015;12:35.
[126] Lin YT, Wu YH, Tseng CK, Lin CK, Chen WC, Hsu YC, Lee JC. Green tea phenolic epicatechins inhibit hepatitis C virus replication via cycloxygenase-2 and attenuate virus-induced inflammation. PloSONE. 2013;8:1–10.
[127] Yue X, Yang F, Yang Y, Mu Y, Sun W, Li W, Zhu Y. Induction of cyclooxygenase-2 expression by hepatitis B virus depends on demethylation-associated recruitment of transcription factors to the promoter. Virol J. 2011;8:118.
[128] Zhu H, Cong JP, Yu D, Bresnahan WA, Shenk TE. Inhibition of cyclooxygenase 2 blocks human cytomegalovirus replication. Proc Natl Acad Sci. 2002;99:3932–3937.
[129] Dumais N, Barbeau B, Olivier M, Tremblay MJ. Prostaglandin E_2 up-regulates HIV-1 long terminal repeat-driven gene activity in T cells via NF-κB-dependent and-independent signaling pathways. J Biol Chem. 1998;273:27306–27314.
[130] Gebhardt BM, Varnell ED, Kaufman HE. Inhibition of cyclooxygenase 2 synthesis suppresses Herpes simplex virus type 1 reactivation. J Ocul Pharmacol Ther. 2005;21:114–120.
[131] Pyeon D, Diaz FJ, Splitter GA. Prostaglandin E_2 Increases Bovine Leukemia Virustax and pol mRNA Levels via Cyclooxygenase 2: Regulation by Interleukin-2, Interleukin-10, and Bovine Leukemia Virus. J Virol. 2000;74:5740–5745.
[132] Moriuchi M, Inoue H, Moriuchi H. Reciprocal interactions between human T-lymphotropic virus type 1 and prostaglandins: Implications for viral transmission. J Virol. 2001;75:192–198.
[133] Lee SM, Gai WW, Cheung TK, Peiris JSM. Antiviral effect of a selective COX-2 inhibitor on H5N1 infection in vitro. Antiviral Res. 2011;91:330–334.
[134] Wang W, Ning Y, Wang Y, Deng G, Pace S, Barth SA, Menge C, Zhang K, Dai Y, Cai Y, Chen X. Mycobacterium tuberculosis-Induced Upregulation of the COX-2/mPGES-1 Pathway in Human Macrophages Is Abrogated by Sulfasalazine. Front Immunol. 2022;13.
[135] Marnett LJ. Aspirin and the potential role of prostaglandin in colon cancer. Cancer Res. 1992;52:5575–5589.
[136] Subbaramaiah K, Dannenberg AJ. Cyclooxygenase 2: A molecular target for cancer prevention and treatment. Trends Pharmacol Sci. 2003;24:96–102.
[137] Xu L, Stevens J, Hilton MB, Seaman S, Conrads TP, Veenstra TD, Kalen J. COX-2 inhibition potentiates antiangiogenic cancer therapy and prevents metastasis in preclinical models. Sci Transl Med. 2014;6:1–12.
[138] Jaime LM, Kathleen ML, Alane TK, Ben SZ, Steven LS, Woerner BM, Dorothy AE, Flickinger A, Moore RJ, Seibert K. Antiangiogenic and Antitumor Activities of Cyclooxygenase-2 Inhibitors. Cancer Res. 2000;60:1306–1311.
[139] Bazan N, Botting J, Vane JR. New Targets in Inflammation: Inhibitors of COX-2 or Adhesion Molecules Proceedings of a Conference Held on April 15–16, 1996, in New Orleans, USA, Supported by an Educational Grant from Boehringer Ingelheim. Springer Science & Business Media, 2012.
[140] Niranjan R. The role of inflammatory and oxidative stress mechanisms in the pathogenesis of Parkinson's disease: Focus on astrocytes. Mol Neurobiol. 2014:28–38.
[141] Mitchell JA, Warner TD. COX isoforms in the cardiovascular system: Understanding the activities of non-steroidal anti-inflammatory drugs. Nat Rev Drug Discov. 2006;5:75–86.
[142] Van Dam PS, Cotter MA, Bravenboer B, Cameron NE. Pathogenesis of diabetic neuropathy: Focus on neurovascular mechanisms. Eur J Pharmacol. 2013;719:180–186.

[143] Fondel E, Eilis JO, Fitzgerald KC, Falcone GJ, Mccullough ML, Thun MJ, Park Y, Colonel LN, Ascheri A. Non-steroidal antiinflammatory drugs and amyotrophic lateral sclerosis: Results from five prospective cohort studies. Amyo Lat Sclerosis. 2012;13:573–579.
[144] Palumbo S, Bosetti F. Alterations of brain eicosanoid synthetic pathway in multiple sclerosis and in animal models of demyelination: Role of cyclooxygenase-2. Prostaglandins Leukot Essent Fatty Acids. 2013;89:273–278.
[145] Morales I, Guzmán-Martínez L, Cerda-Troncoso C, Farías GA, Maccioni RB. Neuroinflammation in the pathogenesis of Alzheimer's disease. A rational framework for the search of novel therapeutic approaches. Front Cell Neurosci. 2014;8:1–12.
[146] Jiang J, Yang M, Quan Y, Gueorguieva P, Ganesh T, Dingledine R. Therapeutic window for cyclooxygenase-2 related anti-inflammatory therapy after status epilepticus. Neurobiol Dis. 2015;76:126–136.
[147] McCoy MK, Tansey MG. TNF signaling inhibition in the CNS: Implications for normal brain function and neurodegenerative disease. J Neuroinflammation. 2008;5:45.
[148] Simi A, Tsakiri N, Wang P, Rothwell NJ. Interleukin-1 and inflammatory neurodegeneration. Biochem Soc Trans. 2007;35:1122–1126.
[149] Saijo K, Winner B, Carson CT, Collier JG, Boyer L, Rosenfeld MG, Glass CK. A Nurr1/CoREST pathway in microglia and astrocytes protects dopaminergic neurons from inflammation-induced death. Cell. 2009;137:47–59.
[150] Akiyama H, Barger S, Barnum S, Bradt B, Baucer J, Cole GM. Inflammation and Alzheimer's disease. Neurobiol Aging. 2000;21:383–421.
[151] FitzGerald GA. COX-2 and beyond: Approaches to prostaglandin inhibition in human disease. Nat Rev Drug Discov. 2003;2:879–890.
[152] Sommer IE, De WL, Begemann M, Kahn RS. Nonsteroidal anti-inflammatory drugs in schizophrenia: Ready for practice or a good start? A meta-analysis. J Clin Psychiatry. 2012;73:414–419.
[153] Citraro R, Leo A, Marra R, Sarro GD, Russo E. Antiepileptogenic effects of the selective COX-2 Inhibitor etoricoxib, on the development of spontaneous absence seizures in WAG/Rij rats. Brain Res Bull. 2015;113:1–7.
[154] Zago M, Souza ARD, Hecht E, Rousseau S, Hamid Q, Eidelman DH, Baglole CJ. The NF-B family member RelB regulates microRNA miR-146a to suppress cigarette smoke-induced COX-2 Protein expression in lung fibroblasts. Toxicol Lett. 2014;26:107–116.
[155] Chiles NS, Phillips CL, Volpato S, Bandinelli SP, Ferrucci L, Guralnik JM, Patel KV. Diabetes, peripheral neuropathy, and lower-extremity function. J Diabetes Complicat. 2014;28:91–95.
[156] Stavniichuk R, Obrosov AA, Drel VR, Nadler JL, Obrosova IG, Yorek MA. 12/15-Lipoxygenase inhibition counteracts MAPK phosphorylation in mouse and cell culture models of diabetic peripheral neuropathy. J Diabetes Mellit. 2013;3:1–15.
[157] Kellogg AP, Cheng HT, Pop-Busui R. Cyclooxygenase-2 pathway as a potential therapeutic target in diabetic peripheral neuropathy. Curr Drug Targets. 2008;1:68–76.
[158] Tung WH, Hsieh HL, Yang CM. Enterovirus 71 induces COX-2 expression via MAPKs, NF-κB, and AP-1 in SK–N–SH cells: Role of PGE_2 in viral replication. Cell Signal. 2010;22:234–246.
[159] Tung WH, Lee IT, Hsieh HL, Yang CM. EV71 induces COX-2 expression via c-Src/PDGFR/PI3K/Akt/p42/p44 MAPK/AP-1 and NF-κB in rat brain astrocytes. J Cell Physiol. 2010;224:376–386.
[160] Hashimoto NAOAKI, Watanabe TSUYOSHI, Ikeda Y, Yamada HARUKI, Taniguchi SHIGEO, Mitsui HIROSHI, Kurokawa K. Prostaglandins induce proliferation of rat hepatocytes through a prostaglandin E2 receptor EP3 subtype. Am. J Physiol Gastrointest Liver Physiol. 1997;272:G597–G604.
[161] Kumar S, Bansal K, Holla S, Verma-Kumar S, Sharma P, Balaji KN. ESAT-6 induced COX-2 expression involves coordinated interplay between PI3K and MAPK signaling. Mol Immunol. 2012;49:655–663.

[162] Zhao JW, Sun ZQ, Zhang XY, Zhang Y, Liu J, Ye J, Zhang SL. Mycobacterial 3-hydroxyacyl-l-thioester dehydratase Y derived from Mycobacterium tuberculosis induces COX-2 expression in mouse macrophages through MAPK-NF-κB pathway. Immunology Lett. 2014;161:125–132.
[163] Phipps RP, Stein SH, Roper RL. A new view of prostaglandin E regulation of the immune response. Immunol Today. 1991;12:349–352.
[164] Betz M, Fox BS. Prostaglandin E2 inhibits production of Th1 lymphokines but not of Th2 lymphokines. J Immunol. 1991;146:108–113.
[165] Wu CY, Wang K, McDyer JF, Seder RA. Prostaglandin E2 and dexamethasone inhibit IL-12 receptor expression and IL-12 responsiveness. J Immunol. 1998;161:2723–2730.
[166] Van der Pouw Kraan TC, Boeije LC, Smeenk RJ, Wijdenes J, Aarden LA. Prostaglandin-E2 is a potent inhibitor of human interleukin 12 production. J Exp Med. 1995;181:775–779.
[167] Kuroda E, Sugiura T, Zeki K, Yoshida Y, Yamashita U. Sensitivity difference to the suppressive effect of prostaglandin E2 among mouse strains: A possible mechanism to polarize Th2 type response in BALB/c mice. J Immunol. 2000;164:2386–2395.
[168] Goto T, Herberman RB, Maluish A, Strong DM. Cyclic AMP as a mediator of prostaglandin E-induced suppression of human natural killer cell activity. J Immunol. 1983;130:1350–1355.
[169] Snyder DS, Beller DI, Unanue ER. Prostaglandins modulate macrophage Ia expression. Nature. 1982;299:163–165.
[170] Schultz RM, Pavlidis NA, Stylos WA, Chirigos MA. Regulation of macrophage tumoricidal function: A role for prostaglandins of the E series. Science. 1978;202:320–321.
[171] Alqahtani AM, Chidambaram K, Pino-Figueroa A, Chandrasekaran B, Dhanaraj P, Venkatesan K. Curcumin-Celecoxib: A synergistic and rationale combination chemotherapy for breast cancer. Eur Rev Med Pharmacol Sci. 2021;25:1916–1927.
[172] Cho HY, Thomas S, Golden EB, Gaffney KJ, Hofman FM, Chen TC, Louie SG, Petasis NA, Schönthal AH. Enhanced killing of chemo-resistant breast cancer cells via controlled aggravation of ER stress. Cancer Lett. 2009;282:87–97.
[173] Mishra RK, Ahmad A, Kumar A, Vyawahare A, Raza SS, Khan R. Lipid-based nanocarrier-mediated targeted delivery of celecoxib attenuate severity of ulcerative colitis. Mater Sci Eng. 2020;116:111103.
[174] Li J, Tian X, Wang K, Jia Y, Ma F. Transdermal delivery of celecoxib and α-linolenic acid from microemulsion-incorporated dissolving microneedles for enhanced osteoarthritis therapy. J Drug Target. 2023;31(2):206–216.
[175] Bruno MC, Cristiano MC, Celia C, d'Avanzo N, Mancuso A, Paolino D, Wolfram J, Fresta M. Injectable Drug Delivery Systems for Osteoarthritis and Rheumatoid Arthritis. ACS Nano. 2022;16 (12):19665–19690.
[176] AlSawaftah NM, Awad NS, Pitt WG, Husseini GA. pH-responsive nanocarriers in cancer therapy. Polymers. 2022;14:936.
[177] Singh V, Md S, Alhakamy NA, Kesharwani P. Taxanes loaded polymersomes as an emerging polymeric nanocarrier for cancer therapy. Eur Polymer J. 2022;162:110883.
[178] Üner M, Yener G, Ergüven M. Design of colloidal drug carriers of celecoxib for use in treatment of breast cancer and leukemia. Mater Sci Eng. 2019;103:109874.
[179] Elzoghby AO, Mostafa SK, Helmy MW, ElDemellawy MA, Sheweita SA. Multi-reservoir phospholipid shell encapsulating protamine nanocapsules for co-delivery of letrozole and celecoxib in breast cancer therapy. Pharm Res. 2017;34:1956–1969.
[180] Robledo-Cadena DX, Gallardo-Pérez JC, Dávila-Borja V, Pacheco-Velázquez SC, Belmont-Díaz JA, Ralph SJ, Blanco-Carpintero BA, Moreno-Sánchez R, Rodríguez-Enríquez S. Non-steroidal anti-inflammatory drugs increase cisplatin, paclitaxel, and doxorubicin efficacy against human cervix cancer cells. Pharmaceuticals. 2020;13:463.
[181] Warnakulasuriya S, Greenspan JS. Epidemiology of oral and oropharyngeal cancers. In Warnakulasuriya S, Greenspan J, editors. Textbook of Oral Cancer. Berlin, Springer, 2020, 5–22.

[182] Tołoczko-Iwaniuk N, Dziemiańczyk-Pakieła D, Celińska-Janowicz K, Zaręba I, Klupczyńska A, Kokot ZJ, Nowaszewska BK, Reszeć J, Borys J, Miltyk W. Proline-dependent induction of apoptosis in oral squamous cell carcinoma (OSCC) – The effect of celecoxib. Cancers. 2020;12:136.
[183] Chiang SL, Velmurugan BK, Chung CM, Lin SH, Wang ZH, Hua CH, Tsai MH, Kuo TM, Yeh KT, Chang PY, Yang YH. Preventive effect of celecoxib use against cancer progression and occurrence of oral squamous cell carcinoma. Sci Rep. 2017;7:1–1.
[184] Wróbel K, Wołowiec S, Markowicz J, Wałajtys-Rode E, Uram Ł. Synthesis of biotinylated PAMAM G3 dendrimers substituted with R-glycidol and celecoxib/simvastatin as repurposed drugs and evaluation of their increased additive cytotoxicity for cancer cell lines. Cancers. 2022;14:714.
[185] Yamaguchi I, Nakajima K, Shono K, Mizobuchi Y, Fujihara T, Shikata E, Yamaguchi T, Kitazato K, Sampetrean O, Saya H, Takagi Y. Downregulation of PD-L1 via FKBP5 by celecoxib augments antitumor effects of PD-1 blockade in a malignant glioma model. Neuro-oncol Adv. 2020;2:058.
[186] Xiong S, Huang W, Liu X, Chen Q, Ding Y, Huang H, Zhang R, Guo J. Celecoxib Synergistically Enhances MLN4924-Induced Cytotoxicity and EMT Inhibition Via AKT and ERK Pathways in Human Urothelial Carcinoma. Cell Transplant. 2022;31:09636897221077921.
[187] Suri A, Sheng X, Schuler KM, Zhong Y, Han X, Jones HM, Gehrig PA, Zhou C, Bae-Jump VL. The effect of celecoxib on tumor growth in ovarian cancer cells and a genetically engineered mouse model of serous ovarian cancer. Oncotarget. 2016;7:39582.
[188] Wang YP, Wang QY, Li CH, Li XW. COX-2 inhibition by celecoxib in epithelial ovarian cancer attenuates E-cadherin suppression through reduced Snail nuclear translocation. Chem Biol Interact. 2018;292:24–29.
[189] Patel MI, Subbaramaiah K, Du B, Chang M, Yang P, Newman RA, Cordon-Cardo C, Thaler HT, Dannenberg AJ. Celecoxib inhibits prostate cancer growth: Evidence of a cyclooxygenase-2-independent mechanism. Clin Cancer Investig J. 2005;11:1999–2007.
[190] Wang W, Kardosh A, Su YS, Schonthal AH, Chen TC. Efficacy of celecoxib in the treatment of CNS lymphomas: An in vivo model. Neurosurg Focus. 2006;21:1–8.
[191] Srivastava S, Dewangan J, Mishra S, Divakar A, Chaturvedi S, Wahajuddin M, Kumar S, Rath SK. Piperine and Celecoxib synergistically inhibit colon cancer cell proliferation via modulating Wnt/β-catenin signaling pathway. Phytomedicine. 2021;84:153484.
[192] Lin YM, Lu CC, Hsiang YP, Pi SC, Chen CI, Cheng KC, Pan HL, Chien PH, Chen YJ. c-Met inhibition is required for the celecoxib-attenuated stemness property of human colorectal cancer cells. J Cell Physiol. 2019;234:10336–10344.
[193] Chen WC, Lin MS, Ye YL, Gao HJ, Song ZY, Shen XY. microRNA expression pattern and its alteration following celecoxib intervention in human colorectal cancer. Exp Ther Med. 2012;3:1039–1048.
[194] Lerdwanangkun P, Wonganan P, Storer RJ, Limpanasithikul W. Combined effects of celecoxib and cepharanthine on human colorectal cancer cells in vitro. J Appl Pharm Sci. 2019;9:117–125.
[195] Dai H, Zhang S, Ma R, Pan L. Celecoxib inhibits hepatocellular carcinoma cell growth and migration by targeting PNO1. Med Sci Monit. 2019;25:7351.
[196] Zhang C, Zhou J, Hu J, Lei S, Yuan M, Chen L, Wang G, Qiu Z. Celecoxib attenuates hepatocellular proliferative capacity during hepatocarcinogenesis by modulating a PTEN/NF-κB/PRL-3 pathway. RSC Adv. 2019;9:20624–20632.
[197] Yeh CC, Liao PY, Pandey S, Yung SY, Lai HC, Jeng LB, Chang WC, Ma WL. Metronomic celecoxib therapy in clinically available dosage ablates hepatocellular carcinoma via suppressing cell invasion, growth, and stemness in pre-clinical models. Front Oncol. 2020:2300.
[198] Xie H, Gao L, Chai N, Song J, Wang J, Song Z, Chen C, Pan Y, Zhao L, Sun S, Wu K. Potent cell growth inhibitory effects in hepatitis B virus X protein positive hepatocellular carcinoma cells by the selective cyclooxygenase-2 inhibitor celecoxib. Mol Carcinog. 2009;48:56–65.

[199] Cao N, Lu Y, Liu J, Cai F, Xu H, Chen J, Zhang X, Hua ZC, Zhuang H. Metformin synergistically enhanced the antitumor activity of celecoxib in human non-small cell lung cancer cells. Front Pharmacol. 2020;11:1094.

[200] Kim J, Noh MH, Hur DY, Kim B, Kim YS, Lee HK. Celecoxib upregulates ULBP-1 expression in lung cancer cells via the JNK/PI3K signaling pathway and increases susceptibility to natural killer cell cytotoxicity. Oncol Lett. 2020;20:1.

[201] Zhang W, Yi L, Shen J, Zhang H, Luo P, Zhang J. Comparison of the benefits of celecoxib combined with anticancer therapy in advanced non-small cell lung cancer: A meta-analysis. J Cancer. 2020;11:1816.

[202] Zhang P, He D, Song E, Jiang M, Song Y. Celecoxib enhances the sensitivity of non-small-cell lung cancer cells to radiation-induced apoptosis through downregulation of the Akt/mTOR signaling pathway and COX-2 expression. PLoS One. 2019;14:e0223760.

[203] Liu X, Wu Y, Zhou Z, Huang M, Deng W, Wang Y, Zhou X, Chen L, Li Y, Zeng T, Wang G. Celecoxib inhibits the epithelial-to-mesenchymal transition in bladder cancer via the miRNA-145/TGFBR2/Smad3 axis. Int J Mol Med. 2019;44:683–693.

[204] Saito Y, Suzuki H, Imaeda H, Matsuzaki J, Hirata K, Tsugawa H, Hibino S, Kanai Y, Saito H, Hibi T. The tumor suppressor microRNA-29c is downregulated and restored by celecoxib in human gastric cancer cells. Int J Cancer. 2013;132:1751–1760.

[205] Guo Q, Li Q, Wang J, Liu M, Wang Y, Chen Z, Ye Y, Guan Q, Zhou Y. A comprehensive evaluation of clinical efficacy and safety of celecoxib in combination with chemotherapy in metastatic or postoperative recurrent gastric cancer patients: A preliminary, three-center, clinical trial study. Medicine. 2019;98.

[206] Choi SM, Cho YS, Park G, Lee SK, Chun KS. Celecoxib induces apoptosis through Akt inhibition in 5-fluorouracil-resistant gastric cancer cells. Toxicol Res. 2021;37:25–33.

[207] Zuo C, Hong Y, Qiu X, Yang D, Liu N, Sheng X, Zhou K, Tang B, Xiong S, Ma M, Liu Z. Celecoxib suppresses proliferation and metastasis of pancreatic cancer cells by down-regulating STAT3/NF-kB and L1CAM activities. Pancreatol. 2018;18:328–333.

[208] Chen JS, Chou CH, Wu YH, Yang MH, Chu SH, Chao YS, Chen CN. CC-01 (chidamide plus celecoxib) modifies the tumor immune microenvironment and reduces tumor progression combined with immune checkpoint inhibitor. Sci Rep. 2022;12:1–8.

[209] Limasale YD, Tezcaner A, Özen C, Keskin D, Banerjee S. Epidermal growth factor receptor-targeted immunoliposomes for delivery of celecoxib to cancer cells. Int J Pharm. 2015;479:364–373.

[210] Ahmed KS, Changling S, Shan X, Mao J, Qiu L, Chen J. Liposome-based codelivery of celecoxib and doxorubicin hydrochloride as a synergistic dual-drug delivery system for enhancing the anticancer effect. J Liposome Res. 2020;30:285–296.

[211] Alajami HN, Fouad EA, Ashour AE, Kumar A, Yassin AE. Celecoxib-loaded solid lipid nanoparticles for colon delivery: Formulation optimization and in vitro assessment of anti-cancer activity. Pharmaceutics. 2022;14:131.

[212] Chandel D, Uppal S, Mehta SK, Shukla G. Preparation and characterization of celecoxib entrapped guar gum nanoparticles targeted for oral drug delivery against colon cancer: An in-vitro study. J Drug Deliv Ther. 2020;10:14–21.

[213] Karai E, Szebényi K, Windt T, Fehér S, Szendi E, Dékay V, Vajdovich P, Szakács G, Füredi A. Celecoxib prevents doxorubicin-induced multidrug resistance in canine and mouse lymphoma cell lines. Cancers. 2020;12:1117.

[214] https://clinicaltrials.gov (A service of the U.S. National Institutes of Health)

[215] Tanwar L, Vaish V, Sanyal SN. Chemopreventive role of etoricoxib (MK-0663) in experimental colon cancer. Eur J Cancer Prev. 2010;19:280–287.

[216] Ali G, Omar H, Hersi F, Abo-Youssef A, Ahmed O, Mohamed W. The protective role of etoricoxib against diethylnitrosamine/2-acetylaminofluorene-induced hepatocarcinogenesis in Wistar rats: The impact of NF-κB/COX-2/PGE2 signaling. Curr Mol Pharmacol. 2022;15:252–262.
[217] Md S, Alhakamy NA, Alharbi WS, Ahmad J, Shaik RA, Ibrahim IM, Ali J. Development and evaluation of repurposed etoricoxib loaded nanoemulsion for improving anticancer activities against lung cancer cells. Int J Mol Sci. 2021;22:13284.
[218] Nadda N, Vaish V, Setia S, Sanyal SN. Angiostatic role of the selective cyclooxygenase-2 inhibitor etoricoxib (MK0663) in experimental lung cancer. Biomed Pharmacother. 2012;66:474–483.
[219] Nadda N, Setia S, Vaish V, Sanyal SN. Role of cytokines in experimentally induced lung cancer and chemoprevention by COX-2 selective inhibitor, etoricoxib. Mol Cell Biochem. 2013;372:101–112.
[220] Xiong W, Li WH, Jiang YX, Liu S, Ai YQ, Liu R, Chang L, Zhang M, Wang XL, Bai H, Wang H. Parecoxib: An enhancer of radiation therapy for colorectal cancer. Asian Pac J Cancer. 2015;16:627–633.
[221] O'Byrne KJ, Danson S, Dunlop D, Botwood N, Taguchi F, Carbone D, Ranson M. Combination therapy with gefitinib and rofecoxib in patients with platinum-pretreated relapsed non–small-cell lung cancer. J Clin Oncol. 2007;25:3266–3273.
[222] Gridelli C, Gallo C, Ceribelli A, Gebbia V, Gamucci T, Ciardiello F, Carozza F, Favaretto A, Daniele B, Galetta D, Barbera S. Factorial phase III randomised trial of rofecoxib and prolonged constant infusion of gemcitabine in advanced non-small-cell lung cancer: The GEmcitabine-COxib in NSCLC (GECO) study. Oncology. 2007;8:500–512.
[223] Dicker AP, Williams TL, Grant DS. Targeting angiogenic processes by combination rofecoxib and ionizing radiation. Am J Clin Oncol. 2001;24:438–442.
[224] Mohseni H, Zaslau S, McFadden DW, Riggs DR, Jackson BJ, Kandzari SJ. Cox-2 inhibition demonstrates potent anti-proliferative effects on bladder cancer in vitro. J Surg Res. 2003;114:276–277.
[225] Zhu FS, Chen XM, Huang ZG, Wang ZR, Zhang DW, Zhang X. Rofecoxib augments anticancer effects by reversing intrinsic multidrug resistance gene expression in BGC-823 gastric cancer cells. J Dig Dis. 2010;11:34–42.
[226] Tseng WW, Deganutti A, Chen MN, Saxton RE, Liu CD. Selective cyclooxygenase-2 inhibitor rofecoxib (Vioxx) induces expression of cell cycle arrest genes and slows tumor growth in human pancreatic cancer. J Gastrointest Surg. 2002;6:838–844.
[227] Vona-Davis L, Riggs DR, Jackson BJ, McFadden DW. Anti-proliferative and apoptotic effects of rofecoxib on esophageal cancer in vitro. J Surg Res. 2003;114:276.
[228] Hochstrasser T, Hohsfield LA, Sperner-Unterweger B, Humpel C. β-Amyloid induced effects on cholinergic, serotonergic, and dopaminergic neurons is differentially counteracted by anti-inflammatory drugs. J Neurosci Res. 2013;91:83–94.
[229] Guo JW, Guan PP, Ding WY, Wang SL, Huang XS, Wang ZY, Wang P. Erythrocyte membrane-encapsulated celecoxib improves the cognitive decline of Alzheimer's disease by concurrently inducing neurogenesis and reducing apoptosis in APP/PS1 transgenic mice. Biomaterials. 2017;145:106–127.
[230] Kalra J, Kumar P, Majeed AB, Prakash A. Modulation of LOX and COX pathways via inhibition of amyloidogenesis contributes to mitoprotection against β-amyloid oligomer-induced toxicity in an animal model of Alzheimer's disease in rats. Pharmacol Biochem Behav. 2016;146:1–2.
[231] Dassati S, Schweigreiter R, Buechner S, Waldner A. Celecoxib promotes survival and upregulates the expression of neuroprotective marker genes in two different in vitro models of Parkinson's disease. Neuropharmacol. 2021;194:108378.
[232] Reksidler AB, Lima MM, Zanata SM, Machado HB, Da Cunha C, Andreatini R, Tufik S, Vital MA. The COX-2 inhibitor parecoxib produces neuroprotective effects in MPTP-lesioned rats. Eur J Pharmacol. 2007;560:163–175.

[233] Yan H, Zhao H, Kang Y, Ji X, Zhang T, Wang Y, Cui R, Zhang G, Shi G. Parecoxib alleviates the motor behavioral decline of aged rats by ameliorating mitochondrial dysfunction in the substantia nigra via COX-2/PGE2 pathway inhibition. Neuropharmacol. 2021;194:108627.

[234] Klivenyi P, Gardian G, Calingasan NY, Yang L, Beal MF. Additive neuroprotective effects of creatine and a cyclooxygenase 2 inhibitor against dopamine depletion in the 1-methyl-4-phenyl-1, 2, 3, 6-tetrahydropyridine (MPTP) mouse model of Parkinson's disease. J Mol Neurosci. 2003;21:191–198.

[235] Gupta A, Kumar A, Kulkarni SK. Targeting oxidative stress, mitochondrial dysfunction and neuroinflammatory signaling by selective cyclooxygenase (COX)-2 inhibitors mitigates MPTP-induced neurotoxicity in mice. Prog Neuropsychopharmacol Biol Psychiatry. 2011;35:974–981.

[236] Vijitruth R, Liu M, Choi DY, Nguyen XV, Hunter RL, Bing G. Cyclooxygenase-2 mediates microglial activation and secondary dopaminergic cell death in the mouse MPTP model of Parkinson's disease. J Neuroinflam. 2006;3:1–6.

[237] Gordon PH, Cheung YK, Levin B, Andrews H, Doorish C, Macarthur RB, Montes J, Bednarz K, Florence J, Rowin J, Boylan K. A novel, efficient, randomized selection trial comparing combinations of drug therapy for ALS. Amyotrophic Lateral Sclerosis. 2008;9:212–222.

[238] Goldshtein H, Muhire A, Petel Legare V, Pushett A, Rotkopf R, Shefner JM, Peterson RT, Armstrong GA, Russek-Blum N. Efficacy of Ciprofloxacin/Celecoxib combination in zebrafish models of amyotrophic lateral sclerosis. Ann Clin Transl Neurol. 2020;7:1883–1897.

[239] Cheng X, Zhao L, Ke T, Wang X, Cao L, Liu S, He J, Rong W. Celecoxib ameliorates diabetic neuropathy by decreasing apoptosis and oxidative stress in dorsal root ganglion neurons via the miR-155/COX-2 axis. Exp Ther Med. 2021;22:1–1.

[240] Suarez-Mendez S, Tovilla-Zarate CA, Ortega-Varela LF, Bermudez-Ocaña DY, Blé-Castillo JL, González-Castro TB, Zetina-Esquivel AM, Diaz-Zagoya JC, Esther Juárez-Rojop I. Isobolographic analyses of proglumide–celecoxib interaction in rats with painful diabetic neuropathy. Drug Dev Res. 2017;78:116–123.

[241] Bujalska-Zadrożny M, De Cordé A, Pawlik K. Influence of nitric oxide synthase or cyclooxygenase inhibitors on cannabinoids activity in streptozotocin-induced neuropathy. Pharmacol Rep. 2015;67:209–216.

[242] Rawat C, Kukal S, Dahiya UR, Kukreti R. Cyclooxygenase-2 (COX-2) inhibitors: Future therapeutic strategies for epilepsy management. J Neuroinflammation. 2019;16:1–5.

[243] Jain R, Khare N, Jain A. Anticonvulsant property of celecoxib in maximal electroshock seizure: Correlating the role of inflammation in pathophysiology of epileptic seizures. Natl J Physiol Pharm Pharmacol. 2022;13.

[244] Schlichtiger J, Pekcec A, Bartmann H, Winter P, Fuest C, Soerensen J, Potschka H. Celecoxib treatment restores pharmacosensitivity in a rat model of pharmacoresistant epilepsy. Br J Pharmacol. 2010;160:1062–1071.

[245] Jung KH, Chu K, Lee ST, Kim J, Sinn DI, Kim JM, Park DK, Lee JJ, Kim SU, Kim M, Lee SK. Cyclooxygenase-2 inhibitor, celecoxib, inhibits the altered hippocampal neurogenesis with attenuation of spontaneous recurrent seizures following pilocarpine-induced status epilepticus. Neurobiol Dis. 2006;23:237–246.

[246] Haiju Z, Ruopeng S, Gefei L, Lu Y, Chunxi L. Cyclooxygenase-2 inhibitor inhibits the hippocampal synaptic reorganization by inhibiting MAPK/ERK activity and modulating GABAergic transmission in pilocarpine-induced status epilepticus rats. Med Chem Res. 2009;18:71–90.

[247] Zhou L, Zhou L, Cao SL, Xie YJ, Wang N, Shao CY, Wang YN, Zhou JH, Cowell JK, Shen Y. Celecoxib ameliorates seizure susceptibility in autosomal dominant lateral temporal epilepsy. J Neurosci. 2018;38:3346–3357.

[248] Alsaegh H, Eweis H, Kamel F, Alrafiah A. Celecoxib decrease seizures Susceptibility in a rat model of inflammation by inhibiting HMGB1 translocation. Pharmaceuticals. 2021;14:380.

[249] El-Bahrawy H, Hegazy S, Farrag W, Werida R. Targeting inflammation using celecoxib with glimepiride in the treatment of obese type 2 diabetic Egyptian patients. Int J Diabetes Dev Ctries. 2017;37:97–102.

[250] Amior L, Srivastava R, Nano R, Bertuzzi F, Melloul D. The role of Cox-2 and prostaglandin E2 receptor EP3 in pancreatic β-cell death. FASEB J. 2019;33:4975–4986.

[251] Pattaragarn A, Alon US. Treatment of congenital nephrogenic diabetes insipidus by hydrochlorothiazide and cyclooxygenase-2 inhibitor. Pediatr Nephrol. 2003;18:1073–1076.

[252] Kim TJ, Lee HJ, Pyun DH, Abd El-Aty AM, Jeong JH, Jung TW. Valdecoxib improves lipid-induced skeletal muscle insulin resistance via simultaneous suppression of inflammation and endoplasmic reticulum stress. Biochem Pharmacol. 2021;188:114557.

[253] Knudsen JF, Carlsson U, Hammarström P, Sokol GH, Cantilena LR. The cyclooxygenase-2 inhibitor celecoxib is a potent inhibitor of human carbonic anhydrase II. Inflammation. 2004;28:285–290.

[254] Di Fiore A, Pedone C, D'Ambrosio K, Scozzafava A, De Simone G, Supuran CT. Carbonic anhydrase inhibitors: Valdecoxib binds to a different active site region of the human isoform II as compared to the structurally related cyclooxygenase II 'selective' inhibitor celecoxib. Bioorg Med Chem Lett. 2006;16:437–442.

[255] Zimmerman S, Innocenti A, Casini A, Ferry JG, Scozzafava A, Supuran CT. Carbonic anhydrase inhibitors. Inhibition of the prokariotic beta and gamma-class enzymes from Archaea with sulfonamides. Bioorg Med Chem Lett. 2004;14:6001–6006.

[256] Lehtonen JM, Parkkila S, Vullo D, Casini A, Scozzafava A, Supuran CT. Carbonic anhydrase inhibitors. Inhibition of cytosolic isozyme XIII with aromatic and heterocyclic sulfonamides: A novel target for the drug design. Bioorg Med Chem Lett. 2004;14:3757–3762.

[257] Vullo D, Del Prete S, Osman SM, De Luca V, Scozzafava A, AlOthman Z, Supuran CT, Capasso C. Sulfonamide inhibition studies of the δ-carbonic anhydrase from the diatom Thalassiosira weissflogii. Bioorg Med Chem Lett. 2014;24:275–279.

[258] Baghaki S, Yalcin CE, Baghaki HS, Aydin SY, Daghan B, Yavuz E. COX2 inhibition in the treatment of COVID-19: Review of literature to propose repositioning of celecoxib for randomized controlled studies. Int J Infect Dis. 2020;101:29–32.

[259] Wang R. Etoricoxib may inhibit cytokine storm to treat COVID-19. Med Hypotheses. 2021;150:110557.

[260] Pettersen FO, Torheim EA, Dahm AE, Aaberge IS, Lind A, Holm M, Aandahl EM, Sandset PM, Taskén K, Kvale D. An exploratory trial of cyclooxygenase type 2 inhibitor in HIV-1 infection: Downregulated immune activation and improved T cell-dependent vaccine responses. J Virol. 2011;85:6557–6566.

[261] Winston A, Pozniak A, Mandalia S, Gazzard B, Pillay D, Nelson M. Which nucleoside and nucleotide backbone combinations select for the K65R mutation in HIV-1 reverse transcriptase. Aids. 2004;18:949–951.

[262] Kvale D, Ormaasen V, Kran AM, Johansson CC, Aukrust P, Aandahl EM, Frøland SS, Taskén K. Immune modulatory effects of cyclooxygenase type 2 inhibitors in HIV patients on combination antiretroviral treatment. Aids. 2006;20:813–820.

[263] World Health Organization (WHO), Global Tuberculosis Report, 2021. https://www.who.int/tb/publications/global_report/en/.

[264] Mohr E, Ferlazzo G, Hewison C, Azevedo VD, Isaakidis P. Bedaquiline and delamanid in combination for treatment of drug-resistant tuberculosis. Lancet Infect Dis. 2019;19:470.

[265] Jøntvedt Jørgensen M, Nore KG, Aass HC, Layre E, Nigou J, Mortensen R, Tasken K, Kvale D, Jenum S, Tonby K, Dyrhol-Riise AM. Plasma LOX-Products and Monocyte Signaling Is Reduced by Adjunctive Cyclooxygenase-2 Inhibitor in a Phase I Clinical Trial of Tuberculosis Patients. Front Cell Infect Microbiol. 2021;11:669623.

Kirti Sharma, Himanshu Shrivastava, Asim Kumar

13 Advantages, opportunities, and challenges to drug repurposing

Abstract: Drug repurposing, also known as drug repositioning or drug reprofiling, involves establishing new therapeutic indications for a clinically well-established drug. The drug repurposing strategy has shown great promise over the years owing to some inherent limitations of the de novo drug discovery and development approach. Its significance could be understood in a way that approximately one-third of the new approvals in recent years are of repurposed drugs and have helped the pharmaceutical industry to generate 25% of their annual revenue. However, this strategy has its own advantages, opportunities, and bottlenecks. It could be a great savior, particularly when an epidemic or pandemic breaks out in the society. As it is well known that the classical drug development and discovery approach of orphan diseases requires high input cost, longer gestational time, and extremely high rate of drug attrition, it is high time that the drug repurposing and repositioning strategy should be streamlined. The current chapter encompasses the advantages offered and the challenges faced in drug repurposing.

Keywords: Drug repurposing, advantages, drug discovery, challenges

13.1 Introduction: opportunities and advantages of drug repurposing

The science and practice of drug repurposing are known to medicinal chemists; however, it's application and research has witnessed a paradigm shift during the last two-three decades. It offers a plethora of advantages such as the reduction of latency period for drug discovery, significant reduction in the input cost for the development of a newer drug analogue, relatively less stringent regulatory aspects, devising newer therapy for some of the rare and orphan diseases [1–4]. Medicinal chemistry has received a tremendous impetus due to the accumulation of accurate biological data, evolution of computational chemical biology, and by the advent of AI-based software, which have revolutionized the drug discovery and development. Value addition in the drug repurposing strategy is observed due to multiparameter optimization, diversity-oriented synthesis, structure-based drug discovery, and biological system-mediated drug delivery [5]. Figure 13.1 outlines the timeline of the drug repurposing approach.

Acknowledgments: Dr. A.K. wants to thank Amity University Haryana (AUH) for providing necessary academic support

https://doi.org/10.1515/9783110791150-013

Figure 13.1: Drug repurposing approaches.

In silico techniques such as docking, molecular simulation, and molecular dynamics have provided a great impetus for drug repurposing. One of the most recent successful case studies of drug repurposing is pembrolizumab, a monoclonal antibody (MAB) developed and marketed by Merck, originally indicated for advanced melanoma. It was studied for other cancer indications and was later approved for the treatment of 14 different types of cancers [6]. Other notable clinically successful examples of MABs being repurposed are Bristol-Myers Squibb's (BMS) Opdivo (nivolumab), Novartis' Arzerra (ofatumumab), and Pfizer's immunomodulator Rapamune (sirolimus). Rapamune was originally approved to treat organ transplant rejection; it was later approved for lymphangioleiomyomatosis (LAM). Interestingly, it was found that some of the noncancerous drugs are showing great results in treating cancer [7, 8]. Apart from above the mentioned MABs success story, some of the earlier examples of chemical drug repurposing include valproic acid for HDAC inhibitor, zidovudine for cancer, amphotericin B for black fungus, minoxidil for alopecia or male pattern baldness, methotrexate for immunomodulator action, protease inhibitors for SARS CoV-2, bromocriptine for Type 2 diabetes, Thalidomide for multiple myeloma, and sildenafil for erectile dysfunction [9, 10]. Figure 13.2 encompasses

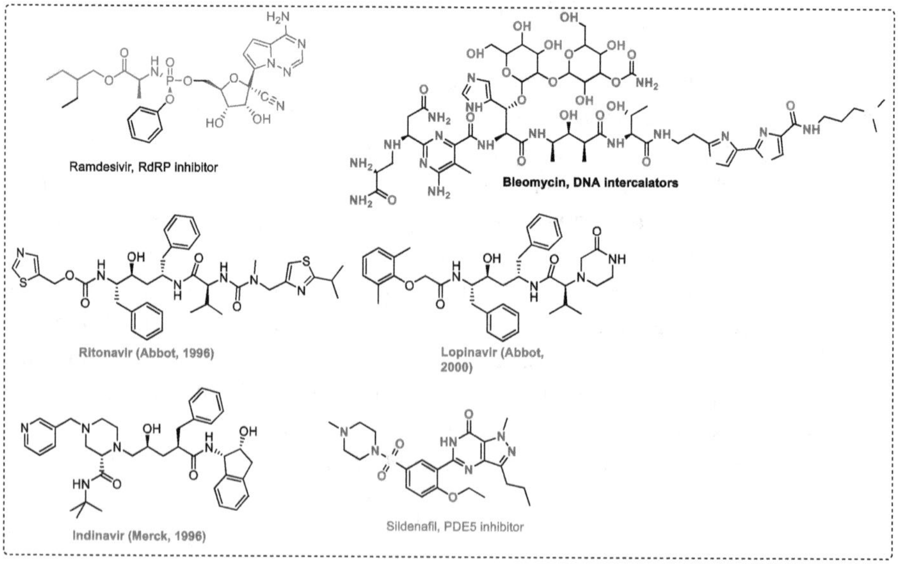

Ramdesivir, RdRP inhibitor

Bleomycin, DNA intercalators

Ritonavir (Abbot, 1996)

Lopinavir (Abbot, 2000)

Indinavir (Merck, 1996)

Sildenafil, PDE5 inhibitor

Figure 13.2: Some examples of repurposed drugs.

some of the examples of repurposed drugs. Suberanilohydroxamic acid (SAHA), vorinostat, originally indicated for cutaneous T cell lymphoma was repurposed for pulmonary fibrosis and acute lung injury. It was found that SAHA was inhibiting LTB4 biosynthesis, thereby reducing neutrophilic inflammation [11].

13.2 Application of drug repurposing via computational tools, pertaining to rare and neglected diseases

Drug repurposing strategy is highly indispensable in the hunt for newer therapies for rare, neglected, and orphan diseases. In US, a rare disease is defined as one affecting less than 200,000 people, for which an orphan drug is utilized, i.e., a disease affecting approximately 61 people per one lakh of population. In Japan, the definition is if the disease affects <40/100,000 population, and in EU it is <50/100,000 population. It is also estimated that approximately 30 million individuals are suffering from rare diseases in the US [12]. It is clear from the above data that the target group for big pharma giants is not commercially lucrative; therefore, a meaningful collaboration of the academics and the pharma industry, adequately incentivized by government agencies, is the need of hour, especially for rare and neglected diseases. The prevalence of these neglected diseases is mainly in the African continent and in other subtropical low-income countries where the per capita income is very low and hence pharmaceutical companies felt that developing and marketing orphan drugs would not be beneficial for them. To start with this drug repurposing strategy, a robust database of clinically established drugs is required, having both the chemical and biological data [13]. For better data mining of chemical 3D structures, there are various databases available, such as the Cambridge Structural Database (CSD) containing about 500,000 chemical structures, [14] the MDL drug data report (MDDR) containing structures of 1,80,000 bioactive compounds [15], and the World of molecular bioactivity (WOMBAT) registry [16]. As bioinformatics and cheminformatics have witnessed significant progress during the last decade, the translational potentials of existing drug therapies have been explored, leading to newer bioactivities of existing drugs. A novel initiative, named Drug for Neglected Disease Initiative (DNDi), is being practiced that comprises performing clinical trials of various drugs and utilizing in-silico studies [17, 18]. Rare and neglected diseases often suffer from vague pathophysiology and unknown transducer pathways, and computational techniques for predictive repurposing, genome sequencing, proteomics, high throughput screening (HTS), metabolic profiling by GCMS have offered a great respite for devising pharmacotherapy for them. African sleeping sickness is one such rare and neglected disease against which various drugs have been repurposed. Various studies suggested that Network- and metabolic control-based analysis could show great promise in designing multitarget therapeutics and also in developing a synergistic drug combination for rare tropical diseases. Nifurtimox-eflornithine combination is one such combination used to treat advanced stage African trypanosomiasis or sleeping sickness. WHO has included this combination therapy in its model list of essential medicines [19]. The original indication of Nifurtimox is to treat American trypanosomiasis. Various literature reports have indicated a possible repurposing of antitrypanosomial drugs for chagas disease [20]. Figure 13.3 outlines some of the inputs required for the academic aspect of drug repurposing.

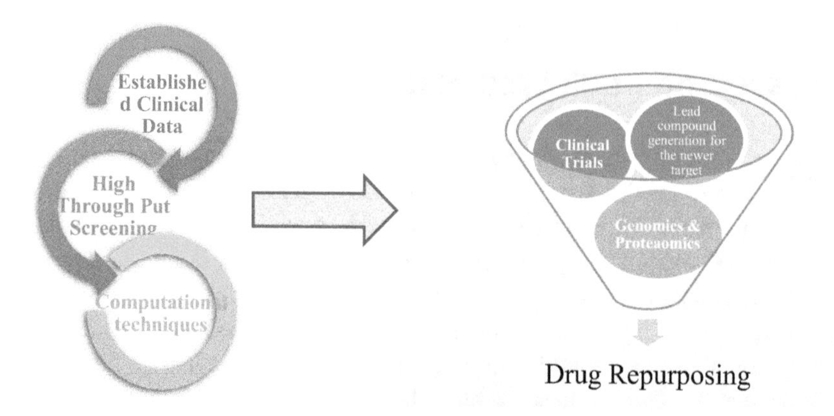

Figure 13.3: Academic aspects of drug repurposing.

Some of the seminal contributions in drug repurposing via advanced in silico techniques could be understood by the following examples. Li and coworkers utilized multiple ligand simultaneous docking (MLSD) with the fragment-based drug design study and found the inhibition of STAT pathway by celecoxib [21]. Utilizing the same MLSD technique, Li and coworkers reported raloxifene, an anticancer drug, to be IL-6/GP130 inhibitor [22]. Astolfi and coworkers developed novel NorA EPIs (Efflux pump inhibitors) by constructing a ligand-based pharmacophore model. NorA is an efflux pump expressed in staphylococcus aureus. The drugs that were found to exhibit this NorA EPI action were dasatinib, gefitinib and nicardipine [23]. Turk and coworkers modified the structure of tozasertib, originally developed as AurA inhibitor for cancer therapy, via automated multiobjective in silico design and shifted its selectivity for a pain-inducing target, TrkA (Figure 13.4) [24].

Some other notable approaches for computational-based drug repurposing are Similarity Ensemble Approach (SEA), Google's DeepMind AI-based software, DRviaSPCN software package, SAveRUNNER (Searching off-lAbel dRUg aNd NEtwoRk), which is a network-based software tool, and DrugRep database and NOD web server [25–28]. Figure 13.5 depicts the network representation of the in silico method [29].

Recently, Yang and coworkers outlined the application of machine learning in drug repurposing. They emphasized the exploiting of molecular fingerprints in terms of their static physicochemical and dynamic properties [30].

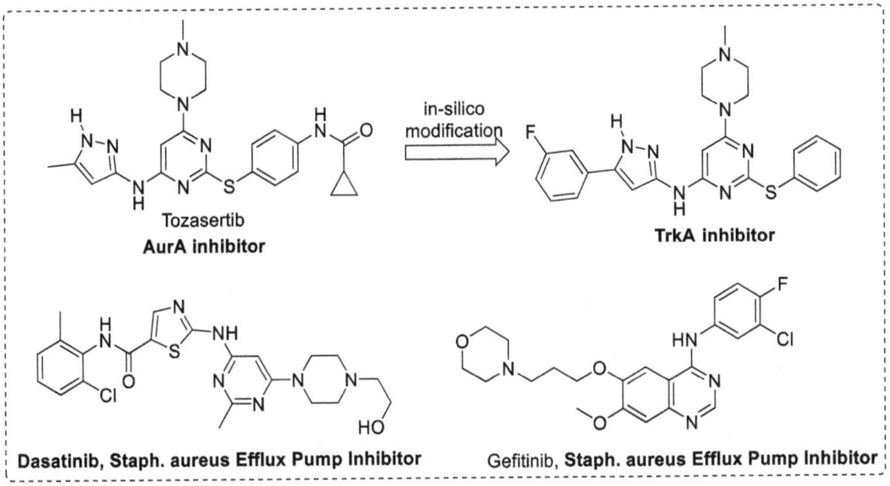

Tozasertib
AurA inhibitor
in-silico modification
TrkA inhibitor
Dasatinib, Staph. aureus Efflux Pump Inhibitor
Gefitinib, **Staph. aureus Efflux Pump Inhibitor**

Figure 13.4: Repurposing of Tozasertib, Dasatinib, and Gefitinib.

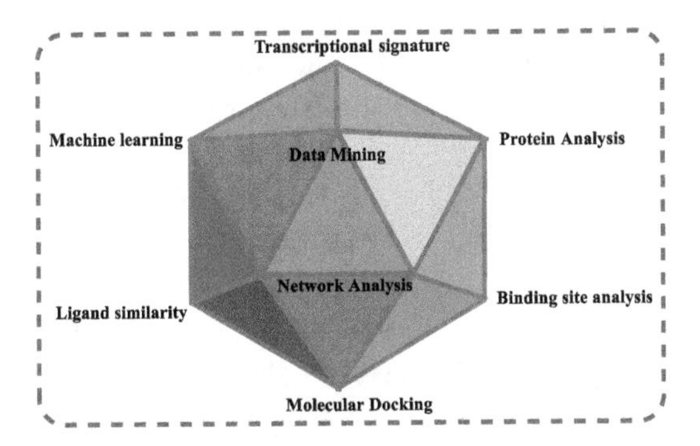

Figure 13.5: Integration of different in silico methods.

13.3 Cost-benefit ratio

The most exciting aspect of drug repurposing strategy is the less amount of cost incurred as compared to the classical or de-novo drug discovery approach. It is estimated that the drug repurposing strategy lowers down the overall cost of drug development by at least $300 million. Currently, it is estimated that almost one-fourth of the revenue generated by the pharmaceutical industry is by repurposed drugs [31]. The attrition rate of new chemical entities is so much that it becomes nonviable for companies. Given the stringent IP rules of various drug administration agencies, the de novo drug discovery faces various uncalled for challenges. In a recent study published in JAMA, it is estimated that de novo drug discovery method for a new chemical entity costs somewhere between $314 million and $2.8 billion. The study also highlights that during the study period of 2009 to 2018 in the US, the estimated median capitalized cost per product on research and develop was $985 million [32]. Drug repurposing not only saves time, resources, and input cost but also reduces the incubation period. The near future would certainly witness further expedited, cost effective, and streamlined newer therapeutics due to the rampant evolution of AI-based machine learning.

13.4 Enrichment of translational domains for drug discovery

The de novo drug discovery approach has experienced a tremendous shift, with the enrichment of translational domains such as precision medicine, system medicine, proteomics, genomics, HTS, metabolomics, and in silico techniques. Precision medicine focuses on the individual aspect of disease prognosis, i.e., genetic polymorphism, phenotypic effects by the surrounding environment, and food habits [33]. After the successful completion of the human genome project in 2003, a huge impetus was given to genomics in the drug discovery and development process. Gene-wide association studies (GWAS), a systemic study of whole genome sequencing of a disease, catapulted the drug repositioning strategy. It was further found out that the single nucleotide polymorphism (SNP), a small sequence variant, is responsible for various physiological imbalances. SNPs were found to be in linkage disequilibrium with the disease. The GWAS study was utilized to repurpose Denosumab, an anticancer drug, and it was found that it is effective against osteoporosis [34]. De novo drug discovery hardly accounts for the off-tract transducer pathway kick-started by the drug molecule when it binds at its designated receptor site. Often, it is seen that a drug activates many of the accessory cell signaling pathways in a cell, leading to severe adverse drug effects. Microarray transcriptomic profiling, coupled with genome sequencing, leads to identifying the transcriptional factors affecting a disease process. Transcriptome is an array of mRNA transcripts produced in a cell. A connectivity map database (cMAP) outlines transcriptional fingerprints of some 1,600 compounds against various diseases [35]. Further, to speed up, phenotyping at a relatively larger scale electronic medical records and genomics (eMERGE) netwok is used [12].

13.5 Bottlenecks and disadvantages of drug repurposing

The drug repurposing strategy, despite looking so promising and indispensable, suffers from several bottlenecks. The main apprehension and reluctance of pharmaceutical companies is to establish a safety profile of the repurposed drug for a newer indication. This apart, the cost effectiveness of the repurposed therapy, regulatory framework for claimed efficacy and intellectual property aspects, and their market access are some of the most intriguing aspects. The legal frameworks of different countries pertaining to the drug intellectual property rights and patenting have become quite stringent, post GATT. Therefore, the actual incentive for the repurposing of a drug is not very fruitful and is often penal. It is also being observed in many cases that the off-level indications of the repurposed drug that a research group explores, is already reported in various literature or are already explored clinically. Exclusivity regimens are provided by many drug regulatory bodies across the globe to encourage the research and development for drug repurposing. "Regulatory data exclusivity" of the generic drug gives some space for undertaking "second medical use patents" [36]. However, it should be noted that the enforcement of a patent could be difficult as there would always be a risk of disclosing the clinical data/indications/off-level uses to the public domain, which could further put patentees in a disadvantaged state. Therefore, it is pertinent that for generic drugs, repurposing offers no incentive to developers to obtain their invested cost. For the off-patent drugs, the patentability for the second use could be difficult unless the strength of the dose for the newer indication would not be less than the already marketed strength of the drug. The data protection regimen differs in various regulatory domains. In the US, it is five years of data protection along with three years of extension for the new indication and EU gives eight years of data protection along with two years of market exclusivity. At the same time, if the original developer found a second indication of the drug during this eight years span of data exclusivity, EU may provide one additional year of data protection [37].

In addition to above mentioned aspects, the lack of coordination between the various stakeholders in the pharmaceutical industry is also a bottleneck in the drug repurposing strategy. Starting with the R & D laboratory of a research lab to the pharmaceutical industry and finally at the retail chain end, there seems to be a disconnect many a time. This disconnect could also be possible since physicians in different countries prescribe medicine by their own laid down norms, for example in the US, the physician follows the international nonproprietary name (INN) or the United States Adopted Name (USAN) for the active pharmaceutical ingredient. This system might vary in different countries. Pharmacists are being encouraged by governments to fill the generic version of any API and it provides financial incentives for it. Therefore, it becomes difficult for the inventor to recoup their investment. It is also being observed that pharmaceutical companies often show reluctance in releasing their chemical libraries or share their clinical data with other academic collaborators or inventors, for a possible repurposing. Availability of these compounds also becomes challenging, sometimes, due to their company policies or when the compound is exhausted during the clinical study or has gone to the international market [38]. Given the current pace of this drug repurposing strategy, it is said that soon the chemical space also would be exhausted. Recently, due to the outbreak of the coronavirus pandemic, medicinal chemists were compelled to scan almost every bioactive chemical structures, be it antiviral, anticancer, monoclonal antibodies, immunomodulators, or antibiotics for a possible drug therapy against SARS CoV-2 [39, 40].

13.6 Conclusion

Drug discovery and development is never ending and is a perpetually evolving domain of science. With the advent of modern and sophisticated equipment and the evolution of AI, it is being felt that drug repurposing could enrich the medicinal chemist's toolbox like never before. The recent outbreak of SARS CoV-2 has left the drug discovery scientists with no choice but to explore the newer and advance techniques of drug repurposing. Various drugs, such as hydroxychloroquine, doxycycline, ivermectin, remdesivir (RdRp inhibitor), tocilizumab and related MAbs, protease inhibitors and immunomodulators, were repurposed for possible respite against the corona virus. However, it was felt over the years that the bioactive chemical libraries should be enriched further. For this, it is highly imperative to have a robust unison between the academia and the industry or possibly between all the stakeholders of this drug discovery domain. A well-trained and research-oriented pharmacy graduate, a physician with an interdisciplinary bend of mind, and a well-informed patient could really transform the process of drug discovery and repurposing. The coming decade or so is going to experience a paradigm shift in the de novo drug discovery approach as well as the drug repurposing strategy. A summary of the advantages of drug repurposing is represented in Figure 13.6.

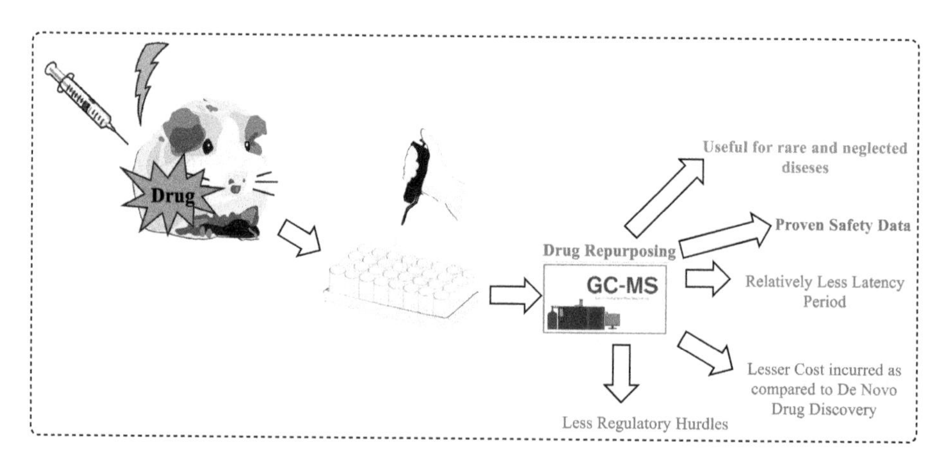

Figure 13.6: Illustration of the advantages that drug repurposing brings in drug discovery.

References

[1] Krishnamurthy N, Grimshaw AA, Axson SA, Choe SH, Miller JE. Drug repurposing: A systematic review on root causes, barriers and facilitators. BMC Health Serv Res. 2022;22:970.

[2] Pushpakom S, Iorio F, Eyers PA, Escott KJ, Hopper S, Wells A, Doig A, Guilliams T, Latimer J, McNamee C, Norris A, Sanseau P, Cavalla D, Pirmohamed M. Drug repurposing: Progress, challenges and recommendations. Nat Rev Drug Discov. 2018;18(1):41–58. doi: https://doi.org/10.1038/nrd.2018.168.

[3] Xue H, Li J, Xie H, Wang Y. Review of drug repositioning approaches and resources. Int J Biol Sci. 2018;14(10):1232.

[4] Ashburn TT, Thor KB. Drug repositioning: Identifying and developing new uses for existing drugs. Nat Rev Drug Discov. 2004;3(8):673–683.

[5] Wu G, Zhao T, Kang D, Zhang J, Song Y, Namasivayam V, Kongsted J, Pannecouque C, Clercq ED, Poongavanam V, Liu X, Zhan P. J Med Chem. 2019;62(21):9375–9414.

[6] Thuru X, Magnez R, El-Bouazzati H, Vergoten G, Quesnel B, Bailly C. Drug repurposing to enhance antitumor response to PD-1/PD-L1 immune checkpoint inhibitors. Cancers. 2022;14(14):3368.

[7] https://www.technologynetworks.com/drug-discovery/articles/drug-repurposing-advantages-and-key-approaches-344261

[8] Costa B, Vale N. A review of repurposed cancer drugs in clinical trials for potential treatment of COVID-19. Pharmaceutics. 2021;13(6):815.

[9] Ashburn TT, Thor KB. Drug repositioning: Identifying and developing new uses for existing drugs. Nat Rev Drug Discov. 2004;3:673–683.

[10] Nosengo N. Can you teach old drugs new tricks?. Nature. 2016;534:314–316.

[11] Lu W, Yao X, Ouyang P, Dong N, Wu D, Jiang X, Wu Z, Zhang C, Xu Z, Tang Y. Drug repurposing of histone deacetylase inhibitors that alleviate neutrophilic inflammation in acute lung injury and idiopathic pulmonary fibrosis via inhibiting leukotriene A4 hydrolase and blocking LTB4 biosynthesis. J Med Chem. 2017;60:1817–1828.

[12] Kort E, Jovinge S. Drug repurposing: Claiming the full benefit from drug development. Cur Cardiol Rep. 2021;23:62.

[13] Lo Y-C, Senese S, Damoiseaux R, Torres JZ. 3D chemical similarity networks for structure-based target prediction and scaffold hopping. ACS Chem Biol. 2016;11(8):2244–2253.

[14] Allen FH, Kennard OG, Motherwell WDS, Town W. Cambridge crystallographic data centre. 2. structural data file. J Chem Doc. 1973;13(3):119–123.

[15] Sheridan RP, Shpungin J. Calculating similarities between biological activities in the MDL Drug Data Report database. J Chem Inf Comp Sci. 2004;44(2):727–740.

[16] (www.cbs.dtu.dk/services/ChemProt/ChemProt-2.0/) or Carlsbad (http://carlsbad.health.unm.edu/carlsbad)

[17] Ferreira LG, Andricopulo AD. Drug repositioning approaches to parasitic diseases: A medicinal chemistry perspective. Drug Discov Today. 2016;21:1699–1710.

[18] Pollastri MP. Fexinidazole: A new drug for African sleeping sickness on the horizon. Trends Parasitol. 2018;34:178–179.

[19] Sun W, Sanderson PE, Zheng W. Drug combination therapy increases successful drug repositioning. Drug Discov Today. 2016;21(7):1189–1195.

[20] Trindade J, Freire-de-lima DS, Côrte-Real CG, Decote-Ricardo S, Lima MEF. Drug repurposing for Chagas disease: In vitro assessment of nimesulide against Trypanosoma cruzi and insights on its mechanisms of action. PLOS ONE. 2021. doi: https://doi.org/10.1371/journal.pone.0258292O.

[21] Li H, Liu A, Zhao Z, Xu Y, Lin J, Jou D, Li C. Fragment-based drug design and drug repositioning using multiple ligand simultaneous docking (MLSD): Identifying celecoxib and template compounds as

novel inhibitors of signal transducer and activator of transcription 3 (STAT3). J Med Chem. 2011;54:5592–5596.

[22] Li H, Xiao H, Lin L, Jou D, Kumari V, Lin J, Li C. Drug design targeting protein-protein interactions (PPIs) using multiple ligand simultaneous docking (MLSD) and drug repositioning: Discovery of raloxifene and bazedoxifene as novel inhibitors of IL-6/ GP130 interface. J Med Chem. 2014;57:632–641.

[23] Astolfi A, Felicetti T, Iraci N, Manfroni G, Massari S, Pietrella D, Tabarrini O, Kaatz GW, Barreca ML, Sabatini S, Cecchetti V. Pharmacophore-based repositioning of approved drugs as novel staphylococcus aureus NorA efflux pump inhibitors. J Med Chem. 2017;60:1598–1604.

[24] Turk S, Merget B, Eid S, Fulle S. From cancer to pain target by automated selectivity inversion of a clinical candidate. J Med Chem. 2018;61:4851–4859.

[25] Paul D, Sanap G, Shenoy S, Kalyane D, Kalia K, Tekade RK. Artificial intelligence in drug discovery and development. Drug Discov Today. 2021;26(1):80–93.

[26] Fiscon G, Paci P. SAveRUNNER: An R-based tool for drug repurposing. BMC Bioinform. 2021;22:150.

[27] Wu J, Li X, Wang Q, Han J. DRviaSPCN: A software package for drug repurposing in cancer via a subpathway crosstalk network. Bioinformatics. 2022;38(21):4975–4977.

[28] Narwani TJ, Srinivasan N, Chakraborti S. NOD: A web server to predict New use of Old Drugs to facilitate drug repurposing. Sci Rep. 2021;11:13540.

[29] March-Vila E, Pinzi L, Sturm N, Tinivella A, Engkvist O, Chen H, Rastelli G. On the integration of in silico drug design methods for drug repurposing. Front Pharmacol. 2017;8:298.

[30] Yang F, Zhang Q, Ji X, Zhang Y, Li W, Peng S, Xue F. Machine learning applications in drug repurposing. Interdis Sci Comput Life Sci. 2022;14:15–21.

[31] Jourdan J-P, Bureau R, Rochais C, Dallemagne P. Drug repositioning: A brief overview. J Pharm Pharmacol. 2020;72(9):1145–1151.

[32] Wouters OJ, McKee M, Luyten J. Estimated research and development investment needed to bring a new medicine to market, 2009–2018. JAMA. 2020;323(9).

[33] Rojo AC, Heylen D, Aerts J, Thas O, Hooyberghs J, Gökhan E, Valkenborg D. Towards building a quantitative proteomics toolbox in precision medicine: A mini-review. Front Pharmacol. 2021;12:723510.

[34] Sanseau P, Agarwal P, Barnes MR, Pastinen T, Richards JB, Cardon LR, et al. Use of genome-wide association studies for drug repositioning. Nat Biotechnol. 2012;30(4):317–320.

[35] Toro-Domínguez D, Alarcón-Riquelme ME, Carmona-Sáez P. Chapter 11-Drug repurposing from transcriptome data: Methods and applications. In Silico Drug Design Repurposing Techniques and Methodologies, 2019, 303–327. doi: https://doi.org/10.1016/B978-0-12-816125-8.00011-0.

[36] Grabowski H. Follow-on biologics: Data exclusivity and the balance between innovation and competition. Nat Rev Drug Discov. 2008;7:479–488.

[37] Talevi A, Bellera CL. Challenges and opportunities with drug repurposing: Finding strategies to find alternative uses of therapeutics. Expert Opin Drug Discov. 2020;15(4):397–401. doi: https://doi.org/10.1080/17460441.2020.1704729.

[38] Breckenridge A, Jacob R. Overcoming the legal and regulatory barriers to drug repurposing. Nat Rev Drug Discov. 2019;18(1):1–2.

[39] Chakraborty C, Sharma AR, Bhattacharya M, Agoramoorthy G, Lee -S-S. The drug repurposing for COVID-19 clinical trials provide very effective therapeutic combinations: Lessons learned from major clinical studies. Front Pharmacol. 2021;12:704205.

[40] Rodrigues L, Cunha RB, Vassilevskaia T, Viveiros M, Cunha C. Drug repurposing for COVID-19: A review and a novel strategy to identify new targets and potential drug candidates. Molecules. 2022;27(9):2723.

Index

https://doi.org/10.1515/9783110791150-014

Printed in the USA
CPSIA information can be obtained
at www.ICGtesting.com
LVHW080158140624
783111LV00003B/357